LIBRARY
UNIV OF MAINE

LIBRARY
UNIV OF MAINE

ADVANCES IN
CARBOHYDRATE CHEMISTRY

VOLUME 5

LIBRARY
UNIV OF MAINE

ADVANCES IN CARBOHYDRATE CHEMISTRY

Edited by

CLAUDE S. HUDSON
National Institutes of Health
Bethesda, Maryland

SIDNEY M. CANTOR
American Sugar Refining Company
Philadelphia, Pennsylvania

Associate Editors for the British Isles

STANLEY PEAT
University College of North Wales
Bangor, Caernarvonshire, Wales

MAURICE STACEY
The University
Birmingham, England

Board of Advisors

WILLIAM L. EVANS
HERMANN O. L. FISCHER
E. L. HIRST
R. C. HOCKETT
E. G. V. PERCIVAL

W. W. PIGMAN
C. B. PURVES
J. C. SOWDEN
ROY L. WHISTLER
M. L. WOLFROM

VOLUME 5

1950
ACADEMIC PRESS INC., PUBLISHERS
NEW YORK, N. Y.

Copyright 1950, by
ACADEMIC PRESS INC.
125 EAST 23RD STREET
NEW YORK 10, N. Y.

All Rights Reserved

NO PART OF THIS BOOK MAY BE REPRODUCED IN ANY
FORM, BY PHOTOSTAT, MICROFILM, OR ANY OTHER MEANS
WITHOUT WRITTEN PERMISSION FROM THE PUBLISHERS

PRINTED IN THE UNITED STATES OF AMERICA

CONTRIBUTORS TO VOLUME 5

Mildred Adams, *Takamine Laboratory, Clifton, New Jersey.*

E. J. Bourne, *The University, Birmingham, England.*

Mary L. Caldwell, *Columbia University, New York, New York.*

G. R. Dean, *Corn Products Refining Company, Argo, Illinois.*

M. Doudoroff, *Department of Bacteriology, University of California, Berkeley, California.*

Hewitt G. Fletcher, Jr., *Laboratory of Chemistry and Chemotherapy, Experimental Biology and Medicine Institute, National Institutes of Health, Bethesda, Maryland.*

J. B. Gottfried, *Corn Products Refining Company, Argo, Illinois.*

Alfred Gottschalk, *The Walter and Eliza Hall Institute of Medical Research, Melbourne, Australia.*

W. Z. Hassid, *Division of Plant Nutrition, College of Agriculture, University of California, Berkeley, California.*

Z. I. Kertesz, *New York State Agricultural Experiment Station, Cornell University, Geneva, New York.*

R. J. McColloch, *New York State Agricultural Experiment Station, Cornell University, Geneva, New York.*

R. F. Nickerson, *Textile Research Laboratory, Monsanto Chemical Company, Everett, Massachusetts.*

Stanley Peat, *The University College of North Wales, Bangor, Caernarvonshire, Wales.*

Nelson K. Richtmyer, *Laboratory of Chemistry and Chemotherapy, Experimental Biology and Medicine Institute, National Institutes of Health, Bethesda, Maryland.*

Roy L. Whistler, *Department of Agricultural Chemistry, Purdue University, Lafayette, Indiana.*

L. F. Wiggins, *The Imperial College of Tropical Agriculture, Trinidad, British West Indies.*

EDITORS' PREFACE

The editors of Volume V wish to express their appreciation of the excellent service that has been rendered by Professor W. W. Pigman and Professor M. L. Wolfrom in the editing of the earlier volumes of this series. They have generously consented to assist in the publication of later volumes in the capacity of members of a "Board of Advisors," which now takes the place of the "Executive Committee" of the earlier volumes.

During the past year we have suffered a great loss through the death of Sir Norman Haworth. His assistance was invaluable in the organization of the group that initiated the publication of the Advances and he gave much of his time to the details of the solicitation of articles during the remaining years of his life. We pay homage to the memory of this great man and genial friend.

The editorial policy of the publication will continue in its past form. To quote from Volume I: "It is our plan to have the individual contributors furnish critical, integrating reviews rather than mere literature surveys, and to have the articles presented in such a form as to be intelligible to the average chemist rather than only to the specialist." Invitations will be extended to selected research workers to prepare critical reviews of special topics in the broad field of the carbohydrates, including the sugars, polysaccharides and glycosides. It is also the intention to include biochemical and analytical developments in the carbohydrate field as well as critical reviews of important industrial advances.

Criticisms and suggestions of any kind are respectfully solicited.

THE EDITORS
C. S. H.
S. M. C.

Bethesda, Maryland
Philadelphia, Pennsylvania

CONTENTS

	Page
Contributors to Volume 5	v
Editor's Preface	vii

Applications in the Carbohydrate Field of Reductive Desulfurization by Raney Nickel

By Hewitt G. Fletcher, Jr. and Nelson K. Richtmyer, *Laboratory of Chemistry and Chemotherapy, Experimental Biology and Medicine Institute, National Institutes of Health, Bethesda, Maryland*

I. Introduction	1
II. Hydrogenolysis of Various Classes of Sulfur-Containing Sugar Derivatives	5
III. Tabular Survey of Reductive Desulfurizations in the Sugar Series	26

Enzymatic Synthesis of Sucrose and Other Disaccharides

By W. Z. Hassid and M. Doudoroff, *Division of Plant Nutrition, College of Agriculture, and Department of Bacteriology, University of California, Berkeley, California*

I. Introduction	29
II. Structure of Sucrose	31
III. Synthesis of Sucrose through the Mechanism of Phosphorolysis	33
IV. Specificity of Sucrose Phosphorylase	35
V. Synthetic Non-reducing Disaccharides	36
VI. Synthetic Reducing Disaccharide	43
VII. Formation of Sucrose and Other Disaccharides through Exchange of Glycosidic Linkages	46

Principles Underlying Enzyme Specificity in the Domain of Carbohydrates

By Alfred Gottschalk, *The Walter and Eliza Hall Institute of Medical Research, Melbourne, Australia*

I. Introduction	49
II. The Enzyme-Substrate Compound as Chemical Basis of Enzyme Specificity	51
III. The Effect of Configurational and Substitutional Changes in the Sugar Molecule on the Rate of Various Enzyme Reactions and Its Application in Marking the Contacting Groups in the Substrate	60
IV. Discussion of the Principles Controlling and of the Factors Impairing the Formation of the Intermediate Compound between Carbohydrates and Their Specific Enzymes	76

Enzymes Acting on Pectic Substances

By Z. I. Kertesz and R. J. McColloch, *New York State Agricultural Experiment Station, Cornell University, Geneva, New York*

I. Introduction	79
II. Pectin Chemistry and Nomenclature	80

	Page
III. Possible Changes Brought About in Pectic Substances by Enzymes	82
IV. Protopectinase	84
V. Pectin-polygalacturonase (PG)	85
VI. Pectin-methylesterase (PM)	92
VII. Role and Application of Pectic Enzymes	98

The Relative Crystallinity of Celluloses

By R. F. Nickerson, *Textile Research Laboratory, Monsanto Chemical Company, Everett, Massachusetts*

I. Introduction	103
II. Purpose	104
III. Historical	105
IV. X-ray Diffraction Studies	106
V. Chemical Studies	109
VI. Discussion	124
VII. Implications	125

The Commercial Production of Crystalline Dextrose

By G. R. Dean and J. B. Gottfried, *Corn Products Refining Company, Argo, Illinois*

I. Introduction	127
II. History	127
III. Manufacture	131
IV. Application of Ion Exchange Refining to the Commercial Manufacture of Crystalline Dextrose	137

The Methyl Ethers of D-Glucose

By E. J. Bourne, *The University, Birmingham, England*, and Stanley Peat, *The University College of North Wales, Bangor, Caernarvonshire, Wales*

I. Introduction	145
II. Monomethyl Ethers	148
III. Dimethyl Ethers	160
IV. Trimethyl Ethers	172
V. Tetramethyl Ethers	186

Anhydrides of the Pentitols and Hexitols

By L. F. Wiggins, *The Imperial College of Tropical Agriculture, Trinidad, British West Indies*

I. Introduction	191
II. Anhydrides of Hexitols	192
III. Anhydrides of Pentitols	220
IV. Some Uses of Hexitol Anhydrides	222
V. Tables of Properties of Pentitol and Hexitol Anhydrides and Their Derivatives	225

Action of Certain Alpha Amylases

By Mary L. Caldwell, *Columbia University, New York, New York* and Mildred Adams, *Takamine Laboratory, Clifton, New Jersey*

I. Foreword	229
II. Introduction	229
III. Beta Amylases	231
IV. Alpha Amylases	234
V. Discussion and Summary	266

Xylan

By Roy L. Whistler, *Department of Agricultural Chemistry, Purdue University, Lafayette, Indiana*

I. Introduction	269
II. Occurrence	270
III. Pretreatment of Plant Material for Polysaccharide Isolation	272
IV. Removal of Lignin	274
V. Extractive Isolation of Xylan	274
VI. Purification	276
VII. Composition and Structure	278
VIII. Oxidation	284
IX. Degree of Polymerization	285
X. Derivatives	286
XI. Biological Decomposition of Xylan	288
XII. Industrial Uses	288
XIII. Addendum	289

Author Index ... 291

Subject Index ... 304

APPLICATIONS IN THE CARBOHYDRATE FIELD OF REDUCTIVE DESULFURIZATION BY RANEY NICKEL

By Hewitt G. Fletcher, Jr., and Nelson K. Richtmyer

Laboratory of Chemistry and Chemotherapy, Experimental Biology and Medicine Institute, National Institutes of Health, Bethesda, Maryland

Contents

I. Introduction 1
II. Hydrogenolysis of Various Classes of Sulfur-Containing Sugar Derivatives . 5
    1. Tetraacetyl-1-thioaldose and Octaacetyldialdosyl Disulfide 5
    2. Thioacetals (Mercaptals) 5
    3. 1-Thioglycosides 14
    4. Thioethers 19
    5. Thiol Esters 22
    6. S-Substituted Xanthates 23
    7. S-Substituted Isothioureas 24
    8. Thiocyanates 24
III. Tabular Survey of Reductive Desulfurizations in the Sugar Series 26

I. Introduction

The hydrogenolysis of the carbon–sulfur bond by Raney nickel con-

$$RSR' + 2H \rightarrow RH + R'H + (S)$$

taining hydrogen was apparently first observed by Bougault, Cattelain and Chabrier[1] and reported by these authors in 1939. A few years later the reaction played an important part in the elucidation of the structure of biotin[2] and was shown by Mozingo and his coworkers[3] to be a tool of general applicability and of preparative importance.

Reductive desulfurization with Raney nickel is a reaction which requires but the simplest of techniques, is carried out under rather mild conditions and affords relatively good yields; it is now a standard and well-recognized reaction, having found use in practically all the various

---

(1) J. Bougault, E. Cattelain and P. Chabrier, *Compt. rend.*, **208**, 657 (1939); *Bull. soc. chim. France*, [5] **7**, 780, 781 (1940).

(2) V. du Vigneaud, D. B. Melville, K. Folkers, D. E. Wolf, R. Mozingo, J. C. Keresztesy and S. A. Harris, *J. Biol. Chem.*, **146**, 475 (1942).

(3) R. Mozingo, D. E. Wolf, S. A. Harris and K. Folkers, *J. Am. Chem. Soc.*, **65**, 1013 (1943).

fields of organic chemistry. The present review will deal solely with its applications in the sugar field.[4]

Concerning the essential nature of the reaction almost nothing is known. The hydrogen which is present in the nickel in quantities varying greatly according to the method of preparation and age of the catalyst appears to displace the sulfur in the organic compound; the sulfur, in turn, becomes bound to the nickel whence it may subsequently be released as hydrogen sulfide by treatment with acid.[5] Desulfurization carried out in aqueous alcohol has been observed[5] to give rise to acetaldehyde; in view of the well-known dehydrogenating action of Raney nickel[6] it seems likely that the solvent may play a part as a hydrogen donor in the reaction.

In practice the sulfur-containing derivative is dissolved or suspended in a suitable solvent, treated with a sufficient quantity of Raney nickel, and either left at room temperature or heated under reflux, recovery of the product being effected by concentration of the liquid phase. The following example may be taken as typical.[7]

Heptaacetyl-1,5-anhydrocellobiitol from Phenyl 1-Thio-β-cellobioside Heptaacetate. Ten grams of pure phenyl 1-thio-β-cellobioside heptaacetate was suspended in 100 ml. of absolute alcohol, treated with approximately 100 g. of freshly prepared Raney nickel in absolute alcohol and then boiled gently for one hour. After cooling, the supernatant solution was decanted and the nickel washed by decantation with three successive 100-ml. portions of boiling absolute alcohol. The combined decantates, after filtration through a fine sintered glass plate, were concentrated at 80° (bath) under a slight vacuum to a volume of 125 ml.; crystallization of the product as a mass of fine needles was spontaneous. Concentration of the mother liquor afforded a very small quantity of additional material; total yield 5.9 g. or 69%. Two recrystallizations from warm 95% ethanol furnished with little loss material (heptaacetyl-1,5-anhydrocellobiitol) melting at 194–195° (corr.) and showing $[\alpha]^{20}_D$ +4.0° in chloroform (c, 1.39).

In order to ascertain whether sufficient nickel to complete a given reaction has been used, the liquid phase may be tested for starting material before an attempt is made to isolate a product. In general the sodium fusion test for sulfur is satisfactory for this purpose but in certain individual cases a specific test may be more convenient. Thus, in the desulfurization of thioacetals (mercaptals), unreacted material

(4) For a brief review of the use of Raney nickel reductive desulfurization in organic chemistry in general see J. F. W. McOmie, *Ann. Repts. on Progress Chem. (Chem. Soc. London),* **45,** 198 (1948).

(5) M. L. Wolfrom and J. V. Karabinos, *J. Am. Chem. Soc.,* **66,** 909 (1944).

(6) W. Reeve and H. Adkins, *J. Am. Chem. Soc.,* **62,** 2874 (1940); E. C. Kleiderer and E. C. Kornfeld, *J. Org. Chem.,* **13,** 455 (1948).

(7) H. G. Fletcher, Jr., and C. S. Hudson, *J. Am. Chem. Soc.,* **70,** 310 (1948).

may be detected through the odor of the thiol liberated on acidification of a sample of the solution.[5]

Reductive desulfurizations with Raney nickel have been conducted in absolute alcohol, aqueous alcohol, dioxane, and ether, the first two solvents being most frequently used. Presumably any neutral solvent, containing neither halogen nor sulfur and not readily reduced, could be employed.

A variety of methods have been reported for the preparation of Raney nickel catalyst. The original procedure of Covert and Adkins,[8] based on the patent of Raney,[9] was modified by Mozingo[10] in 1941. In this modified procedure the nickel-aluminum alloy was finally heated in strong alkali on the steam-bath for eight to twelve hours. Mozingo and his coworkers[3] later found that this prolonged heating resulted in a considerable loss of hydrogen from the nickel. Analysis showed that when the final treatment with alkali was carried out at 80° for one hour the resultant catalyst contained 40–43 ml. of hydrogen per gram while even milder conditions, 50° for one hour, afforded a catalyst containing 118 ml. of hydrogen per gram. This catalyst, prepared under mild conditions (either at 50° or 80° for one hour), has been used in the majority of the reductive desulfurizations which have been reported. Other very active preparations have been made by Adkins and his coworkers[11] and recent work indicates that some of these are superior for desulfurization purposes.[12]

The ratio of catalyst to compound in desulfurization has been varied greatly and it seems likely that unnecessarily large excesses have frequently been employed. As will be seen later, the quantity of the catalyst used may affect the nature of the reaction; with certain methyl methylthio-benzylidenehexosides a relatively small quantity of catalyst suffices to cleave the carbon–sulfur bond while the use of a larger quantity results in the removal of the benzylidene residue as well. The proportion of nickel necessary for a given reaction depends to a certain extent upon the quality and age of the catalyst for, on standing, the catalyst becomes less active, both through loss of its hydrogen as gas and through air oxidation. Measurement of catalyst, seldom particularly accurate, is usually made

(8) L. W. Covert and H. Adkins, *J. Am. Chem. Soc.*, **54**, 4116 (1932).
(9) M. Raney, U. S. Pat. 1,628,190 (1927).
(10) R. Mozingo, *Org. Syntheses*, **21**, 15 (1941).
(11) A. A. Pavlic and H. Adkins, *J. Am. Chem. Soc.*, **68**, 1471 (1946); H. Adkins and A. A. Pavlic, *ibid.*, **69**, 3039 (1947); H. Adkins and H. R. Billica, *ibid.*, **70**, 695 (1948); H. R. Billica and H. Adkins, *Org. Syntheses*, **29**, 24 (1949).
(12) G. B. Spero, A. V. McIntosh, Jr., and R. H. Levin, *J. Am. Chem. Soc.*, **70**, 1907 (1948).

volumetrically or by converting a weighed quantity of the approximately 1:1 nickel-aluminum alloy to catalyst and using the latter entire.

No definite statement regarding the temperature or duration of reaction may be made. It seems probable, however, that, as is the case with most relatively new reactions, the conditions reported are in general much more rigorous than is necessary.

As will be seen through the specific examples discussed later, the yields in Raney nickel desulfurization are quite variable, rarely approaching quantitative; doubtless some of the reported yields would have been higher had the spent nickel, which tends to adsorb the products, been extracted more exhaustively.

Side reactions which may be encountered in the course of reductive desulfurization have been studied by Mozingo, Spencer and Folkers.[13] These authors showed that carbon-to-carbon double bonds, carbonyl groups, nitro groups, hydrazo and azoxy linkages are reduced. Benzyl alcohol gave toluene and an amine was alkylated by ethyl alcohol in the presence of Raney nickel at 78°; benzene rings, acids and esters are, however, stable. As mentioned above, benzylidene groups may be cleaved from sugar residues. In 70% aqueous solution an aged sample of Raney nickel was found[14] to cause the hydrolysis of an ethyl thioglycoside to the corresponding sulfur-free sugar.

That the desulfurizing action of Raney nickel is retained even in the absence of hydrogen has been demonstrated by Hauptmann, Wladislaw and Camargo[15] who found that Raney nickel which had been deprived of its hydrogen through heating in vacuum at 200° converted, for example, benzaldehyde dibenzyl thioacetal into a mixture of stilbene and bibenzyl. That this type of reaction proceeds by a free-radical mechanism appears

$$C_6H_5CH(SCH_2C_6H_5)_2 \rightarrow C_6H_5CH=CHC_6H_5 + C_6H_5CH_2CH_2C_6H_5$$

most likely; indeed, Kenner, Lythgoe and Todd[16] have remarked that ordinary reductive desulfurization may involve an attack by atomic hydrogen with the formation of free radicals. Regardless of mechanism it appears possible that some desulfurization unaccompanied by hydrogenation may take place in the course of ordinary reductive desulfurization.

(13) R. Mozingo, C. Spencer and K. Folkers, *J. Am. Chem. Soc.*, **66**, 1859 (1944).

(14) N. G. Brink, F. A. Kuehl, Jr., E. H. Flynn and K. Folkers, *J. Am. Chem. Soc.*, **70**, 2085 (1948).

(15) H. Hauptmann, B. Wladislaw and P. F. Camargo, *Experientia*, **4**, 385 (1948); H. Hauptmann and B. Wladislaw, *J. Am. Chem. Soc.*, **72**, 707, 710 (1950).

(16) G. W. Kenner, B. Lythgoe and A. R. Todd, *J. Chem. Soc.*, 957 (1948).

II. Hydrogenolysis of Various Classes of Sulfur-Containing Sugar Derivatives

1. *Tetraacetyl-1-thioaldose and Octaacetyldialdosyl Disulfide*

The first application of Raney nickel reductive desulfurization in the sugar field was made by Richtmyer, Carr and Hudson[17] in 1943. At that time the structure and configuration of the naturally occurring hexitol anhydride, polygalitol, had been but recently proved by Zervas and Papadimitriou[18] and additional evidence was desirable, particularly in view of the conflicting conclusions of earlier authors.[19] By synthesizing authentic tetraacetyl-1,5-anhydro-D-glucitol (II) through the reductive desulfurization of octaacetyl-β,β-di-D-glucopyranosyl disulfide (I) and of tetraacetyl-1-thio-β-D-glucose (III) and showing it to be identical with the tetraacetate of polygalitol, they obtained the desired additional proof for the configuration of the natural product as 1,5-anhydro-D-glucitol.

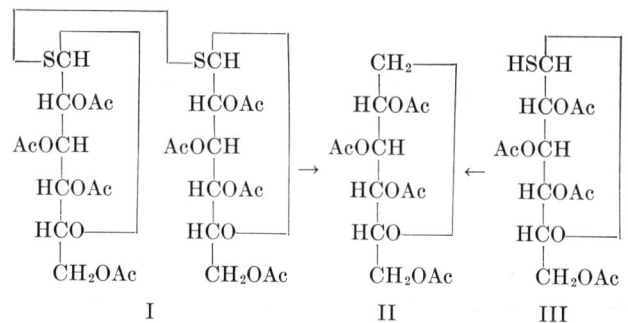

2. *Thioacetals (Mercaptals)*

Wolfrom and Karabinos[5] have shown that a carbonyl group may be reduced to the hydrocarbon stage *via* Raney nickel hydrogenolysis of its thioacetal, the net effect being the same as that accomplished by the older Clemmensen and Wolff–Kishner procedures. Since sugar diethyl thioacetals (mercaptals) have long been known and are relatively easily prepared, these offer synthetic access to the 1- and 2-desoxy derivatives of the sugar alcohols. Thus Wolfrom and Karabinos[5] desulfurized

(17) N. K. Richtmyer, C. J. Carr and C. S. Hudson, *J. Am. Chem. Soc.*, **65**, 1477 (1943).

(18) L. Zervas and Irene Papadimitriou, *Ber.*, **73**, 174 (1940).

(19) Y. Asahina and H. Takimoto, *Ber.*, **64**, 1803 (1931) [*cf.* R. C. Hockett and Maryalice Conley, *J. Am. Chem. Soc.*, **66**, 464 (1944)]; W. Freudenberg and J. T. Sheehan, *J. Am. Chem. Soc.*, **62**, 558 (1940).

D-galactose diethyl thioacetal pentaacetate (IV) to obtain 1-desoxy-D-galactitol (*syn.*, L-fucitol) pentaacetate (V) in 66% yield. Likewise the desulfurization of D-fructose diethyl thioacetal pentaacetate (VI) gave the pentaacetate of 2-desoxy-D-mannitol (*syn.*, 2-desoxy-D-glucitol) (VII) in 20% yield. 1-Desoxy-D-glucitol (*syn.*, L-gulomethylitol) pentaacetate was synthesized from D-glucose diethyl thioacetal pentaacetate and the general nature of the method demonstrated by desoxidation of a variety of simple aldehydes and ketones. That the course of

$$\begin{array}{ccc}
CH(SC_2H_5)_2 & & CH_3 \\
| & & | \\
HCOAc & & HCOAc \\
| & & | \\
AcOCH & & AcOCH \\
| & \rightarrow & | \\
AcOCH & & AcOCH \\
| & & | \\
HCOAc & & HCOAc \\
| & & | \\
CH_2OAc & & CH_2OAc \\
IV & & V
\end{array}$$

$$\begin{array}{ccc}
CH_2OAc & & CH_2OAc \\
| & & | \\
C(SC_2H_5)_2 & & CH_2 \\
| & & | \\
AcOCH & & AcOCH \\
| & \rightarrow & | \\
HCOAc & & HCOAc \\
| & & | \\
HCOAc & & HCOAc \\
| & & | \\
CH_2OAc & & CH_2OAc \\
VI & & VII
\end{array}$$

this synthesis may not always be the simple one indicated here was found by Wolfrom, Lew and Goepp[20] in 1946; in an attempt to synthesize 2-desoxy-D-allitol, *keto*-D-psicose pentaacetate (VIII) was treated with ethanethiol and zinc chloride in the same manner as that used for the synthesis of the D-fructose diethyl thioacetal pentaacetate (VI) mentioned above. The amorphous product obtained was hydrogenolysed and then deacetylated with barium hydroxide and there resulted a crystalline substance having the composition of a didesoxyhexitol. Since an aqueous solution of this tetrol was devoid of optical activity throughout the visible spectrum it was inferred to have the *meso* structure X which was named 1,6-(*erythro*-3,4)-hexanetetrol. If this structure is indeed the actual one, it is evident that during the treatment of *keto*-D-

(20) M. L. Wolfrom, B. W. Lew and R. M. Goepp, Jr., *J. Am. Chem. Soc.*, **68**, 1443 (1946).

psicose pentaacetate (VIII) with ethanethiol and zinc chloride the acetoxy group at position 5 was removed and a sulfur linkage introduced. Replacement of acetoxy groups by ethylthio groups, while rare, is not unknown[21] and structure IX may represent the intermediate involved here.

$$
\begin{array}{ccc}
\text{CH}_2\text{OAc} & \text{CH}_2\text{OAc} & \text{CH}_2\text{OH} \\
| & | & | \\
\text{C}=\text{O} & \text{C}(\text{SC}_2\text{H}_5)_2 & \text{CH}_2 \\
| & | & | \\
\text{HCOAc} & \text{HCOAc} & \text{HCOH} \\
| \rightarrow & | \rightarrow & | \\
\text{HCOAc} & \text{HCOAc} & \text{HCOH} \\
| & | & | \\
\text{HCOAc} & \text{CHSC}_2\text{H}_5 & \text{CH}_2 \\
| & | & | \\
\text{CH}_2\text{OAc} & \text{CH}_2\text{OAc} & \text{CH}_2\text{OH} \\
\text{VIII} & \text{IX} & \text{X}
\end{array}
$$

Extension of this procedure for the desulfurization of sugar mercaptals has been made by Bollenback and Underkofler.[22] These authors, who sought various desoxy sugar alcohols as substrates in a study of the specificity of the action of *Acetobacter suboxydans*, consider that the method is valuable only for preparations of the order of five grams, excessive adsorption of the product upon the nickel making larger scale work in their opinion impracticable. Using Raney nickel prepared according to the directions of Pavlic and Adkins[11] these authors reduced D- and L-arabinose diethyl thioacetal tetraacetates to 1-desoxy-D- and 1-desoxy-L-arabitol (*syn.*, D- and L-lyxomethylitol) tetraacetates, respectively. In like fashion they have desulfurized D-mannose diethyl thioacetal pentaacetate, hydrolysis of the intermediate 1-desoxy-D-mannitol pentaacetate (not isolated) giving 1-desoxy-D-mannitol (*syn.*, D-rhamnitol) in 36% over-all yield. Hydrogenolysis of the unacetylated D-mannose diethyl thioacetal[23] (XI) has been found to give 1-desoxy-D-

$$
\begin{array}{cccc}
\text{CH}(\text{SC}_2\text{H}_5)_2 & \text{CH}_3 & \text{CH}_2\text{OH} & \text{CHO} \\
| & | & | & | \\
\text{HOCH} & \text{HOCH} & \text{HOCH} & \text{HOCH} \\
| & | & | & | \\
\text{HOCH} & \text{HOCH} & \text{HOCH} & \text{HOCH} \\
| \rightarrow & | = & | \leftarrow & | \\
\text{HCOH} & \text{HCOH} & \text{HCOH} & \text{HCOH} \\
| & | & | & | \\
\text{HCOH} & \text{HCOH} & \text{HCOH} & \text{HCOH} \\
| & | & | & | \\
\text{CH}_2\text{OH} & \text{CH}_2\text{OH} & \text{CH}_3 & \text{CH}_3 \\
\text{XI} & \text{XIIa} & \text{XIIb} & \text{XIII}
\end{array}
$$

(21) M. L. Wolfrom and A. Thompson, *J. Am. Chem. Soc.*, **56**, 1804 (1934).
(22) G. N. Bollenback and L. A. Underkofler, *J. Am. Chem. Soc.*, **72**, 741 (1950).
(23) N. K. Richtmyer and C. S. Hudson, *J. Am. Chem. Soc.*, in press.

mannitol (XIIa) in 64% yield; that this substance is identical with the 6-desoxy-D-mannitol (XIIb) obtained by Haskins, Hann and Hudson[24] through the reduction of 6-desoxy-D-mannose (D-rhamnose) (XIII) demonstrates the equivalency of carbon atoms 1 and 6 (as well as the equivalence of carbon atoms 2 and 5, and 3 and 4) in D-mannitol.

Digressing from reductive desulfurization into stereochemistry, we may use this experimental proof of the equivalent symmetry of D-mannitol as a basis for an independent proof of the configurations of D-mannitol and D-arabitol. The reduction of D-arabinose yields the optically active pentitol, D-arabitol; application of the Sowden–Fischer synthesis to D-arabinose yields D-mannose[25] which upon reduction gives D-mannitol.

In the formula for D-mannitol (XIV), by Emil Fischer's second convention,[26] the hydroxyl on carbon atom 5 is placed on the right. Since D-arabitol (XV) is optically active, the hydroxyl on carbon atom 3 must then be on the left, for regardless of the configuration at carbon atom 4, arabitol would otherwise be an optically inactive *meso* form. Finally, by reason of the equivalent symmetry of D-mannitol, the

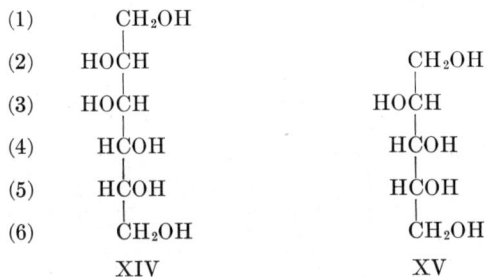

hydroxyl on carbon atom 2 of this hexitol must be on the left, that is, opposite to that on carbon atom 5, and the hydroxyl on carbon atom 4 must be on the right, opposite to that on carbon atom 3. Thus the complete configurations of D-mannitol and D-arabitol, as well as those of the corresponding sugars, may be derived.

Another proof of the configuration of D-mannitol and also of D-manno-L-*manno*-octitol (XVI), which is likewise dependent on the experimental proof of the equivalent symmetry of D-mannitol is the following. D-Mannose has been converted, by successive cyanohydrin syntheses, first to a mannoheptose and then to a mannooctose which on reduction yielded a mannooctitol whose octaacetyl derivative is optically inactive. (It was not possible to examine the octitol itself because of its very low solubility in water.)[27] The *meso* character of the octaacetate shows that the mannooctitol must possess a *meso* configuration, with a plane of symmetry between carbon atoms 4 and 5. To write its formula, the hydroxyl at carbon atom 7 is placed on the

(24) W. T. Haskins, R. M. Hann and C. S. Hudson, *J. Am. Chem. Soc.*, **68**, 628 (1946).

(25) J. C. Sowden and H. O. L. Fischer, *J. Am. Chem. Soc.*, **69**, 1963 (1947); J. C. Sowden, *J. Biol. Chem.*, **180**, 55 (1949).

(26) See C. S. Hudson, *Advances in Carbohydrate Chem.*, **3**, 8 (1948).

(27) E. Fischer and F. Passmore, *Ber.*, **23**, 2226 (1890); R. M. Hann, W. D. Maclay, A. E. Knauf and C. S. Hudson, *J. Am. Chem. Soc.*, **61**, 1268 (1939).

right in accordance with Fischer's second convention. By reason of the equivalent symmetry of D-mannitol, the hydroxyl on carbon atom 4 must be opposite to that on carbon atom 7, that is, on the left. Then, because of the *meso* character of the mannooctitol, the hydroxyl on carbon atom 5 must be on the same side as that on carbon atom 4, that is, on the left. Again, because of the equivalent symmetry of D-mannitol, the hydroxyl on carbon atom 6 must be opposite to that on carbon atom 5, that is, on

(1) CH$_2$OH
(2) HCOH
(3) HCOH
(4) HOCH
(5) HOCH
(6) HCOH
(7) HCOH
(8) CH$_2$OH

XVI

the right. Finally, because of the *meso* character of the mannooctitol, the hydroxyls on carbon atoms 2 and 3 must be on the same side as those on carbon atoms 6 and 7, that is, both on the right.

The utility of Raney nickel desulfurization in solving problems of configuration in the sugar series has recently been demonstrated[28] in the ω-desoxyglycitol series. The addition of hydrocyanic acid to 6-desoxy-L-galactose (L-fucose), first effected by Mayer and Tollens,[29] and later studied more extensively by Votoček,[30] yields two epimeric heptonic acids which were originally called "fucohexonic" acids. With the aid of the amide rule of rotation, Votoček assigned to his "α-fucohexonic" acid the configuration of 7-desoxy-L-gala-D-*manno*-heptonic acid; the rotations of the barium salts of the two acids, as well as the rotation of the benzylphenylhydrazone of the "α-fucohexose," also agreed with that conclusion. The correctness of Votoček's views has recently been established in an absolute manner by Zissis, Richtmyer and Hudson.[28] By the reductive desulfurization with Raney nickel of the diethyl thioacetal of D-manno-D-*gala*-heptose (XVII), whose configuration is well established, they prepared 1-desoxy-D-manno-D-*gala*-heptitol (XVIIIa), and showed it to be identical with the 7-desoxy-L-gala-D-*manno*-heptitol ("α-fucohexitol") (XVIIIb) described by Votoček as the reduction product of his "α-fucohexose" (XIX). Thus the configurations of the

(28) E. Zissis, N. K. Richtmyer and C. S. Hudson, *J. Am. Chem. Soc.*, in press.
(29) W. Mayer and B. Tollens, *Ber.*, **40**, 2434 (1907).
(30) E. Votoček, *Collection Czech. Chem. Communs.*, **6**, 528 (1934).

"fucohexoses" are proved, and the rules of rotation relating to amides, salts and benzylphenylhydrazones are given additional support.

```
CH(SC₂H₅)₂      CH₃           CH₂OH         CHO
   |             |              |             |
  HCOH          HCOH           HOCH          HOCH
   |             |              |             |
  HOCH      →   HOCH      =    HCOH     ←    HCOH
   |             |              |             |
  HCOH          HCOH           HCOH          HCOH
   |             |              |             |
  HCOH          HCOH           HOCH          HOCH
   |             |              |             |
  CH₂OH         CH₂OH          CH₃           CH₃
  XVII          XVIIIa         XVIIIb         XIX
```

In a similar fashion it should be a relatively simple task to relate the well-known D-gala-L-*manno*-heptose (XX) to "α-rhamnohexose" (XXII)[31] through the known 1-desoxy-D-gala-L-*manno*-heptitol (*syn.*, 7-desoxy-L-manno-L-*gala*-heptitol) (XXI).[31] In this particular case, however, there is already ample proof of the configuration of the latter sugar (XXII) because nitric acid has been found to oxidize the corresponding aldonic acid to mucic acid.[32]

```
   CHO           CH₃           CHO
    |             |             |
   HCOH          HCOH          HOCH
    |             |             |
   HCOH          HCOH          HCOH
    |             |             |
   HOCH          HOCH          HCOH
    |             |             |
   HOCH          HOCH          HOCH
    |             |             |
   HCOH          HCOH          HOCH
    |             |             |
   CH₂OH         CH₂OH         CH₃
    XX            XXI           XXII
```

The same scheme may be extended to the octoses, *e.g.*, in the correlation of D-manno-L-*manno*-octose (whose configuration is proved by its reduction to an optically inactive *meso* octitol)[27] and the "L-rhamno-

$$\text{HOH}_2\text{C} \cdot \overset{\text{H}}{\underset{\text{OH}}{\text{C}}} \cdot \overset{\text{H}}{\underset{\text{OH}}{\text{C}}} \cdot \overset{\text{OH}}{\underset{\text{H}}{\text{C}}} \cdot \overset{\text{OH}}{\underset{\text{H}}{\text{C}}} \cdot \overset{\text{H}}{\underset{\text{OH}}{\text{C}}} \cdot \overset{\text{H}}{\underset{\text{OH}}{\text{C}}} \cdot \text{CHO}$$

heptose" of Fischer and Piloty.[31] The latter sugar, according to Hud-

(31) E. Fischer and O. Piloty, *Ber.*, **23**, 3102 (1890).
(32) E. Fischer and R. S. Morrell, *Ber.*, **27**, 382 (1894).

son,[33] is probably 8-desoxy-L-manno-D-*manno*-octose, with the following configuration.

$$H_3C \cdot \underset{H}{\overset{OH}{C}} \cdot \underset{H}{\overset{OH}{C}} \cdot \underset{OH}{\overset{H}{C}} \cdot \underset{OH}{\overset{H}{C}} \cdot \underset{H}{\overset{OH}{C}} \cdot \underset{H}{\overset{OH}{C}} \cdot CHO$$

Wolfrom, Lemieux and Olin[34] have recently employed reductive desulfurization as one step in a direct experimental correlation of the configurations of L-(*levo*)-glyceraldehyde with natural (*dextro*)-alanine. Pentaacetyl-D-glucosamine diethyl thioacetal (XXIII), whose configuration is known through the work of Haworth, Lake and Peat[35] on the configuration of D-glucosamine, was desulfurized to give pentaacetyl-2-amino-1,2-didesoxy-D-glucitol (XXIV). Partial deacetylation of the product to XXV, followed by oxidation, first with lead tetraacetate and then with bromine water, gave an N-acetyl-2-amino-propionic acid

```
 CH(SC2H5)2       CH3            CH3
     |             |              |
   HCNHAc        HCNHAc         HCNHAc
     |             |              |              CH3              COOH
   AcOCH         AcOCH          HOCH              |                |
     |      →      |       →     |       →     HCNHAc  =  AcHNCH
   HCOAc         HCOAc          HCOH             |                |
     |             |              |             COOH             CH3
   HCOAc         HCOAc          HCOH
     |             |              |
   CH2OAc        CH2OAc         CH2OH

    XXIII         XXIV           XXV              XXVI
```

(XXVI) identical with the N-acetyl derivative of natural (*dextro*)-alanine. The configuration of carbon 2 in D-glucosamine is related to D-(*dextro*)-glyceraldehyde; in this synthetic process, however, carbon 1 is reduced to the hydrocarbon stage and carbon 3 oxidized to a carboxyl group. Since the preferred orientation of the resulting compound is with the carboxyl group written uppermost, the net effect of these changes is the same as would be produced by interchanging the groups attached to carbon 2; as a result carbon 2 in natural (*dextro*)-alanine is configurationally related to L-(*levo*)-glyceraldehyde.

A somewhat similar configurational correlation between (*levo*)-glyceraldehyde and (*dextro*)-lactic acid has been made by Wolfrom, Lemieux, Olin and Weisblat.[36] Reductive desulfurization of tetraacetyl-2-methyl-D-glucose diethyl thioacetal (XXVII) and hydrolysis of the product gave 2-methyl-1-desoxy-D-glucitol (XXVIII); oxidation

(33) C. S. Hudson, *Advances in Carbohydrate Chem.*, **1**, 28 (1945).
(34) M. L. Wolfrom, R. U. Lemieux and S. M. Olin, *J. Am. Chem. Soc.*, **71**, 2870 (1949).
(35) W. N. Haworth, W. H. G. Lake and S. Peat, *J. Chem. Soc.*, 271 (1939).
(36) M. L. Wolfrom, R. U. Lemieux, S. M. Olin and D. I. Weisblat, *J. Am. Chem. Soc.*, **71**, 4057 (1949).

of this latter substance, first with periodic acid and then with aqueous bromine, produced O-methyl-L-lactic acid (XXIX), isolated in a pure state through chromatography of its *p*-phenylphenacyl ester. Methylation of the *p*-phenylphenacyl ester of authentic L-(*dextro*)-lactic acid, obtained through a fermentation procedure, gave the *p*-phenylphenacyl ester of O-methyl-L-lactic acid, identical with the specimen produced indirectly from 2-methyl-1-desoxy-D-glucitol (XXVIII) and afforded direct chemical proof of the relationship between (*levo*)-glyceraldehyde and (*dextro*)-lactic acid. As in the preceding example from the same laboratory, convention again requires that the carboxyl group be written

$$\begin{array}{cccc}
CH(SC_2H_5)_2 & CH_3 & & \\
| & | & & \\
HCOCH_3 & HCOCH_3 & CH_3 & COOH \\
| & | & | & | \\
AcOCH & HOCH & HCOCH_3 = H_3COCH \\
| & | & | & | \\
HCOAc & HCOH & COOH & CH_3 \\
| & | & & \\
HCOAc & HCOH & & \\
| & | & & \\
CH_2OAc & CH_2OH & & \\
XXVII & XXVIII & & XXIX
\end{array}$$

uppermost in the formula for lactic acid and carbon 2, originally written in the glucose derivative XXVII as a D-carbon, becomes in the lactic acid derivative an L-carbon.

As more progress is made on the synthesis of other higher-carbon ω-desoxy sugars, especially those from sugars less readily available than 6-desoxy-L-galactose (L-fucose) and 6-desoxy-L-mannose (L-rhamnose), additional interesting and useful correlations will undoubtedly be made through Raney nickel hydrogenolysis.

Reductive desulfurization with Raney nickel of the mercaptolysis products of streptomycin has supplied one of the keys to the elucidation of the structure and configuration of that antibiotic. Since the chemistry of streptomycin has recently been reviewed[37] the reactions discussed here will be considered solely as examples of hydrogenolysis and their bearing upon the problem of the structure and configuration of streptomycin will be ignored.

Kuehl, Flynn, Brink and Folkers[38] in 1946 showed that mercaptolysis of streptomycin leads to the formation of an ethyl thiostreptobiosaminide diethyl thioacetal hydrochloride (XXX). The same substance

(37) R. U. Lemieux and M. L. Wolfrom, *Advances in Carbohydrate Chem.*, **3**, 337 (1948).

(38) F. A. Kuehl, Jr., E. H. Flynn, N. G. Brink and K. Folkers, *J. Am. Chem. Soc.*, **68**, 2096 (1946).

could be obtained through mercaptolysis of methyl streptobiosaminide dimethyl acetal hydrochloride which had, in turn, been obtained by methanolysis of streptomycin. Acetylation of XXX gave a tetraacetate (XXXI), m. p. 81–82°, $[\alpha]^{25}_D$ −178°, which, upon treatment with Raney nickel, was converted to didesoxydihydrostreptobiosamine tetraacetate (XXXII), all three ethylthio groups being hydrogenolyzed. Wolfrom

and his coworkers[39] performed a similar series of changes but found not only the form of XXXI which had previously been reported but also an isomeric form melting at 111–111.5° and showing $[\alpha]^{20}_D$ −29°. Since this second form gave on hydrogenolysis the same didesoxydihydrostreptobiosamine tetraacetate (XXXII) as the first isomer it is probable that the two are anomers. Brink, Kuehl, Flynn and Folkers[14] have reported more recently that, while hydrogenolysis of XXXI to XXXII proceeded normally with fresh Raney nickel, one hydrogenolysis conducted with an aged sample of catalyst gave also desoxydihydrostreptobiosamine tetraacetate (XXXIII). Thus it would appear that 1-thioglycosides may be hydrolyzed to the sulfur-free sugar under the usual conditions of Raney nickel hydrogenolysis.

(39) I. R. Hooper, L. H. Klemm, W. J. Polglase and M. L. Wolfrom, *J. Am. Chem. Soc.*, **68**, 2120 (1946); **69**, 1052 (1947).

3. 1-Thioglycosides

In the preceding section the reductive desulfurization of some substances which are both thioglycosides and thioacetals has been mentioned; here the hydrogenolysis of the simple 1-thioglycosides will be discussed.

Numerous alkyl and aryl 1-thioglycosides have been hydrogenolyzed with Raney nickel; the ease with which many 1-thioglycosides may be prepared[40] makes these substances attractive intermediates in the synthesis of anhydrides of the sugar alcohols.[41] Thus phenyl 1-thio-β-D-xylopyranoside triacetate (XXXIV), which may be prepared from D-xylose tetraacetate in 80% yield,[40] gave on hydrogenolysis a 69% yield of triacetyl-1,5-anhydroxylitol (XXXV),[42] a derivative of one of the two possible *meso* 1,5-anhydropentitols. In an analogous fashion both

```
    C₆H₅SCH          CH₂——
       |              |
     HCOAc          HCOAc
       |              |
     AcOCH   →      AcOCH
       |              |
     HCOAc          HCOAc
       |              |
     CH₂O——         CH₂O——

     XXXIV           XXXV
```

the 2′-naphthyl and the phenyl 1-thio-α-D-arabinopyranoside triacetates (XXXVI)[43] gave, after deacetylation of the intermediary triacetate, 1,5-anhydro-D-arabitol (XXXVII).

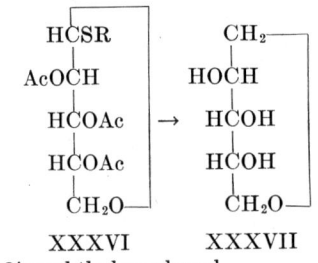

R = 2′-naphthyl or phenyl

(40) C. B. Purves, *J. Am. Chem. Soc.*, **51**, 3619 (1929); W. T. Haskins, R. M. Hann and C. S. Hudson, *ibid.*, **69**, 1668 (1947).

(41) For a review of the anhydrides of the sugar alcohols, see L. F. Wiggins, *Advances in Carbohydrate Chem.*, **5**, 191 (1950).

(42) H. G. Fletcher, Jr., and C. S. Hudson, *J. Am. Chem. Soc.*, **69**, 921 (1947).

(43) H. G. Fletcher, Jr., and C. S. Hudson, *J. Am. Chem. Soc.*, **69**, 1672 (1947).

The second of the two possible *meso* 1,5-anhydropentitols, 1,5-anhydroribitol (XXXIX), has been prepared by Jeanloz, Fletcher and Hudson[44] through reductive desulfurization of the amorphous 2′-naphthyl 1-thio-β-D-ribopyranoside tribenzoate (XXXVIII).

Styracitol, or 1,5-anhydro-D-mannitol (XLI), discovered in *Styrax Obassia* Sieb. et Zucc. by Asahina[45] and subsequently synthesized by

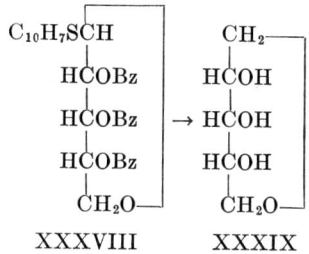

XXXVIII   XXXIX

Zervas[46] through the catalytic reduction of 2,3,4,6-tetraacetyl-2-hydroxy-D-glucal, has recently been prepared by the reductive desulfurization of ethyl tetraacetyl-1-thio-β-D-mannopyranoside (XL),[47] a substance which has been obtained through the mercaptolysis of mannosidostreptomycin[48] and may be prepared through the prolonged action of ethyl mercaptan and hydrochloric acid on D-mannose.

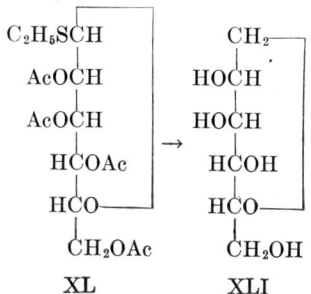

XL   XLI

The sugar alcohol anhydride polygalitol, 1,5-anhydro-D-glucitol, may be prepared from the root of *Polygala Senega* L.[49] The process is, however, a time-consuming one and chemical synthesis of the substance by Raney

(44) R. Jeanloz, H. G. Fletcher, Jr., and C. S. Hudson, *J. Am. Chem. Soc.*, **70**, 4052 (1948).
(45) Y. Asahina, *Arch. Pharm.*, **245**, 325 (1907); **247**, 157 (1909).
(46) L. Zervas, *Ber.*, **63**, 1689 (1930).
(47) J. Fried and Doris E. Walz, *J. Am. Chem. Soc.*, **71**, 140 (1949).
(48) J. Fried and H. E. Stavely, *J. Am. Chem. Soc.*, **69**, 1549 (1947); H. E. Stavely and J. Fried, *ibid.*, **71**, 135 (1949).
(49) N. K. Richtmyer and C. S. Hudson, *J. Am. Chem. Soc.*, **65**, 64 (1943).

nickel reductive desulfurization, even on a relatively large laboratory scale, has proved wholly practicable.[50] Thus a total of 165 g. of p-tolyl 1-thio-β-D-glucoside tetraacetate was desulfurized in 15-g. batches, the product deacetylated, and 48.5 g. or 81% of the theory of 1,5-anhydro-D-glucitol obtained; in a similar fashion phenyl 1-thio-β-D-glucoside tetraacetate gave a 78% yield of polygalitol.

This general type of process has proved useful in the synthesis of sugar alcohol anhydrides of authentic structure and configuration for comparison with similar anhydrides of uncertain structure or configuration which had previously been reported in the literature. Thus Maurer and Plötner[51] reduced both heptaacetyl-2-hydroxycellobial (XLII) and heptaacetyl-2-hydroxygentiobial (XLV) to crystalline substituted anhydrohexitols in yields, respectively, of 62% and 53%. The free anhydrides were termed "1,4-glucosidostyracitol" and "1,6-glucosidostyracitol," respectively, that is, as derivatives of an anhydride (styracitol) which is now known to have the D-mannitol configuration.

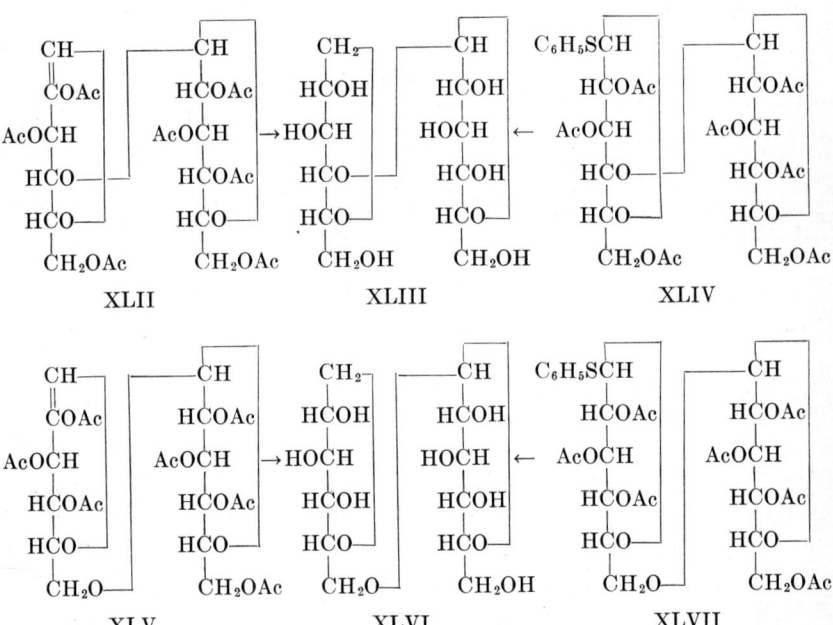

Apparently these names were chosen solely upon the assumption that the course of the reduction of the acetates of the two substituted 2-hydroxyglucals had been similar to that of 2,3,4,6-tetraacetyl-2-hydroxy-D-glucal

(50) N. K. Richtmyer and C. S. Hudson, unpublished work.
(51) K. Maurer and K. Plötner, *Ber.*, **64**, 281 (1931).

which, when reduced in the presence of palladium, appears to give predominantly 1,5-anhydro-D-mannitol.[17,46,52] Definite proof of the configuration of the second carbon atom in these two 1,5-anhydro-β-D-glucopyranosyl-D-hexitols was lacking. The synthesis of authentic 1,5-anhydro-4-(β-D-glucopyranosyl)-D-glucitol (XLIII) through the reductive desulfurization of phenyl 1-thio-β-cellobioside heptaacetate (XLIV)[7] and of authentic 1,5-anhydro-6-(β-D-glucopyranosyl)-D-glucitol (XLVI) through the similar desulfurization of phenyl 1-thio-β-gentiobioside heptaacetate (XLVII) was a comparatively simple matter; the authentic anhydrides proved to have the same physical constants as those reported by Maurer and Plötner for their products which were, therefore, not styracitol derivatives as these authors indicated, but derivatives of polygalitol or 1,5-anhydro-D-glucitol.

A similar situation arose in the galactose series: Freudenberg and Rogers[53] reduced tetraacetyl-2-hydroxy-D-galactal (XLVIII) to obtain in 70% yield the tetraacetate of an anhydrohexitol whose structure as either 1,5-anhydro-D-galactitol (L) or 1,5-anhydro-D-talitol (XLIX) remained uncertain. Reductive desulfurization of 2'-naphthyl 1-thio-β-D-galactopyranoside tetraacetate (LI)[7] gave 1,5-anhydro-D-galactitol tetraacetate. The physical constants of this latter substance as well as

CH—	CH$_2$—	CH$_2$—	C$_{10}$H$_7$SCH—
‖			
COAc	HOCH	HCOH	HCOAc
AcOCH	HOCH	HOCH	AcOCH
	→		←
AcOCH	HOCH	HOCH	AcOCH
HCO—	HCO—	HCO—	HCO—
CH$_2$OAc	CH$_2$OH	CH$_2$OH	CH$_2$OAc
XLVIII	XLIX	L	LI

of the free crystalline 1,5-anhydro-D-galactitol (L) were found to differ markedly from those which Freudenberg and Rogers[53] had reported for the comparable substance which they had obtained by reduction of tetraacetyl-2-hydroxy-D-galactal (XLVIII) and led to the conclusion that the product of these authors was 1,5-anhydro-D-talitol (XLIX). An unequivocal synthesis of 1,5-anhydro-D-talitol[54] has since confirmed this conclusion.

(52) R. C. Hockett and Maryalice Conley, *J. Am. Chem. Soc.*, **66**, 464 (1944).
(53) W. Freudenberg and E. F. Rogers, *J. Am. Chem. Soc.*, **59**, 1602 (1937).
(54) D. A. Rosenfeld, N. K. Richtmyer and C. S. Hudson, *J. Am. Chem. Soc.*, **70**, 2201 (1948).

Two new anhydrides, 1,5-anhydro-4-(β-D-galactopyranosyl)-D-glucitol (1,5-anhydrolactitol) (LIII) and 1,5-anhydro-4-(α-D-glucopyranosyl)-D-glucitol (1,5-anhydromaltitol) (LV) have recently been synthesized[55] through the reductive desulfurization of 2′-naphthyl 1-thio-β-lactopyranoside (LII) and phenyl 1-thio-β-maltopyranoside heptaacetate (LIV), respectively.

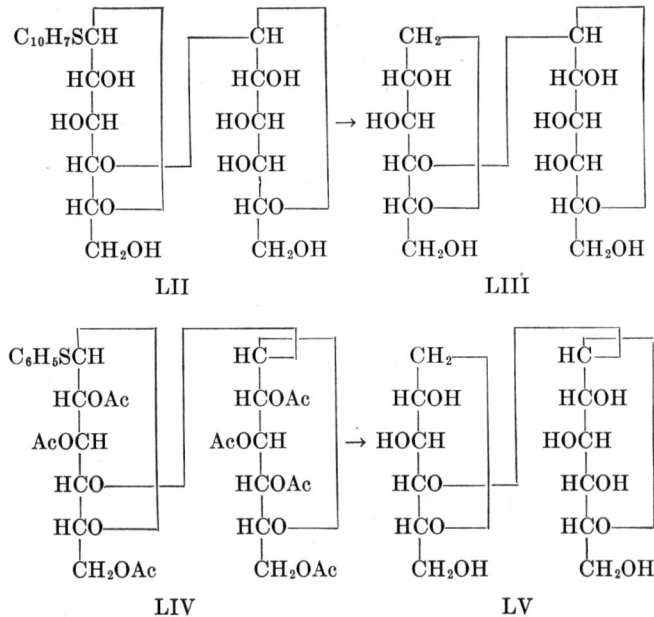

The facility with which 1,5-anhydroglycitols of known structure and configuration may be synthesized through reductive desulfurization has resulted in the accumulation of sufficient new data on compounds of this class so that some generalizations have already been made concerning the relationship of configuration to rotatory power in this interesting series.[56]

The recent discovery that acetylated glycopyranosyl bromides may be reduced in high yields with lithium aluminum hydride to the corresponding 1,5-anhydroglycitols[56a] represents an improvement over the reductive desulfurization process and will doubtless make this class of compound even more readily available in the future.

(55) H. G. Fletcher, Jr., Leonore H. Koehler and C. S. Hudson, *J. Am. Chem. Soc.*, **71**, 3679 (1949).

(56) H. G. Fletcher, Jr., and C. S. Hudson, *J. Am. Chem. Soc.*, **71**, 3682 (1949).

(56a) R. K. Ness, H. G. Fletcher, Jr., and C. S. Hudson, *J. Am. Chem. Soc.*, in press.

Methanolysis of dihydrostreptomycin with subsequent acetylation of the products was found by Brink, Kuehl, Flynn and Folkers[57] to give both anomeric forms of methyl pentaacetyldihydrostreptobiosaminide (LVI). Mercaptolysis of either of these anomers, followed by reacetylation, converted them into ethyl pentaacetylthiodihydrostreptobiosaminide (LVII) which was not isolated but immediately desulfurized to the acetylated anhydride LVIII, pentaacetyldihydrodesoxystreptobiosamine. Starting with a crystalline form of LVII, Lemieux, Polglase, Dewalt and Wolfrom[58] performed a similar desulfurization.

$$
\begin{array}{ccc}
\text{LVI} & \rightarrow & \text{LVII} \\
\end{array}
$$

LVIII

4. Thioethers

A novel approach to the synthesis of desoxy sugars was made by Jeanloz, Prins and Reichstein[59] in 1945. In general terms, a 1,2-epoxide was cleaved with sodium thiomethoxide to give a β-hydroxy thioether which, in turn, was hydrogenolyzed with Raney nickel. It is apparent

(57) N. G. Brink, F. A. Kuehl, Jr., E. H. Flynn and K. Folkers, *J. Am. Chem. Soc.*, **68**, 2557 (1946).
(58) R. U. Lemieux, W. J. Polglase, C. W. Dewalt and M. L. Wolfrom, *J. Am. Chem. Soc.*, **68**, 2747 (1946).
(59) R. Jeanloz, D. A. Prins and T. Reichstein, *Experientia*, **1**, 336 (1945); *Helv. Chim. Acta*, **29**, 371 (1946).

that with an unsymmetrical ethylene oxide two possible products may be expected, depending upon which of the two carbon–oxygen bonds is broken by the sodium thiomethoxide; an isomer instead of the desired desoxy sugar is therefore sometimes obtained. As an example of the method, the synthesis of 2-desoxyallose derivatives by Jeanloz, Prins

$$\begin{array}{c} HC \\ \diagdown \\ O + NaSCH_3 \rightarrow \\ \diagup \\ HC \end{array} \quad \begin{array}{c} CH_3SCH \\ | \\ HCOH \end{array} \rightarrow \begin{array}{c} CH_2 \\ | \\ HCOH \end{array}$$

and Reichstein[59] may be cited. These authors treated methyl 2,3-anhydro-4,6-benzylidene-α-D-alloside (LIX) with sodium thiomethoxide to obtain methyl 2-methylthio-4,6-benzylidene-α-D-altroside (LX) which was desulfurized with concomitant loss of its benzylidene residue to give methyl 2-desoxy-α-D-alloside (LXI). In like fashion methyl 2-methylthio-3-methyl-4,6-benzylidene-α-D-altroside (LXII) was hydrogenolyzed to methyl 2-desoxy-3-methyl-α-D-alloside (LXIII).

| LIX | LX, R = H
LXII, R = CH₃ | LXI, R = H
LXIII, R = CH₃ |

Shortly after the above work appeared, Bolliger and Prins[60] reported a similar synthesis leading to a 3- rather than a 2-desoxy derivative. Methyl 2,3-anhydro-4,6-benzylidene-α-D-mannoside (LXIV), upon treatment with sodium thiomethoxide, was converted into methyl 3-methylthio-4,6-benzylidene-α-D-altroside (LXV) which, after desulfurization and rebenzylidenation, gave methyl 3-desoxy-4,6-benzylidene-α-D-mannoside (LXVI). Methylation of LXV with methyl iodide and silver oxide followed by desulfurization gave methyl 2-methyl-3-desoxy-4,6-benzylidene-α-D-mannoside in poor yield.

Maehly and Reichstein[61] have found that treatment of methyl 2-methylthio-4,6-benzylidene-α-D-idoside (LXVII) with Raney nickel

(60) H. R. Bolliger and D. A. Prins, *Helv. Chim. Acta*, **29**, 1061 (1946).
(61) A. C. Maehly and T. Reichstein, *Helv. Chim. Acta*, **30**, 496 (1947).

prepared from a four-fold quantity of Raney alloy (approximately 50% nickel) caused only desulfurization and gave methyl 2-desoxy-4,6-benzylidene-α-D-guloside (LXVIII) in 57% yield. When the Raney nickel prepared from a thirty-fold quantity of alloy was used the benzylidene group also was removed and the product was methyl 2-desoxy-α-D-guloside (LXIX). Other idose derivatives were hydrogenated by

```
        HCOCH₃                    HCOCH₃                    HCOCH₃
         |                         |                         |
    ┌──CH                         HOCH                      HOCH
  O │   |                          |                         |
    └──CH              →         HCSCH₃          →          CH₂
         |                         |                         |
        HCO──┐                   HCO──┐                    HCO──┐
         |   │                    |   │                     |   │
        HCO──┼─CH·C₆H₅           HCO──┼─CH·C₆H₅            HCO──┼─CH·C₆H₅
         |   │                    |   │                     |   │
        H₂CO─┘                   H₂CO─┘                    H₂CO─┘
          LXIV                     LXV                       LXVI
```

```
        HCOCH₃                    HCOCH₃                    HCOCH₃
         |                         |                         |
       CH₃SCH                     CH₂                       CH₂
         |                         |                         |
        HCOH            →         HCOH             →        HCOH
         |                         |                         |
      ┌──OCH                    ┌──OCH                      HOCH
      │  |                      │  |                         |
 C₆H₅·HC HCO──┐            C₆H₅·HC HCO──┐                   HCO
      │  |   │                 │  |   │                      |
      └──OCH₂┘                 └──OCH₂┘                     CH₂OH
          LXVII                    LXVIII                    LXIX
```

```
        CH₃OCH                    CH₃OCH
         |                         |
        HOCH                      HOCH
         |                         |
        HCSCH₃          →         CH₂
         |                         |
      ┌──OCH                    ┌──OCH
      │  |                      │  |
 C₆H₅·HC HCO──┐            C₆H₅·HC HCO──┐
      │  |   │                 │  |   │
      └──OCH₂┘                 └──OCH₂┘
          LXX                      LXXI
```

Gut, Prins and Reichstein:[62] methyl 3-methylthio-4,6-benzylidene-β-D-idoside (LXX) with a four-fold quantity of Raney nickel (prepared from an eight-fold quantity of alloy) gave methyl 3-desoxy-4,6-benzylidene-β-D-idoside (LXXI) in 92% yield; use of a large excess of nickel resulted in simultaneous debenzylidenation.

(62) M. Gut, D. A. Prins and T. Reichstein, *Helv. Chim. Acta*, **30**, 743 (1947).

In an attempt to synthesize 2-desoxy-L-ribose, Mukherjee and Todd[63] treated methyl 2,3-anhydro-β-L-ribopyranoside (LXXII) with sodium thiomethoxide and then reduced the amorphous product. The sulfur-free sirup thus obtained was inert to the action of periodate and is therefore methyl 3-desoxy-β-L-riboside (*syn.*, methyl 3-desoxy-β-L-xyloside) (LXXIV) rather than the desired 2-desoxy derivative. The methylthio intermediate (LXXIII) was consequently a 3-methylthio-L-xylose derivative rather than a 2-methylthio-L-arabinose compound. The presence of a trace of the latter, however, is not wholly excluded since the sirupy desoxypentoside (LXXIV) gave a feeble green coloration in the Keller–Kiliani test, a reaction generally accepted as being specific for 2-desoxy-sugars.

```
      HCOCH₃         HCOCH₃         HCOCH₃
       |              |              |
     ┌─CH            HOCH           HOCH
   O │              |              |
     └─CH    →      HCSCH₃   →     CH₂
       |              |              |
      HOCH           HOCH           HOCH
       |              |              |
      CH₂O──┘         CH₂O──┘        CH₂O──┘
      LXXII          LXXIII         LXXIV
```

Prins[64] has recently discovered that lithium aluminum hydride hydrogenates 1,2-epoxides smoothly in high yield; this single-step reaction would appear to outmode the Raney nickel desulfurization process for this purpose.

5. *Thiol Esters*

The hydrogenolysis of the thiol esters of carboxylic acids offers a delicate method for reducing an acid to the corresponding aldehyde or alcohol; application of this method to the sugar series has been made by Wolfrom and Karabinos[65] who reduced D-ribonyl chloride tetraacetate (LXXV) to *aldehydo*-D-ribose tetraacetate (LXXVII) in 22% yield *via* ethyl thiol-D-ribonate tetraacetate (LXXVI). The desulfurization of the thiol ester (LXXVI) was carried out with a somewhat aged sample of Raney nickel[66] which doubtless had less activity than the

(63) S. Mukherjee and A. R. Todd, *J. Chem. Soc.*, 969 (1947).

(64) D. A. Prins, *J. Am. Chem. Soc.*, **70**, 3955 (1948); H. Hauenstein and T. Reichstein, *Helv. Chim. Acta*, **32**, 22 (1949).

(65) M. L. Wolfrom and J. V. Karabinos, *J. Am. Chem. Soc.*, **68**, 724, 1455 (1946).

(66) Personal communication from Dr. J. V. Karabinos. Subsequent work on the reduction of steroid acids by Spero, McIntosh and Levin (ref. 12) has shown that Raney nickel partially deactivated by acetone is particularly suitable for the desulfurization of thiol esters to aldehydes.

freshly prepared material; almost simultaneous work by Jeger, Norymberski, Szpilfogel and Prelog[67] showed that fresh Raney nickel readily carries the thiol ester over to the alcohol stage. Methyl thiol-D-gluconate

$\overset{O}{\underset{}{\overset{\|}{C}}}$—Cl	$\overset{O}{\underset{}{\overset{\|}{C}}}$—SC$_2H_5$	$\overset{O}{\underset{}{\overset{\|}{C}}}$—H
HCOAc	HCOAc	HCOAc
HCOAc	HSC$_2$H$_5$ → HCOAc	→ HCOAc
HCOAc	HCOAc	HCOAc
CH$_2$OAc	CH$_2$OAc	CH$_2$OAc
LXXV	LXXVI	LXXVII

pentaacetate (LXXVIII) gave amorphous 2,3,4,5,6-pentaacetyl-D-glucitol (LXXIX) which, on acetylation, was converted to D-glucitol hexaacetate (LXXX) in an over-all yield of 72%.

$\overset{O}{\underset{}{\overset{\|}{C}}}$—SCH$_3$	CH$_2$OH	CH$_2$OAc
HCOAc	HCOAc	HCOAc
AcOCH	AcOCH	AcOCH
HCOAc →	HCOAc →	HCOAc
HCOAc	HCOAc	HCOAc
CH$_2$OAc	CH$_2$OAc	CH$_2$OAc
LXXVIII	LXXIX	LXXX

6. S-Substituted Xanthates

The S-glycosyl xanthates are readily prepared through the condensation of acetohalogeno sugars with potassium ethyl xanthate[68] and offer a convenient route to the synthesis of sugar alcohol anhydrides. Thus, crystalline ethyl tetraacetyl-β-D-glucopyranosyl xanthate (LXXXI) gave 2,3,4,6-tetraacetyl-1,5-anhydro-D-glucitol (*syn.*, tetraacetylpolygalitol) (II)[69] in 81% yield. In order to obtain 1,5-anhydro-D-arabitol,[43] crystalline 2,3,4-triacetyl-β-D-arabinosyl bromide was converted to the sirupy xanthate which was then desulfurized and hydrolyzed, the yield of 1,5-anhydro-D-arabitol, based on the original bromide, being 24%.

(67) O. Jeger, J. Norymberski, S. Szpilfogel and V. Prelog, *Helv. Chim. Acta*, **29**, 684 (1946).
(68) W. Schneider, R. Gille and K. Eisfeld, *Ber.*, **61**, 1244 (1928).
(69) H. G. Fletcher, Jr., *J. Am. Chem. Soc.*, **69**, 706 (1947).

```
        S
        ‖
   C₂H₅OCSCH              CH₂─┐
        |                     |
      HCOAc               HCOAc
        |                     |
      AcOCH      →         AcOCH
        |                     |
      HCOAc               HCOAc
        |                     |
      HCO─┘               HCO─┘
        |                     |
      CH₂OAc              CH₂OAc
      LXXXI                  II
```

7. S-Substituted Isothioureas

A critical study of the preparation of the two anomeric 6-desoxy-D-glucose tetraacetates (LXXXIV) from D-glucose has been made by Hardegger and Montavon.[70] The final step, namely, the reduction of the two 6-iodo-6-desoxy-D-glucose tetraacetates (LXXXII) in high yield, proved to be particularly difficult and a variety of methods were studied. Highest yields were obtained when LXXXII was allowed to react with thiourea to form the S-substituted isothiuronium iodide (LXXXIII) which was then desulfurized with Raney nickel. In actual practice the 6-iodo derivative was refluxed briefly in amyl alcohol with thiourea and

```
   CHOAc              CHOAc              CHOAc
     |                  |                  |
   HCOAc              HCOAc              HCOAc
     |                  |                  |
   AcOCH      →       AcOCH      →       AcOCH
     |                  |                  |
   HCOAc              HCOAc              HCOAc
     |                  |                  |
   HCO─┐              HCO─┐              HCO─┐
     |                  |                  |
   CH₂I              CH₂SCNH₃I           CH₃
                        ‖
                        NH
   LXXXII             LXXXIII            LXXXIV
```

then desulfurized directly; under these conditions the yields of 6-desoxy compounds from the α- and β-isomers were 55 and 70% respectively.

8. Thiocyanates

Reaction of 1-tosyl-2,4:3,5-dimethylene-D,L-xylitol (LXXXV) with sodium thiocyanate affords the 1-thiocyano-1-desoxy-2,4:3,5-dimethyl-

(70) E. Hardegger and R. M. Montavon, *Helv. Chim. Acta*, **29**, 1199 (1946).

ene-D,L-xylitol (LXXXVI); reductive desulfurization[71] converts this latter into the same 1-desoxy derivative (LXXXVII) which had previously been prepared through the reduction of the corresponding iodo derivative.[72]

$$\begin{array}{ccc}
CH_2OSO_2C_7H_7 & CH_2SCN & CH_3 \\
| & | & | \\
HCO\text{---} & HCO\text{---} & HCO\text{---} \\
\text{---}OCH\ CH_2 & \rightarrow \quad \text{---}OCH\ CH_2 \rightarrow & \text{---}OCH\ CH_2 \\
H_2C\ HCO\text{---} & H_2C\ HCO\text{---} & H_2C\ HCO\text{---} \\
\text{---}OCH_2 & \text{---}OCH_2 & \text{---}OCH_2 \\
\text{LXXXV} & \text{LXXXVI} & \text{LXXXVII}
\end{array}$$

(71) R. M. Hann, N. K. Richtmyer, H. W. Diehl and C. S. Hudson, *J. Am. Chem. Soc.*, **72**, 561 (1950).

(72) R. M. Hann, A. T. Ness and C. S. Hudson, *J. Am. Chem. Soc.*, **66**, 670 (1944)

III. Tabular Survey of Reductive Desulfurizations in the Sugar Series

Conversion of	Ratio of Ni to compound	Solvent	Temp.	Time	Yield %	Reference
Octaacetyl-β,β-di-D-glucopyranosyl disulfide to tetraacetyl-1,5-anhydro-D-glucitol	3	Abs. EtOH	Room	10 days	27	17
Tetraacetyl-1-thio-β-D-glucose to tetraacetyl-1,5-anhydro-D-glucitol	2.4	"	"	"	—	17
D-Galactose diethyl thioacetal pentaacetate to 1-desoxy-D-galactitol pentaacetate	15	70% EtOH	Reflux	5 hrs.	66	5
Same		"In usual manner"			48	22
D-Galactose diethyl thioacetal to 1-desoxy-D-galactitol	15	70% EtOH	Reflux	1 hr.	24	5
D-Glucose diethyl thioacetal pentaacetate to 1-desoxy-D-glucitol pentaacetate	15	70% EtOH	Reflux	5 hrs.	60	5
Same		"In usual manner"			40	22
D-Fructose diethyl thioacetal pentaacetate to 2-desoxy-D-glucitol pentaacetate	15	70% EtOH	Reflux	3 hrs.	20	5
D-Arabinose diethyl thioacetal tetraacetate to 1-desoxy-D-arabitol tetraacetate	10	—	—	—	26	22
L-Arabinose diethyl thioacetal tetraacetate to 1-desoxy-L-arabitol tetraacetate	10	—	—	—	50	22
D-Mannose diethyl thioacetal pentaacetate to 1-desoxy-D-mannitol	—	—	—	—	36	22
D-Mannose diethyl thioacetal to 1-desoxy-D-mannitol	9	70% EtOH	Reflux	1 hr.	64	23
D-Manno-D-gala-heptose diethyl thioacetal to 1-desoxy-D-manno-D-gala-heptitol	9	70% EtOH	Reflux	1 hr.	57	28
Pentaacetyl-D-glucosamine diethyl thioacetal to pentaacetyl-2-amino-1,2-didesoxy-D-glucitol	9	70% EtOH	Reflux	5 hrs.	47	34
Tetraacetyl-2-methyl-D-glucose diethyl thioacetal to tetraacetyl-2-methyl-1-desoxy-D-glucitol	5	70% EtOH	Reflux	5 hrs.	92	36
Ethyl tetraacetylthiostreptobiosaminide diethyl thioacetal to tetraacetyldidesoxydihydrostreptobiosamine	14	70% EtOH	Reflux	4 hrs.	60	38
Same	—	—	—	—	78; 70	39
Same	28	70% EtOH	Reflux	1.5 hrs.	33	14

Compound		Solvent	Temp	Time		Ref
Ethyl tetraacetylthiostreptobiosaminide diethyl thioacetal to tetraacetyl-desoxystreptobiose	28	70% EtOH	Reflux	1.5 hrs.	15	14
Phenyl 1-thio-β-D-xylopyranoside triacetate to 1,5-anhydroxylitol triacetate	6	Abs. EtOH	Reflux	2 hrs.	69	42
Phenyl 1-thio-α-D-arabinopyranoside triacetate to 1,5-anhydro-D-arabitol	4	Abs. EtOH	Reflux	0.5 hr.	27	43
2'-Naphthyl 1-thio-α-D-arabinopyranoside triacetate to 1,5-anhydro-D-arabitol	8	Abs. EtOH	Reflux	2 hrs.	44	43
Phenyl 1-thio-β-cellobioside heptaacetate to 1,5-anhydro-4-(β-D-glucopyranosyl)-D-glucitol heptaacetate	10	Abs. EtOH	Reflux	1 hr.	69	7
Phenyl 1-thio-β-gentiobioside heptaacetate to 1,5-anhydro-6-(β-D-glucopyranosyl)-D-glucitol heptaacetate	10	Abs. EtOH	Reflux	1 hr.	72	7
2'-Naphthyl 1-thio-β-D-galactopyranoside tetraacetate to 1,5-anhydro-D-galactitol tetraacetate	10	Abs. EtOH	Reflux	1 hr.	73	7
2'-Naphthyl 1-thio-β-lactopyranoside to 1,5-anhydro-4-(β-D-galactopyranosyl)-D-glucitol	12.5	70% EtOH	Reflux	1 hr.	63	55
Phenyl 1-thio-β-maltopyranoside heptaacetate to 1,5-anhydro-4-(α-D-glucopyranosyl)-D-glucitol heptaacetate	4	Abs. EtOH	Reflux	1 hr.	78	55
2'-Naphthyl 1-thio-β-D-ribopyranoside tribenzoate to 1,5-anhydroribitol tribenzoate	10	Abs. EtOH	Reflux	2 hrs.	59	44
Ethyl tetraacetyl-1-thio-β-D-mannopyranoside to 1,5-anhydro-D-mannitol	8	Abs. EtOH	Reflux	4 hrs.	56	47
Ethyl pentaacetylthiodihydrostreptobiosaminide to pentaacetyldihydrodesoxystreptobiosamine	—	—	—	—	—	57, 58
p-Tolyl 1-thio-β-D-glucopyranoside tetraacetate to 1,5-anhydro-D-glucitol	6	95% EtOH	Reflux	1 hr.	81	50
Phenyl 1-thio-β-D-glucopyranoside tetraacetate to 1,5-anhydro-D-glucitol	6	95% EtOH	Reflux	1 hr.	78	50
Methyl 2-methylthio-4,6-benzylidene-α-D-altroside to methyl 2-desoxy-α-D-alloside	15	Aq. EtOH	Reflux	2 hrs.	85	59
Methyl 2-methylthio-3-methyl-4,6-benzylidene-α-D-altroside to methyl 2-desoxy-3-methyl-α-D-alloside	17	Aq. EtOH	Reflux	2 hrs.	95	59
Methyl 3-methylthio-4,6-benzylidene-α-D-altroside to methyl 3-desoxy-4,6-benzylidene-α-D-mannoside	28	80% EtOH	Reflux	2 hrs.	47	60
Methyl 2-methylthio-3-methylthio-4,6-benzylidene-α-D-altroside to methyl 2-methyl-3-desoxy-4,6-benzylidene-α-D-mannoside	7	—	—	—	7.7	60

III. TABULAR SURVEY OF REDUCTIVE DESULFURIZATIONS IN THE SUGAR SERIES (Continued)

Conversion of	Ratio of Ni to compound	Solvent	Temp.	Time	Yield %	Reference
Methyl 2-methylthio-4,6-benzylidene-α-D-idoside to methyl 2-desoxy-4,6-benzylidene-α-D-guloside	2	Aq. EtOH	Reflux	2 hrs.	57	61
Methyl 2-methylthio-4,6-benzylidene-α-D-idoside to methyl 2-desoxy-α-D-guloside	15	Aq. EtOH	Reflux	2 hrs.	96	61
Methyl 3-methylthio-4,6-benzylidene-β-D-idoside to methyl 3-desoxy-4,6-benzylidene-β-D-idoside	2	Aq. EtOH	Reflux	2 hrs.	92	62
Methyl 2-methyl-3-methylthio-4,6-benzylidene-β-D-idoside to methyl 2-methyl-3-desoxy-4,6-benzylidene-β-D-idoside	1.5	Et$_2$O	Reflux	2.5 hrs.	27	62
Methyl 3-methylthio-β-L-xyloside to methyl 3-desoxy-β-L-xyloside	25	Aq. EtOH	Reflux	2 hrs.	79	63
Ethyl thiol-D-ribonate tetraacetate to *aldehydo*-D-ribose tetraacetate	15	80% EtOH	Reflux	6 hrs.	22	65
Methyl thiol-D-gluconate pentaacetate to D-glucitol hexaacetate	10	EtOH	Room	1 hr.	60	67
Ethyl tetraacetyl-β-D-glucopyranosyl xanthate to 1,5-anhydro-D-glucitol tetraacetate	20	Abs. EtOH	Reflux	6 hrs.	81	69
1-Thiocyano-1-desoxy-2,4:3,5-dimethylene-D,L-xylitol to 1-desoxy-2,4:3,5-dimethylene-D,L-xylitol	18	70% EtOH	Reflux	2 hrs.	70	71

ENZYMATIC SYNTHESIS OF SUCROSE AND OTHER DISACCHARIDES

By W. Z. Hassid and M. Doudoroff

Division of Plant Biochemistry, College of Agriculture and Department of Bacteriology, University of California, Berkeley, California

Contents

I. Introduction.	29
II. Structure of Sucrose.	31
III. Synthesis of Sucrose through the Mechanism of Phosphorolysis.	33
IV. Specificity of Sucrose Phosphorylase.	35
V. Synthetic Non-reducing Disaccharides.	36
1. Synthetic Sucrose.	36
a. Preparation of the Enzymatic Extract from *Pseudomonas saccharophila*.	36
b. Enzymatic Synthesis and Isolation of Synthetic Sucrose.	37
c. Properties of Synthetic Sucrose.	38
2. α-D-Glucopyranosyl-β-D-xyloketofuranoside.	39
3. D-Glucosyl-L-araboketoside.	41
4. α-D-Glucopyranosyl-α-L-sorbofuranoside.	41
VI. Synthetic Reducing Disaccharide.	43
3-[α-D-Glucopyranosyl]-L-arabinopyranose.	43
VII. Formation of Sucrose and Other Disaccharides through Exchange of Glycosidic Linkages.	46

I. Introduction

Sucrose is commercially the most important and quantitatively the most abundant sugar in the vegetable kingdom. It is readily formed as the result of the photosynthetic activity of practically all higher plants, constituting a reserve material used in cell respiration. The energy that is liberated in the latter process is utilized for metabolic activities of the plant such as in the synthesis of numerous organic compounds, including especially amino acids, proteins and fats.

The formation of sucrose in plants has occupied the attention of biochemists for the last several decades. The prevailing theory is that the monosaccharides D-glucose and D-fructose, which are present in the plant as products of photosynthesis, are combined by an enzyme or enzymatic system, forming sucrose. Since the existence of such a mechanism is of fundamental importance, investigators at various times have attempted to demonstrate this biochemical process *in vitro* through chemical or enzymatic synthesis.

In 1928, Pictet and Vogel[1] claimed to have accomplished the synthesis of sucrose by coupling tetraacetyl-γ-D-fructose with tetraacetyl-D-glucose in the presence of a dehydrating agent. However, the following year, Zemplén and Gerecs[2] were not successful in achieving this synthesis by Pictet and Vogel's method. Irvine, Oldham and Skinner[3,4] pointed out the difficulties inherent in the synthesis of sucrose by chemical means. Since D-glucose and D-fructose may each exist in the α- and β- form, the following four different disaccharide configurations are possible when the two hexoses are combined: (1) α-D-glucose-β-D-fructose, (2) α-D-glucose-α-D-fructose, (3) β-D-glucose-α-D-fructose, (4) β-D-glucose-β-D-fructose. Thus far it has not been possible to devise conditions which would lead to the production of sucrose by this condensation. In attempting to condense tetraacetyl-γ-D-fructose with tetraacetyl-D-glucose, Irvine and coworkers were unable to obtain sucrose octaacetate, but did produce a disaccharide derivative with a different glycosidic linkage, the so-called isosucrose octaacetate. The acetylated derivative had a different melting point and specific rotation from those of sucrose octaacetate. It may be pointed out that had such chemical synthesis of sucrose been successful, it would not necessarily have contributed to our knowledge of the manner in which this compound is synthesized in living organisms, for it is now well established that synthesis of organic compounds in living cells often takes place through mechanisms entirely different from those that may be successful *in vitro*.

Until recently, claims pertaining to the enzymatic synthesis of sucrose also could not be substantiated. Invertase, known to be widely distributed in plants, was considered for a long time as having the double role of a hydrolytic and a synthetic enzyme, and the formation of sucrose was assumed to occur as a result of reversed inversion. Thus, Oparin and Kurssanov[5] claimed to have synthesized sucrose from invert sugar under the influence of invertase and phosphatase in the presence of inorganic phosphate. However, careful repetition of this work by Lebedew and Dikanowa[6] and other investigators failed to substantiate this claim.

(1) A. Pictet and H. Vogel, *Helv. Chim. Acta*, **11**, 436 (1928); *Ber.*, **62**, 1418 (1929).
(2) G. Zemplén and A. Gerecs, *Ber.*, **62**, 984 (1929).
(3) J. C. Irvine, J. W. H. Oldham and A. F. Skinner, *J. Am. Chem. Soc.*, **51**, 1279 (1929).
(4) J. C. Irvine and J. W. H. Oldham, *J. Am. Chem. Soc.*, **51**, 3609 (1929). See also F. Klages and R. Niemann, *Ann.*, **529**, 185 (1937).
(5) A. Oparin and A. Kurssanov, *Biochem. Z.*, **239**, 1 (1931).
(6) A. Lebedew and A. Dikanowa, *Z. physiol. Chem.*, **231**, 271 (1935).

Within the last few years some progress has been made toward a better understanding of the mechanisms involved in the biochemical synthesis of sucrose and other disaccharides. It is the purpose of the authors to discuss the present state of knowledge of this subject.

II. Structure of Sucrose

A completely satisfactory constitutional formula for sucrose was long a matter of controversy but there is now general agreement that this problem has been solved. Two major problems regarding its structure occupied the attention of carbohydrate chemists: (1) the ring configuration of the constituent D-glucose and D-fructose residues in the disaccharide molecule; (2) the type of glycosidic linkages (α or β) involved in each monosaccharide residue. As early as 1916 Haworth and Law[7] showed that octamethyl sucrose did not undergo inversion of rotation on hydrolysis, the specific rotation in water falling from $[\alpha]_D$ $+66.5°$ to $+56.5°$, whereas an equimolar mixture of methylated normal forms of D-glucose and D-fructose would give a rotation of $[\alpha]_D$ $-18°$. The products of hydrolysis were eventually identified many years later as 2,3,4,6-tetramethyl-D-glucose and 1,3,4,6-tetramethyl-D-fructose by Haworth, Hirst and their collaborators; it was thus shown that sucrose contains D-glucopyranose and D-fructofuranose.[8]

This conclusion depended upon the assumption that the ring configuration of the monosaccharide constituents of sucrose does not change during methylation. Recent data obtained through periodate oxidation[9] has proved this assumption to be correct. It was found that when sucrose is oxidized with sodium metaperiodate, three moles of periodate were consumed and one mole of formic acid was produced in the reaction. These results are consistent with a disaccharide consisting of a glucopyranose and a fructofuranose unit. Any other ring configuration for either the aldose or the ketose component would give different results.

Since both hexose units in sucrose are joined through the carbonyl groups, the disaccharide is at one and the same time a glucoside and a fructoside. Regarding the glucose component there is an abundance of evidence indicating that this constituent exists in sucrose as an α-D-glucoside. This conclusion is drawn from polarimetric data as well as from the behavior of sucrose towards the enzymes, α-glucosidase (maltase) and sucrose phosphorylase.

(7) W. N. Haworth and J. Law, *J. Chem. Soc.*, **109**, 1314 (1916).
(8) W. N. Haworth, "The Constitution of Sugars," Edward Arnold and Co., London (1929).
(9) P. Fleury and J. Courtois, *Bull. soc. chim.*, **10**, 245 (1943); *Compt. rend.*, **214**, 366 (1942).

Hudson[10] showed that the mutarotation of fructose in water at 30° is eleven times faster than that of glucose. He therefore assumed that in a sucrose solution which is undergoing very rapid inversion with invertase at that temperature, practically all of the fructose has reached equilibrium and exists as a mixture of its α and β forms, while the glucose is being liberated in only one form which, however, slowly passes to its α, β equilibrium mixture. The drop in rotation between the apparent and real curves of inversion by invertase must therefore be due almost entirely to the mutarotation of glucose. Hudson thus showed that the D-glucose liberated from sucrose by invertase had a specific rotation between $[\alpha]_D$ $+100°$ and $+125°$ and is thus most likely the α-form.

Data obtained by other workers through different experimental approaches also point to the same conclusion. Weidenhagen[11] demonstrated that at its optimum pH 7, maltase (α-glucosidase), which had

I
Sucrose
α-D-Glucopyranosyl-β-D-fructofuranoside

been freed from invertase, is capable of hydrolyzing sucrose and maltose at about the same rates. More recent evidence that the D-glucose constituent of sucrose is an α-glucoside was furnished by Hassid, Doudoroff and Barker's work[12] on the enzymatic synthesis of sucrose by the enzyme sucrose phosphorylase from *Pseudomonas saccharophila*. Since sucrose is formed as a result of a "dephosphorolytic" condensation involving α-D-glucose-1-phosphate, it is believed that the synthesis is analogous to the known enzymatic syntheses of glycogen and starch from α-D-glucose-1-phosphate; since both of these polysaccharides are known to be of α-D-glucosidic type it is believed that the α-type of the linkage in α-D-glucose-1-phosphate is not altered when the phosphate bond is exchanged for a glycosidic bond to D-fructose or other monosaccharides.

There is also strong evidence for the type of glycosidic linkage existing in the fructose constituent of the sucrose molecule. Schlubach and Rauchalles[13] showed that invertase hydrolyzes sucrose and methyl β-D-fructofuranoside at approximately the same rate, indicating that

(10) C. S. Hudson, *J. Am. Chem. Soc.*, **30**, 1564 (1908).
(11) R. Weidenhagen, *Naturwissenschaften*, **16**, 654 (1928).
(12) W. Z. Hassid, M. Doudoroff and H. A. Barker, *J. Am. Chem. Soc.*, **66**, 1416 (1944).
(13) H. H. Schlubach and G. Rauchalles, *Ber.*, **58**, 1842 (1925).

invertase is a β-D-fructofuranosidase and that therefore fructose probably exists in sucrose as the β-D-fructofuranose form. Purves and Hudson's[14] experiments on the analysis of fructoside mixtures by means of invertase added support to this view. Very strong evidence has come from the fact that they prepared crystalline methyl α-D-fructofuranoside, and found it to be unattacked by invertase. They consider sucrose as "almost certainly α-D-glucopyranosyl-β-D-fructofuranoside." On the basis of this evidence the structural and configurational formula for sucrose is designated as in formula I.[15]

III. Synthesis of Sucrose through the Mechanism of Phosphorolysis

The fact that leaves of sucrose-producing plants such as beets and peas were found to contain considerable amounts of hexose phosphates suggested that sugar phosphates might be involved in the mechanism of sucrose formation and that phosphorylation was an essential step in this process.[16-18] That phosphorylation is essential for sucrose synthesis is also indicated by the fact that iodoacetate, which inhibits phosphorylation, also inhibits sucrose synthesis in plants.[19] While this hypothesis has not as yet been supported by the isolation of an enzyme capable of sucrose synthesis from a plant source, Doudoroff, Kaplan and Hassid[20] prepared such an enzyme from a bacterial source. It was found that dried cells from *Pseudomonas saccharophila* contain an enzyme system which catalyzes the breakdown of sucrose in the presence of inorganic phosphate with the formation of D-glucose-1-phosphate (Cori ester) and D-fructose. It was also demonstrated that this reaction is reversible, i.e., the same sucrose phosphorylase will catalyze the synthesis of sucrose from D-glucose-1-phosphate and D-fructose.

The dried bacteria also contain invertase and this hydrolytic enzyme competes with the sucrose phosphorylase for sucrose. However, it is possible to eliminate most of the invertase from the bacterial preparations by several precipitations with ammonium sulfate. Using a partially purified sucrose phosphorylase preparation and a mixture of

(14) C. B. Purves and C. S. Hudson, *J. Am. Chem. Soc.*, **59**, 1170 (1937).
(15) The foregoing is a very condensed summary of some of the principal advances through which the formula for sucrose has been established. A thorough review of the subject has been published recently by I. Levi and C. B. Purves in Vol. IV of the *Advances in Carbohydrate Chemistry*.
(16) J. Burkard and C. Neuberg, *Biochem. Z.*, **270**, 229 (1934).
(17) W. Z. Hassid, *Plant Physiol.*, **13**, 641 (1938).
(18) A. Kursanov and N. Kriukova, *Biokhimiya*, **4**, 229 (1939).
(19) N. Kriukova, *Biokhimiya*, **5**, 574 (1940).
(20) M. Doudoroff, N. Kaplan and W. Z. Hassid, *J. Biol. Chem.*, **148**, 67 (1943).

D-glucose-1-phosphate and D-fructose, Hassid, Doudoroff and Barker[12] succeeded in crystallizing a non-reducing disaccharide which was indistinguishable from natural sucrose. It was thus that the first laboratory synthesis of sucrose was achieved.

The formation of D-glucose-1-phosphate and D-fructose from sucrose and inorganic phosphate, similar to the phosphorolysis of starch[21] and glycogen,[22] may be considered to occur as the result of phosphorolytic cleavage of D-glucose from the sucrose molecule, the sucrose being disrupted without water entering into the reaction. The reverse reaction, the formation of sucrose from D-glucose-1-phosphate and D-fructose is the result of "de-phosphorolytic" condensation of the two monosaccharides. The reversible phosphorolysis of sucrose can be represented as shown in formula II.

II
Phosphorolysis of Sucrose

A similar enzyme system that would combine D-glucose-1-phosphate and D-fructose to form sucrose and inorganic phosphate has not yet been isolated from the tissues of higher plants. However, biochemical studies on various species of plants support the view that the synthesis of sucrose may involve chemical reactions in which phosphate esters of D-glucose, D-fructose or both hexoses serve as substrates, although the mechanism is probably not identical with that of the bacterial enzyme system. It is also significant that the experimental evidence now available shows that for the synthesis of sucrose from D-glucose and D-fructose in the plant, aerobic metabolism is indispensable.[23,24] Possibly aerobic oxidations are essential to the phosphorylation of one of the substrates involved in the synthesis of sucrose. The problem is complicated by the observation that various substrates other than D-glucose and D-fructose may result in sucrose formation by plant tissues. For example, in experiments with barley shoots, the infiltration of D-galactose as well as of various other carbohydrates was shown to lead to sucrose formation.[24]

(21) C. S. Hanes, *Proc. Roy. Soc. (London)*, **B129**, 174 (1940).
(22) C. F. Cori, *Endocrinology*, **26**, 285 (1940).
(23) C. E. Hart, *Hawaiian Planters' Record*, **47**, 113 (1943).
(24) R. M. McCready and W. Z. Hassid, *Plant Physiol.*, **16**, 599 (1941).

IV. Specificity of Sucrose Phosphorylase

Similar to potato and muscle phosphorylase, the enzyme sucrose phosphorylase is specific with regard to the α-D-glucose portion of its substrates. Potato and muscle phosphorylase do not form polysaccharide from α-maltose-1-phosphate, α-D-galactose-1-phosphate, α-D-mannose-1-phosphate, α-D-xylose-1-phosphate or α-L-glucose-1-phosphate.[25,26] Similarly, these hexose phosphates cannot be substituted for α-D-glucose-1-phosphate when use is made of bacterial sucrose phosphorylase, which has the ability of combining α-D-glucose-1-phosphate with D-fructose to form sucrose. However, this enzyme is not specific with regard to substituents for the second sucrose component, D-fructose. The sucrose phosphorylase will combine the same ester with a number of other ketose monosaccharides, such as D-xyloketose, L-araboketose and L-sorbose to form the corresponding non-reducing disaccharides, D-glucosyl-D-xyloketoside,[27] D-glucosyl-L-araboketoside[28] and D-glucosyl-L-sorboside.[29] Since these disaccharides are non-reducing and their ketose constituents exist in the furanose form, they can be considered as analogs of sucrose.

That the monosaccharide substrate for the enzyme need not necessarily be a ketose is demonstrated by the fact that the same enzyme will catalyze a reaction between D-glucose-1-phosphate and L-arabinose. The product is a reducing disaccharide[30] having no obvious structural relation to sucrose or to any of the previously prepared sucrose analogs. It appears therefore that the same enzyme can catalyze at least two diverse reactions; one involving the carbonyl group on the second carbon atom of a ketose and another involving the secondary alcohol group of the third carbon atom of an aldose.

That the same enzyme,[28] sucrose phosphorylase, is involved in the reaction of L-arabinose is indicated by the following observation. When L-arabinose is added to a mixture containing the enzyme, D-glucose-1-phosphate and D-fructose, of which the last is present in insufficient concentration to give the maximum rate of sucrose formation, an increase

---

(25) W. A. Meagher and W. Z. Hassid, *J. Am. Chem. Soc.*, **68**, 2135 (1946).

(26) A. L. Potter, J. C. Sowden, W. Z. Hassid and M. Doudoroff, *J. Am. Chem. Soc.*, **70**, 1751 (1948).

(27) W. Z. Hassid, M. Doudoroff, H. A. Barker and W. H. Dore, *J. Am. Chem. Soc.*, **68**, 1465 (1946).

(28) M. Doudoroff, W. Z. Hassid and H. A. Barker, *J. Biol. Chem.*, **168**, 733 (1947).

(29) W. Z. Hassid, M. Doudoroff, H. A. Barker and W. H. Dore, *J. Am. Chem. Soc.*, **67**, 1394 (1945).

(30) W. Z. Hassid, M. Doudoroff, A. L. Potter and H. A. Barker, *J. Am. Chem. Soc.*, **70**, 306 (1948).

in the rate of utilization of the D-glucose-1-phosphate is observed. However, when the same amount of L-arabinose is added to a similar mixture, with the exception that the D-fructose concentration is so selected as to give a maximum rate of sucrose formation, the total rate of D-glucose-1-phosphate utilization is decreased. Since it is known that the rate of reducing disaccharide formation from D-glucose-1-phosphate and L-arabinose is considerably slower than that of sucrose formation from the same ester and D-fructose, a decrease in the rate of D-glucose-1-phosphate utilization would be expected if the two sugars were competing for the enzyme.

In addition there is other evidence pointing to the fact that the same enzyme is involved in reactions with both D-fructose and L-arabinose. First, the relative rates of reaction with D-fructose and L-arabinose, respectively, remain constant after partial inactivation of the enzyme by heat. Second, the enzyme catalyzing both reactions is produced to a marked extent when sucrose is used as substrate for the growth of the organisms, but not when D-glucose or L-arabinose is used; sucrose phosphorylase is an "adaptive" enzyme. Third, on fractionation of the enzyme preparation with various concentrations of ammonium sulfate, the relative activities of the fractions are the same for both sugars. These observations indicate not only that the same enzyme is involved in both reactions but also that no additional enzyme is required for the formation of D-glucosyl-L-arabinose.

V. Synthetic Non-reducing Disaccharides

1. Synthetic Sucrose

a. Preparation of the Enzymatic Extract from Pseudomonas saccharophila.—Cultures of the bacterium *Pseudomonas saccharophila* Doudoroff[31] were grown in a liquid medium containing $M/30$ KH_2PO_4–Na_2HPO_4 (Sorensen phosphate buffer at pH 6.64), 0.1% NH_4Cl, 0.05% $MgSO_4$, 0.005% $FeCl_3$, 0.001% $CaCl_2$ and 0.3% sucrose at 29°, with constant agitation to provide ample aeration.[18] Under such conditions, almost 50% of the carbon content of sucrose is converted into cell material, the remainder being oxidized to carbon dioxide. Traces of reducing sugar appear in the medium, as well as occasionally small amounts of pyruvic acid, which disappears in the later stages of development. The cells were harvested by centrifugation, washed twice with distilled water and dried at room temperature *in vacuo* over P_2O_5.

Preparations were made by grinding a sufficient amount of dry bacteria with phosphate buffer at pH 6.64 to result, after various further

(31) M. Doudoroff; *Enzymologia*, **9**, 59 (1940).

additions, in a final suspension containing 2% of dry cells by weight. This extract contains a considerable amount of invertase and a small amount of phosphatase. Subsequent treatment of the extract with ammonium sulfate[32] gave rise to preparations which contained from a third to a half of the original phosphorylase but very little invertase or phosphatase.

After two fractionations between 0.33 and 0.6 saturated ammonium sulfate, several precipitations at 0.6 saturation were made throughout a period of two or three days, during which the preparation was stored in the refrigerator. Any denatured, insoluble proteins were removed by centrifugation, and a solution of the desired activity was prepared by dilution with buffer or distilled water.

The chief loss in phosphorylase activity appeared to occur in the early stages of the fractionation; the precipitated material remained active for several days in 0.6 saturated ammonium sulfate at 4°. The invertase was apparently inactivated by prolonged contact with 0.6 saturated ammonium sulfate, since it could not be recovered from the supernatant liquid after precipitation of the active material.

b. Enzymatic Synthesis and Isolation of Synthetic Sucrose.—The phosphorylase preparation that was made from 6 g. of dry bacteria and freed of invertase was added to a solution containing 15 g. of the potassium salt of α-D-glucose-1-phosphate and 15 g. of D-fructose, and the mixture was adjusted to pH 6.8 with acetic acid. Barium acetate was then added to make the final concentration 0.133 M and the total volume was made up to 300 ml. The pH was then adjusted and maintained at 6.85 during incubation. The reaction mixture was kept at 37° with frequent stirring for 12 hours, after which it was covered with toluene and incubated an additional 12 hours at 29° with constant agitation. A quantitative estimation of the synthetic sucrose carried out with the aid of yeast invertase indicated that approximately 3 g. of this sugar had been formed in this reaction.

After removal of the toluene, the mixture was pasteurized at 80° for 5 minutes, cooled, adjusted to pH 7.8 and 2.5 volumes of 95% alcohol added. The mixture was allowed to remain at 4° for 3 hours and the precipitate, containing most of the inorganic and esterified phosphate, was removed by filtration and the alcohol was removed by distillation *in vacuo* at about 30°. The solution was then made up to 600 ml. and passed through columns of Amberlite IR–100 and Amberlite IR–4. This treatment removed all the electrolytes, including the remaining traces of D-glucose-1-phosphate. After washing the columns with water, the volume had increased to about three liters. The solution was concen-

(32) M. Doudoroff, *J. Biol. Chem.*, **151**, 351 (1943).

trated *in vacuo* at 20° to 300 ml. and the residual fructose removed by fermentation. Approximately 38 g. (wet weight) of washed cells of *Torula monosa* (a yeast capable of fermenting monosaccharides only) was added and the fermentation was allowed to proceed at 37° until no reducing sugar was left. A small quantity of bicarbonate had to be added during the fermentation to counteract the accumulation of acid. The yeast cells were centrifuged off and the supernatant liquid again passed through small absorption columns of Amberlite IR–100 and IR–4.

The liquid was then evaporated *in vacuo* to a small volume and two volumes of 95% ethanol were added. A small amount of flocculent precipitate, probably consisting of yeast polysaccharides, was removed by centrifugation, and the solution was concentrated to a sirup in a vacuum oven at 40°. This colorless sirup, when treated with hot absolute alcohol and stirred, set to a crystalline mass. The crystals were filtered, washed with alcohol and ether and dried *in vacuo* at 70°. The yield was 2.8 g.

c. Properties of Synthetic Sucrose.—The synthetic sucrose[12] obtained from α-D-glucose-1-phosphate and D-fructose through the action of the partially purified sucrose phosphorylase from *Pseudomonas saccharophila* possesses properties identical with those of natural sucrose. Its empirical formula obtained by elementary analysis is $C_{12}H_{22}O_{11}$. The compound does not reduce Fehling solution before hydrolysis. After acid or enzymatic hydrolysis the phenylosazone obtained from the inversion mixture is D-glucosazone; the hydrolyzate also gives a positive Seliwanoff reaction for a ketose. The reducing value and the yield of D-glucose and D-fructose are theoretical for invert sugar. The specific rotation $[\alpha]_D$ +66.5° is changed by inversion to −20°. The synthetic product and natural sucrose give an identical X-ray diffraction pattern and the synthetic sucrose is hydrolyzed with acid at the same rate as the natural sugar. The optical properties of the crystals are also the same as for sucrose. An octaacetate derivative was prepared from the synthetic sugar; it had a specific rotation $[\alpha]_D$ + 60° (in chloroform) and a melting point of 69–70°. These constants agree with those of the octaacetate of natural sucrose.

Natural D-glucose-1-phosphate occurs as the α-form. In the "dephosphorolytic" condensation, resulting in the formation of starch or glycogen, the phosphoric acid linked as the α-form in the ester is exchanged for the same type of glycosidic linkage with another monosaccharide unit. It is therefore reasonable to infer that when sucrose is formed through condensation of D-glucose-1-phosphate and D-fructose, the α-configuration of the former is not altered and the D-glucose in the sucrose molecule remains as the α-type. This inference is identical with the conclusions of other workers[10,11,33] regarding the configuration of the glucose in the sucrose molecule.

The enzymatic synthesis of sucrose also throws light on the formation of the furanose form of fructose in the sucrose molecule. The fact that sucrose is directly formed from D-glucose-1-phosphate and D-fructose supports Isbell and Pigman's[34] and Gottschalk's[35] evidence that the latter monosaccharide occurs in solution in an equilibrium mixture of furanose and pyranose forms. This makes it unnecessary to postulate a special mechanism of stabilization of a five membered (furanose) ring before the formation of compound sugars containing the D-fructose molecule.[36]

2. α-D-*Glucopyranosyl-β-D-xyloketofuranoside*

The D-glucosyl-D-xyloketoside, like synthetic sucrose, was formed through the action of the *P. saccharophila* enzyme on α-D-glucose-1-phosphate and D-xyloketose.[27,37] The empirical formula of the disaccharide, obtained from its elementary analysis is $C_{11}H_{20}O_{10}$. The compound does not reduce Fehling solution or alkaline ferricyanide. It is practically unaffected by invertase, but is easily hydrolyzed with acid. When the disaccharide is hydrolyzed with acid and the D-glucose fermented out, a phenylosazone is obtained which is identical with that of D-xylose. The specific rotation of the disaccharide is $[\alpha]_D$ $+43°$ and its melting point is 156–157°. Its rate of hydrolysis with acid is approximately 30% greater than that of sucrose. The acetylated derivative has a specific rotation $[\alpha]_D$ in chloroform of $+22°$ and a melting point of 180–181°.

Since the disaccharide is non-reducing, the D-glucose and D-xyloketose units are obviously linked through the carbonyl groups. Inasmuch as the carbonyl group in xyloketose occurs on the second carbon atom, the largest possible semiacetal ring for the ketose component is the 2,5-furanose ring and the possibility of a pyranose ring is definitely excluded. The furanose structure of the xyloketose was definitely confirmed experi-

(33) H. S. Isbell and W. W. Pigman, *J. Research Natl. Bur. Standards*, **20**, 792 (1938).

(34) H. S. Isbell and W. W. Pigman, *J. Research Natl. Bur. Standards*, **20**, 773 (1938).

(35) A. Gottschalk, *Australian J. Exptl. Biol. Med. Sci.*, **20**, 139 (1943).

(36) W. Z. Hassid, *Plant Physiol.*, **13**, 641 (1938).

(37) The D-xyloketose was prepared by autoclaving at 120° a 10% solution of D-xylose for 45 minutes in the presence of 0.2 M phosphate buffer, pH 6.8. Most of the unreacted D-xylose was separated by crystallization, leaving a sirup consisting chiefly of D-xyloketose.

mentally by oxidation of the disaccharide with sodium periodate, using the technique of Hudson and coworkers.[38] A disaccharide consisting of glucopyranose and xyloketofuranose glycosidically united through positions 1 and 2 of the aldose and ketose monosaccharides, respectively, would possess three adjacent free hydroxyls on carbon atoms 2, 3, and 4 in the glucose residue and two free hydroxyls on carbon atoms 3 and 4 in the xyloketose residue. When subjected to oxidation, a disaccharide of this structure should consume two moles of periodate and form one mole of formic acid due to the glucose residue, and consume one mole of periodate due to the xyloketose residue. A total of three moles of periodate would thus be consumed and one mole of formic acid should be formed per mole of disaccharide. The actual experimental results for this disaccharide agreed with this expectation. Any other ring structure for either the glucose or the ketose component would have given different results.

The fact that this disaccharide, like sucrose, is formed as a result of "de-phosphorolytic" condensation involving α-D-glucose-1-phosphate supports the view that D-glucose also exists in this disaccharide as the α-form.

This non-reducing disaccharide gives a blue-green color with diazouracil, a reaction shown by Raybin[39] to be specific for sucrose and other

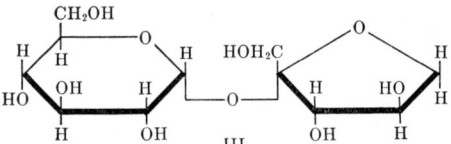

α-D-Glucopyranosyl-β-D-xyloketofuranoside

compounds containing the same type of glycosidic glucose–fructose linkage, such as raffinose, gentianose and stachyose. The analogy of the synthetic non-reducing disaccharide to sucrose in its reaction with diazouracil and with the bacterial sucrose phosphorylase indicates that the local structure about the glycosidic linkage is the same as that of sucrose, that is, the glucose has the α-D- and the ketose component the β-D-configuration. This evidence makes it possible to postulate the structural formula for the disaccharide as shown in formula III and to name it systematically α-D-glucopyranosyl-β-D-xyloketofuranoside. This compound and sucrose are structurally identical, except that sucrose has

(38) E. L. Jackson and C. S. Hudson, *J. Am. Chem. Soc.*, **59**, 994 (1937); **62**, 958 (1940); R. M. Hann, W. D. Maclay and C. S. Hudson, *ibid.*, **61**, 2432 (1939).

(39) H. W. Raybin, *J. Am. Chem. Soc.*, **55**, 2603 (1933); **59**, 1402 (1937).

an additional $-CH_2OH$ group attached to carbon atom 5 of the ring of the ketose moiety.

3. D-*Glucosyl*-L-*araboketoside*

Another non-reducing disaccharide has been recently synthesized as a result of enzymatic action of a *P. saccharophila* preparation on α-D-glucose-1-phosphate and L-araboketose.[28] Like the other two disaccharides, it does not reduce Fehling or alkaline ferricyanide solution and it is not affected by invertase. It is easily hydrolyzed with dilute acid to D-glucose and a pentose sugar. When the glucose is fermented out, a phenylosazone is obtained which is identical with that of L-arabinose. The disaccharide gives Raybin's diazouracil reaction, indicating that it contains the same type of linkages as sucrose. While the structure of this disaccharide has not been definitely determined, there is good reason to believe that it is α-D-glucopyranosyl-α-L-araboketofuranoside.[39a]

4. α-D-*Glucopyranosyl*-α-L-*sorbofuranoside*

The non-reducing disaccharide that was synthesized from D-glucose-1-phosphate and L-sorbose by the same enzyme[29] has an empirical formula of $C_{12}H_{22}O_{11}$. It has a sweet taste and gives a positive Seliwanoff reaction for ketose. It appears to be only very slightly affected by invertase, but it is easily hydrolyzed with acid. The reducing value obtained after acid hydrolysis corresponds to a disaccharide consisting of D-glucose and L-sorbose. After fermenting out the glucose from the hydrolyzate, a pure L-sorbose phenylosazone could be prepared. The melting point of the sugar is 178–180°. The specific rotation is $[\alpha]_D$ +33°. Hydrolysis with acid changes the rotation to +7.5°; this value agrees with the expected rotation for an equimolar mixture of D-glucose and L-sorbose. The rate of acid hydrolysis is approximately twice that of sucrose. The acetylated disaccharide, an octaacetate, has a rotation in chloroform of $[\alpha]_D$ +38°.

The fact that the disaccharide is non-reducing shows that the glucose and sorbose are linked through the carbonyl groups. Evidence that the L-sorbose exists in the disaccharide as sorbofuranose was obtained by oxidizing the compound with sodium periodate. In a disaccharide consisting of glucopyranose and sorbofuranose glycosidically united through positions 1 of the aldose and 2 of the ketose, the glucose residue would possess three adjacent free hydroxyls, on its carbon atoms 2, 3, and 4, and the sorbose residue would possess two adjacent free hydroxyls, on its carbon atoms 3 and 4. On oxidation of such a disaccharide with

---

(39a) The linkage pertaining to the L-araboketoside part of this disaccharide was incorrectly designated in the original publication[28] as the β-type.

periodate, the glucose residue should consume two moles of periodate and form one mole of formic acid, while the sorbose residue should consume one mole of periodate. A total of three moles of periodate would thus be consumed and one mole of formic acid would be formed per mole of disaccharide. If the sorbose residue were to exist in the disaccharide in the pyranose form, it would also contain three adjacent free hydroxyl groups on carbon atoms 3, 4, and 5 and, as in the case of the glucose, it should consume two moles of periodate and give rise to one mole of formic acid. In this case a total of four moles of periodate would be consumed and two moles of formic acid formed per mole of disaccharide.

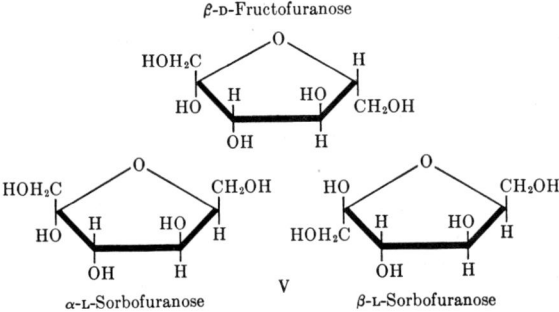

Actually, on oxidation of the carbohydrate with sodium periodate, three moles of periodate are consumed and one mole of formic acid is formed. These data agree with the assumption that the disaccharide contains a pyranose and a furanose ring. The possibility that the disaccharide is made up of glucofuranose and sorbopyranose can also be eliminated on the basis of the periodate oxidation data. Glucofuranose would contain two pairs of adjacent hydroxyls, on carbon atoms 2 and 3 and on 5 and 6, and the sorbopyranose would have three adjacent hydroxyls, on carbon atoms 3, 4 and 5. In oxidizing such a disaccharide, a total of four moles of periodate would thus be used, giving rise to one mole of formic acid. This is inconsistent with the experimental data.

That the glucose exists in the glucosyl-sorboside as the α-form is deduced from the fact that the disaccharide is formed as a result of "de-phosphorolytic" condensation involving α-D-glucose-1-phosphate.

Like the other synthetic disaccharides, it gives the Raybin reaction, which indicates that the local structure about the glycosidic linkage in the glucosyl-sorboside is the same as that of sucrose.

The striking similarity between sucrose and the D-glucosyl-L-sorboside is immediately evident when the structural formula of the latter is constructed in analogy with that of sucrose (see IV). The sorbose component of this disaccharide is an L-sugar in contrast to the D-fructose unit existing in sucrose. Since β-D-fructose and α-L-sorbose have the same configuration for their second carbon atoms[40] (see V), it is necessary to designate the ketose portion of the disaccharide as α-L-sorboside.

It is noteworthy that D-fructose, which has a pyranose structure in the free crystalline state, assumes a furanose configuration whenever it combines with another sugar to form an oligosaccharide or polysaccharide. Apparently the ketohexose L-sorbose shows the same behavior.

VI. Synthetic Reducing Disaccharide

3-[α-D-Glucopyranosyl]-L-*arabinopyranose*

This disaccharide,[30] formed from α-D-glucose-1-phosphate and L-arabinose under the influence of the enzyme from *P. saccharophila*, reduces Fehling and alkaline ferricyanide solutions. It contains two molecules of water of crystallization, and has an empirical formula of $C_{11}H_{20}O_{10} \cdot 2 H_2O$, and a specific rotation $[\alpha]_D$ in water of $+156°$. Unlike sucrose and its analogs, this disaccharide is difficultly hydrolyzable with acid. Upon hydrolysis it yields one mole of D-glucose and one mole of L-arabinose. The phenylosotriazole derivative of the disaccharide, prepared according to Hann and Hudson,[41] is readily hydrolyzed with acid to L-arabinose phenylosotriazole and D-glucose, showing that the L-arabinose constitutes the free reducing unit in the disaccharide. Since α-D-glucose-1-phosphate is involved in the enzymatic synthesis of the disaccharide, it may be assumed that the D-glucose component exists in the α-form.

On oxidation of the phenylosotriazole derivative of the disaccharide with sodium periodate, three moles of periodate are consumed with the formation of one mole each of formic acid and formaldehyde per mole of the phenylosotriazole derivative. If the D-glucose in the D-glucopyranosyl-L-arabinose phenylosotriazole were attached to carbon atom 4 of the L-arabinose derivative, oxidation of this compound with sodium periodate would require two moles of periodate and would liberate one mole of

(40) C. S. Hudson, *J. Am. Chem. Soc.*, **60**, 1537 (1938).
(41) R. M. Hann and C. S. Hudson, *J. Am. Chem. Soc.*, **66**, 735 (1944); W. T. Haskins, R. M. Hann, and C. S. Hudson, *ibid.*, **67**, 939 (1945).

formic acid, with no formaldehyde production; the experimental data exclude the possibility of this union at carbon atom 4. Junction of D-glucopyranose to carbon atom 5 of the L-arabinose phenylosotriazole would require three moles of periodate whereby one mole of formic acid would be produced, and no formaldehyde formed; these expectations likewise disagree with the experimental data and in consequence a union at carbon atom 5 is excluded, and there remains only the possibility of a union at carbon atom 3 as shown in the formula VI. Inspection of it shows that the periodate oxidation of the substance should reduce three

VI
3-[α-D-Glucopyranosyl]-L-arabinose phenylosotriazole

VII   VIII   IX

moles of periodate and generate one mole of formic acid and one mole of formaldehyde, which conforms with the experimental data. The structure of the phenylosotriazole of the disaccharide is therefore 3-[α-D-glucopyranosyl]-L-arabinose phenylosotriazole, in which D-glucose is attached through its carbon atom 1 to the carbon atom 3 of L-arabinose.

On methylation of the disaccharide with dimethyl sulfate and sodium hydroxide a hexamethyl ether of the carbohydrate was obtained. When this fully methylated derivative (VII) was hydrolyzed with acid, 2,3,4,6-tetramethyl-D-glucose (VIII) and dimethyl-L-arabinose (IX) were produced. Since position 3 in the L-arabinose component of VI was shown to be occupied in glycosidic linkage with D-glucose, the dimethyl-L-arabinose could be either the 2,5- or 2,4-dimethyl derivative (IX), depending on whether the L-arabinose unit originally exists in the disaccharide in the furanose or pyranose form. The ring type of the L-arabinose was ascertained through a series of oxidations; by the action of hypoiodite it was converted to the corresponding dimethyl-L-arabonic lactone (X), from which a dimethyl-L-arabonic acid (XI) was obtained.

If this acid were the 2,5-dimethyl-L-arabonic acid, it would possess a pair of adjacent hydroxyls (on positions 3 and 4) which on oxidation with

$$
\text{IX} \xrightarrow{\text{Oxidation NaIO}_4}
\begin{array}{c}
\text{CO} \\
| \\
\text{HCOCH}_3 \\
| \\
\text{HOCH} \\
| \\
\text{CH}_3\text{OCH} \\
| \\
\text{H}_2\text{C}
\end{array}
\!\!\!\!\!\text{O}
\xrightarrow{\text{Hydrolysis}}
\begin{array}{c}
\text{COOH} \\
| \\
\text{HCOCH}_3 \\
| \\
\text{HOCH} \\
| \\
\text{CH}_3\text{OCH} \\
| \\
\text{CH}_2\text{OH}
\end{array}
$$
$$\text{X} \qquad\qquad \text{XI}$$

sodium periodate would consume one mole of periodate in the reaction. On the other hand, a 2,4-dimethyl-L-arabonic acid (XI), lacking a pair of adjacent hydroxyls, would not be attacked by periodate. Actually, no periodate was consumed when the dimethyl-L-arabonic acid was treated with periodate, which shows that the dimethyl derivative is 2,4-dimethyl-L-arabonic acid and accordingly the sugar is 2,4-dimethyl-L-arabinose. A free hydroxyl in position 3 is obviously restored when

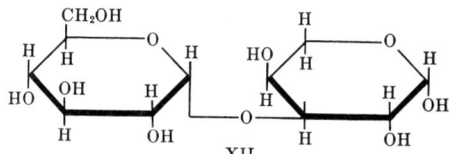

XII

3-[α-D-Glucopyranosyl]-L-arabopyranose

the methylated disaccharide is hydrolyzed and the 2,4-dimethyl-L-arabinose generated. The hydroxyl in position 5 of dimethyl-L-arabonic acid is formed when the ring is broken in the process of hydrolysis of the lactone.

A more direct confirmation that the dimethyl-L-arabinose possesses a pyranose configuration was obtained from the study of the rate with which its lactone derivative is hydrolyzed to the corresponding open chain acid.[8]

When the dimethyl-L-arabonolactone was dissolved in water, it was found to be almost completely hydrolyzed within four hours. This was indicated by a change of its rotation from $[\alpha]_D$ +60° to +24°. A constant value of $[\alpha]_D$ +17° was reached within less than twenty-four hours. Since the rate of change in rotation of this methylated lactone, due to hydrolysis, is a rapid one it strongly indicates that the lactone is of the delta type. This observation confirms the periodate oxidation data, showing that the dimethyl acid is 2,4-dimethyl-L-arabonic acid.

On the basis of these results this reducing disaccharide may be designated as 3-[α-D-glucopyranosyl]-L-arabinopyranose and its structural formula written as in XII.

VII. Formation of Sucrose and Other Disaccharides through Exchange of Glycosidic Linkages

It was found with the aid of the enzyme from *P. saccharophila* that the synthetic disaccharides could also be formed and broken down through a mechanism other than phosphorolysis, namely, through the exchange of one glycosidic linkage for another. The information concerning this new mechanism was first obtained through the use of radioactive phosphate.[42] It was observed that when α-D-glucose-1-phosphate and radioactive inorganic phosphate were added to sucrose phosphorylase preparations in the absence of ketose sugars, a rapid redistribution of the isotope between the organic and inorganic fractions occurred. D-Glucose, which is known to inhibit sucrose phosphorylase, was also found to inhibit the exchange reaction. The presence of D-fructose was also found to decrease the rate of exchange. This would be expected if D-fructose competes with phosphate for the glucose residue of D-glucose-1-phosphate. On the basis of these observations the following reaction was postulated:

D-Glucose-1-phosphate + enzyme ⇌ D-glucose–enzyme + phosphate.

It is to be expected from this equilibrium reaction that the D-glucose–enzyme complex should be capable of donating D-glucose to a suitable acceptor, which could be D-fructose or even some other sugar in accordance with the following scheme:

```
                            D-Glucose-1-fructoside (sucrose)
                                    ↑↓  ± D-fruc-
                         ± phosphate       tose       ± L-sorbose
D-Glucose-1-phosphate ⇌ D-Glucose-enzyme ⇌ D-Glucose-1-sorboside
                                    ↑↓  ± D-xyloketose
                            D-Glucose-1-xyloketoside
```

That the above scheme is actually correct, has been established by the following observations:

(42) M. Doudoroff, H. A. Barker and W. Z. Hassid, *J. Biol. Chem.*, **168**, 725 (1947).

1. In the absence of phosphate or D-glucose-1-phosphate, the enzyme brings about a rapid equilibration between added radioactive free D-fructose and the D-fructose moiety of sucrose.[43]

2. The direct syntheses of D-glucosyl-L-sorboside from sucrose and L-sorbose and of sucrose from D-glucosyl-D-xyloketoside and D-fructose can be obtained in phosphate-free enzyme preparations.[40] The reactions are:

$$\text{D-Glucose-1-D-fructoside} + \text{L-sorbose} \underset{}{\overset{\text{enzyme}}{\rightleftarrows}} \text{D-glucose-1-L-sorboside} + \text{D-fructose}$$
(Sucrose)

$$\text{D-Glucose-1-D-xyloketoside} + \text{D-fructose} \rightleftarrows \text{D-glucose-1-D-fructoside} + \text{D-xyloketose}$$
(Sucrose)

3. Arsenate, which can apparently serve as glucose acceptor to the enzyme–D-glucose complex but which does not form a stable glucose ester, causes the rapid decomposition of both sucrose and D-glucose-1-phosphate in accordance with the following equations:[44]

$$\text{D-Glucose-1-phosphate} \xrightarrow[\text{arsenate}]{\text{enzyme}} \text{D-glucose} + \text{phosphate}$$

$$\text{D-Glucose-1-D-fructoside} \xrightarrow[\text{arsenate}]{\text{enzyme}} \text{D-glucose} + \text{D-fructose}$$
(Sucrose)

Sucrose phosphorylase can then be regarded as a rather versatile "transglucosidase," capable of exchanging glycosidic and ester bonds and of donating D-glucose to a variety of substrates such as ketoses, an aldose, inorganic phosphate and arsenate.

It is of interest that similar enzymes, capable of exchanging glycosidic linkages, appear to be involved in the synthesis of some polysaccharides. Thus; the polysaccharide dextran is formed from sucrose by enzyme preparations from *Leuconostoc*[45] while the polysaccharide levan is produced from sucrose or raffinose by enzymes of other bacteria.[46] The reactions may be written as follows:

$$n\ \text{C}_{12}\text{H}_{22}\text{O}_{11} \rightarrow (\text{C}_6\text{H}_{10}\text{O}_5)_n + n\ \text{C}_6\text{H}_{12}\text{O}_6$$
(D-glucose-1-D-fructoside)   (dextran)   (D-fructose)

$$n\ \text{C}_{12}\text{H}_{22}\text{O}_{11} \rightarrow (\text{C}_6\text{H}_{10}\text{O}_5)_n + n\ \text{C}_6\text{H}_{12}\text{O}_6$$
(D-glucose-1-D-fructoside)   (levan)   (D-glucose)

Although these reactions normally go almost completely from left to right, indirect evidence has been presented for their reversible nature.

(43) H. Wolochow, E. W. Putman, M. Doudoroff, W. Z. Hassid and H. A. Barker, *J. Biol. Chem.*, **180**, 1237 (1949).

(44) M. Doudoroff, H. A. Barker and W. Z. Hassid, *J. Biol. Chem.*, **170**, 147 (1947).

(45) E. J. Hehre and J. Y. Sugg, *J. Exptl. Med.*, **75**, 339 (1942); E. J. Hehre, *Proc. Soc. Exptl. Biol. Med.*, **54**, 240 (1943).

(46) S. Hestrin and S. Avinieri-Shapiro, *Biochem. J.*, **38**, 2 (1944).

Thus when D-glucose was added to a solution containing levan, yeast invertase and an enzyme preparation from *Bacillus subtilis*, it was found that the decomposition of the levan became accelerated, presumably through the synthesis and decomposition of sucrose[47] as follows:

$$n\ C_6H_{12}O_6 + (C_6H_{10}O_5)_n \underset{\longleftarrow}{\overset{\text{enzyme}}{\longrightarrow}} n\ C_{12}H_{22}O_{11} \xrightarrow[H_2O]{\text{invertase}} n\ C_6H_{12}O_6 + n\ C_6H_{12}O_6$$

(D-glucose)   (levan)   (sucrose)   D-glucose   D-fructose

It seems quite possible that not only true phosphorylase but also other "transglycosidases" may play an important role in the biological syntheses, decompositions and interconversions of disaccharides and polysaccharides.

(47) M. Doudoroff and R. O'Neal, *J. Biol. Chem.*, **159**, 585 (1945).

PRINCIPLES UNDERLYING ENZYME SPECIFICITY IN THE DOMAIN OF CARBOHYDRATES

By Alfred Gottschalk

The Walter and Eliza Hall Institute of Medical Research, Melbourne, Australia

Contents

I. Introduction..... 49
II. The Enzyme—Substrate Compound as Chemical Basis of Enzyme Specificity 51
   1. The Intermediate Compound Theory in Homogeneous and Heterogeneous Catalysis..... 51
   2. Identification of Intermediate Enzyme—Substrate Compounds..... 53
   3. The Hypothesis of a Multipoint Contact between Enzyme and Substrate (Carbohydrate)..... 54
   4. The Forces of Attraction between Enzyme and Substrate (Carbohydrate) Groupings..... 55
   5. Orientation by Chemisorption of the Substrate (Carbohydrate) at the Enzyme Surface..... 56
   6. Resolution into Successive Steps of Transfer Reactions between One Enzyme and Two Species of Substrate..... 57
   7. Summary..... 59
III. The Effect of Configurational and Substitutional Changes in the Sugar Molecule on the Rate of Various Enzyme Reactions and Its Application in Marking the Contacting Groups in the Substrate..... 60
   1. β-D-Glucosidase and β-D-Glucosides..... 60
   2. α-D-Mannosidase and α-D-Mannosides..... 66
   3. α-D-Galactosidase and α-D-Galactosides..... 66
   4. β-D-Fructofuranosidase and Sucrose..... 67
   5. Transglucosidase and Sucrose..... 70
   6. Hexokinase (Yeast) and D-Glucose..... 73
IV. Discussion of the Principles Controlling and of the Factors Impairing the Formation of the Intermediate Compound between Carbohydrates and Their Specific Enzymes..... 76

I. Introduction

It would appear that the specific action of an enzyme upon its substrate is conditioned by a definite chemical structure and spatial arrangement of the constituent polar and non-polar groups of the enzyme protein as well as by the constitution and configuration of the substrate. In some cases an enzyme interacts with one chemical compound only. For example, galactokinase extracted from *Saccharomyces fragilis* (grown on whey) catalyzes the transphosphorylation between adenosine triphosphate (ATP) and D-galactose (Fig. 1); the stereoisomers D-glucose and D-mannose are not attacked by the enzyme.[1] More frequently, however, an enzyme reacts with a group of structurally related substances, though the rate of reaction may differ with the various members of the group. Thus, yeast carboxylase splits off carbon dioxide from pyruvic acid, dimethylpyruvic acid, α-ketobutyric acid, the various α-ketovaleric and α-ketocaproic acids, and from aromatic α-ketoacids, but not from

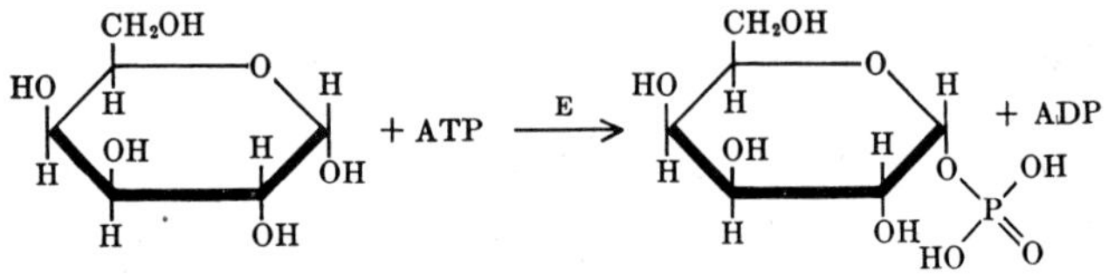

FIG. 1.—Action of galactokinase (E).

β-ketoacids.[2] This suggests that the specificity requirements of the enzyme which breaks the covalent bond between the carbon atom of the carboxylic group and the α-carbon atom are satisfied by the presence in the substrate of the grouping

$$HC-C-COOH.$$

We may conclude from these examples that an enzyme which breaks a chemical bond in its substrate is concerned not only with the electronic structure of the atoms forming this bond but also with the constitution and the configuration of the neighboring groups. Enzyme specificity, then, may be defined as the restriction of the action of an enzyme to the presence in the substrate of a definite atomic structure and spatial arrangement which may comprise either all or part of the constituent groupings of the substrate. The decrease in the reaction rate caused by a change in the constitution or configuration of that (natural) substrate acted upon with the highest reaction velocity is therefore a measure of the degree of specificity.

It is well recognized that specificity is one of the most spectacular aspects of enzymatic action. Thus, the process of alcoholic fermentation of D-glucose by a unicellular organism like yeast has been proved to consist of a sequence of elementary reactions catalyzed by sixteen individual

(1) R. E. Trucco, R. Caputto, L. F. Leloir and N. Mittelman, *Arch. Biochem.* **18**, 137 (1948).

(2) See article by C. Neuberg in "Handbuch der Biochemie der Menschen und der Tiere" by C. Oppenheimer, Verlag Gustav Fischer, Jena, 2nd ed., Vol. II, p. 442 (1925).

enzymes. These individual steps are coordinated in such a manner that each oxidative step is balanced by a reductive one, and the uptake of inorganic phosphate in one reaction synchronizes with the release of phosphate in another. This coordination is inconceivable without a high degree of enzyme specificity. So efficiently do the enzymes, by virtue of their specificity, control this intricate system of chemical reactions that they act even without the support of the macro-molecular structure and the organization in space of the living cell. In cell-free yeast juice, where the enzymes are distributed at random throughout the liquid, the succession of events is only slightly distorted owing to the low concentration or lack of adenosintriphosphatase which is the most sensitive enzyme of the zymase system.[3]

The mechanism of enzyme specificity—like the mechanism of substrate activation by enzymes—is not well understood. It would appear, however, that the accumulated data concerning the effect of substitutional and configurational changes in hexoses and their glycosidic derivatives on the activity of the enzyme concerned will allow the derivation of some principles underlying enzyme specificity in the domain of carbohydrates. Obviously, carbohydrates, due to the possession of many asymmetric centers and to the reactivity of their hydroxyl groups in substitution reactions, lend themselves more than any other group of naturally occurring substances to an investigation of the intimate relationship between an enzyme and those structural features of its substrate which represent the minimum requirements for enzyme action. Perhaps it was not altogether by chance that the highly specific action of enzymes was discovered in work with glucosides.[4]

It may be expedient to recall shortly the current views on catalysis in general and to develop on that foundation the basic assumptions forming the framework of the present article.

II. The Enzyme–Substrate Compound as Chemical Basis of Enzyme Specificity

1. *The Intermediate Compound Theory in Homogeneous and Heterogeneous Catalysis*

Evidently, there does not exist a general theory of catalysis. From a survey of the voluminous literature on catalysis,[5] however, it is apparent

(3) O. Meyerhof, *J. Biol. Chem.*, **157**, 105 (1945).
(4) E. Fischer, *Ber.*, **27**, 2985 (1894).
(5) See W. Langenbeck, "Die organischen Katalysatoren und ihre Beziehungen zu den Fermenten," Verlag Julius Springer, Berlin (1935); W. Langenbeck, "Fermentmodelle" in "Handbuch der Enzymologie" by F. F. Nord and R. Weidenhagen, Akademische Verlagsgesellschaft, Leipzig, Vol. I, p. 325 (1940); W. Frankenburger,

that the following concept is the predominant one: The catalyst takes active part in the chemical reactions by forming with the reactant a labile intermediate compound which readily decomposes into reaction product and regenerated catalyst. The catalyst by its intimate contact with the reactant disturbs the electronic structure within the reactant molecule, resulting in the loosening of a chemical bond. It would appear that part of the energy of reaction between the catalyst and the reactant can be made available to reduce the apparent energy of activation of the catalyzed reaction. Moreover, by attaching itself to a restricted area of the reacting molecule, the catalyst will focus its interfering effect on a small group of atoms or on a single bond, thus steering the chemical reaction into a certain direction. In doing so, the catalyst selects one from several thermodynamically possible reactions.

The concentration of the intermediate compound AC formed by the interaction of reactant A and catalyst C is controlled by the relative

$$A + C \underset{k_2}{\overset{k_1}{\rightleftharpoons}} AC \overset{k_3}{\rightarrow} B + C \tag{1}$$

values for k_1, k_2 and k_3; only in the limiting case where k_3 is less than k_2 and k_1, is AC appreciably present in equilibrium concentration.

It is in the very nature of the catalytic process that the intermediate compound formed between catalyst and reactant is of extreme lability; therefore not many cases are on record where the isolation by chemical means, or identification by physical methods, of intermediate compounds has been achieved concomitant with the evidence that these compounds are true intermediaries and not products of side reactions or artifacts. The formation of ethyl sulfuric acid in ether formation, catalyzed by H_2SO_4, and of alkyl phosphates[6] in olefin polymerization, catalyzed by liquid phosphoric acid, are examples of established intermediate compound formation in homogeneous catalysis. With regard to heterogeneous catalysis, where catalyst and reactant are not in the same

---

"Katalytische Umsetzungen in homogenen und enzymatischen Systemen," Akademische Verlagsgesellschaft, Leipzig (1937); T. P. Hilditch and C. C. Hall, "Catalytic Processes in Applied Chemistry," Chapman and Hall, Ltd., London (1937); S. Berkman, J. C. Morrell and G. Egloff, "Catalysis, Inorganic and Organic," Reinhold Publishing Corporation, New York (1940); C. N. Hinshelwood, "The Kinetics of Chemical Change," Clarendon Press, Oxford (1940); R. P. Bell, "Acid-Base Catalysis," Clarendon Press, Oxford (1941); H. W. Lohse, "Catalytic Chemistry," Chemical Publishing Co., Inc., New York (1945); R. H. Griffith, "The Mechanism of Contact Catalysis," University Press, Oxford (1946).

(6) V. N. Ipatieff, *Ind. Eng. Chem.*, **27**, 1067 (1935). *Cf.* also J. Turkevich and R. K. Smith, *Nature*, **157**, 874 (1946).

physical state, (a) the decomposition of hydrogen peroxide by mercury with the intermediate formation of a film of mercuric peroxide[7] and (b) the production of metallic hydride molecules by hydrogen, chemisorbed on the surface of a metallic lattice, in the ortho-para conversion of hydrogen or deuterium,[8] are cases in point.

Further evidence for the formation of intermediate compounds in catalytic reactions is afforded by the observation (a) that optically active camphor is formed from optically inactive (racemic) camphor carboxylic acid in the presence of the d- or l-forms of quinine, quinidine or nicotine;[9] and (b) that optically active bases, e.g., quinidine, catalyze the synthesis of optically active mandelonitrile from benzaldehyde and hydrocyanic acid.[10] These results hardly admit of any other interpretation than the intermittent production of a catalyst–reactant compound.

2. *Identification of Intermediate Enzyme–Substrate Compounds*

Enzymes are colloids and the reactions catalyzed by them are classified accordingly as microheterogeneous. As in the case of inorganic or organic heterogeneous catalysis, it is assumed that an intermediate compound is formed between enzyme and substrate.[10a]

$$E + S \underset{k_2}{\overset{k_1}{\rightleftharpoons}} ES \overset{k_3}{\to} P + E \tag{2}$$

Concerning the mode of formation of ES, we prefer the concept that the substrate in a monolayer is chemisorbed to the active center of the enzyme protein, just as the experimental evidence pertaining to surface catalysis by inorganic catalysts indicates that in these reactions chemisorbed, not physically adsorbed, reactants are involved. Such a concept is supported by the demonstration of spectroscopically defined unstable intermediate compounds between enzyme and substrate in the decomposition by catalase of ethyl hydroperoxide,[11] and in the interaction between peroxidase and hydrogen peroxide.[12] Recently Chance[13] determined by direct photoelectric measurements the dissociation con-

(7) G. Bredig and A. von Antropoff, *Z. Elektrochem.*, **12**, 585 (1906).
(8) See E. K. Rideal, *Nature*, **161**, 461 (1948).
(9) G. Bredig and K. Fajans, *Ber.*, **41**, 752 (1908); K. Fajans, *Z. physik. Chem.*, **73**, 25 (1910).
(10) G. Bredig and P. S. Fiske, *Biochem. Z.*, **46**, 7 (1912); G. Bredig and M. Minaeff, *ibid.*, **249**, 241 (1932).
(10a) L. Michaelis and M. L. Menten, *Biochem. Z.*, **49**, 333 (1913).
(11) K. G. Stern, *Enzymologia*, **4**, 145 (1937).
(12) D. Keilin and T. Mann, *Proc. Roy. Soc. (London)*, **B122**, 119 (1937).
(13) B. Chance, *J. Biol. Chem.*, **151**, 557 (1943); *Nature*, **161**, 914 (1948).

stant k_2/k_1 and the second order rate constant k_1 of the reaction

$$\text{Peroxidase} + \text{H}_2\text{O}_2 \underset{k_2}{\overset{k_1}{\rightleftharpoons}} \text{Peroxidase} \cdot \text{H}_2\text{O}_2,$$

the values obtained ($k_1 = 1.2 \times 10^7$ liter mole^{-1} sec.$^{-1} \pm 0.4 \times 10^7$, $k_2/k_1 = 2 \times 10^{-8}$ mole liter^{-1}) indicating the extremely quick formation of a relatively tight enzyme–substrate compound. Additional support for the view of a chemical union between enzyme and substrate is provided by the fact that succinic and malonic acid fully protect the functional —SH group of succinic dehydrogenase against oxidation by oxidized glutathione (G—S—S—G).[14]

3. *The Hypothesis of a Multipoint Contact between Enzyme and Substrate (Carbohydrate)*

Turning now to the intermediate compound which, most probably, is formed in the interaction between carbohydrates and their specific enzymes, two qualifications will be made in the following discussion. It is assumed (*a*) that a multipoint contact is established between —OH or substituted —OH groups of the sugar molecule and appropriate groups of the enzyme protein; and (*b*) that this contact involves that atom or group the electronic structure of which will be deformed in the catalytic reaction and one or more —OH groups in *cis*-position to the former, *cis* and *trans* referring to the mean plane of the pyranose or furanose ring.

The idea of a multipoint contact between enzyme and substrate (carbohydrate) is not new. Armstrong[15] was the first to suggest that attachment of a glucoside to its glucosidase takes place through the oxygen atoms of the hydroxyl groups of the sugar molecule, these oxygen atoms possessing "residual affinity." This view was further developed by Josephson[16] and qualified by v. Euler[17] in his theory of "dual affinity." v. Euler emphasized the close agreement between the affinity constant K_M (= reciprocal of the dissociation constant of the enzyme–substrate compound) of the saccharase–sucrose compound and the product $K_M{}^1 \times K_M{}^2$ of the constituent affinity constants, ($K_M{}^1$ and $K_M{}^2$ representing the affinity constants of the saccharase–fructose and of the saccharase–glucose compounds, respectively). He concluded that a

(14) F. G. Hopkins, E. J. Morgan and C. Lutwak-Mann, *Biochem. J.*, **32**, 1829 (1938).

(15) E. F. Armstrong, *Proc. Roy. Soc. (London)*, **73**, 516 (1904); E. F. Armstrong, "The Simple Carbohydrates and the Glucosides," Longmans, Green and Co., London (1910).

(16) K. Josephson, *Z. physiol. Chem.*, **147**, 1 (1925).

(17) H. v. Euler, *Z. physiol. Chem.*, **143**, 79 (1925).

hydrolyzing enzyme like saccharase establishes contact with both moieties of the substrate molecule. In recent years Bergmann,[18] on the basis of his investigations into the specificity requirements of proteolytic enzymes, formed the view that the approach of a peptide to its peptidase results in "a sort of polyaffinity between the polar groups of the enzyme and the substrate." Helferich[19] and Pigman[20] hold a similar opinion for glycosidases and glycosides. The suggestion of a multipoint contact between enzyme and substrate was obviously prompted by the endeavor to explain enzyme specificity. Only a multipoint contact provides for adequate orientation of the substrate molecule at the active center of the enzyme surface, thus correlating the structural features of enzyme and substrate.

4. *The Forces of Attraction between Enzyme and Substrate (Carbohydrate) Groupings*

The most likely forces of attraction between an enzyme protein and a sugar molecule, when they form a transient intermediate compound, are coordinative forces in the form of hydrogen bonds and/or mutual electrostatic attraction of molecular dipoles. Both, the hydroxyl groups of sugars and the hydrophilic groupings of proteins (backbone: $=$O, $=$NH; amino acid residue: —OH, —COOH, —NH$_2$, $=$NH), readily form hydrogen bonds as exemplified by their tendency to act as hydration centers. The ease with which these O and N containing groups form hydrogen bonds, taken together with the low energy value of the hydrogen bond (the energy of the reaction $XH + Y \rightarrow XHY$ being only 5 kcal./mole), would render this bond most suitable for enzyme–substrate coupling. The insolubility in water of furan and of cyclopentene oxide suggests that the ring oxygen in sugars has but little tendency to form a hydrogen bond. The glycosidic oxygen most probably forms a hydrogen bond of less strength than that of the hydroxyl groups, as may be deduced from a comparison of the boiling points and the solubilities in water of hydroquinone and hydroquinone monoethyl ether.

It should be borne in mind, however, that enzyme proteins, like other proteins, are dipolar ionic structures surrounded by an intense electrostatic field. Horse carboxyhemoglobin, for example, has a dipole moment of 480 Debye units (molecular weight, 67,000). Compared with amino acids, sugars have a rather low dipole moment, due mainly to a considerable measure of mutual cancellation of the constituent link moments resulting from their disposition in space (glycine = 15, α- and

(18) M. Bergmann, *Harvey Lectures*, **31,** 37 (1936).
(19) B. Helferich and H. Scheiber, *Z. physiol. Chem.*, **226,** 272 (1934).
(20) W. W. Pigman, *Advances in Enzymol.*, **4,** 41 (1944).

β-pentaacetylglucoses = 3.5 and 2.5 Debye units, respectively). Thus enzyme-substrate association by dipole attraction is a possibility. Regarding the bonds contributing to the dipole moment of a sugar molecule it may be mentioned that the link moments for H—O and C—O are 1.6 and 0.7 Debye units respectively. It would appear, therefore, that the C—O moment in the sugar molecule plays but a small part in the dipole–dipole interaction.

5. *Orientation by Chemisorption of the Substrate (Carbohydrate) at the Enzyme Surface*

Chemisorption by these forces, *e.g.*, of sucrose to saccharase, in such a manner that the mean planes of the pyranose and furanose rings are

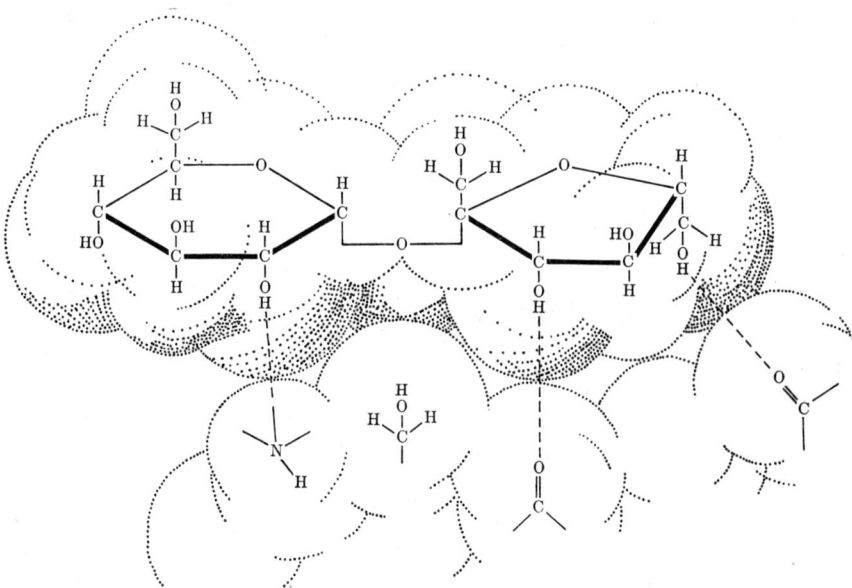

Fig. 2.—Showing the initial stage of interaction between a sucrose molecule and the active center of a saccharase molecule. A hydroxyl group of the enzyme is making an attack on the glucosidic oxygen, whereas three other groups of the enzyme are in the process of forming hydrogen bonds with OH groups of the substrate *cis*-disposed to the glucosidic oxygen. The perspective formulas shown in this article represent planar projections of three-dimensional objects. Though they do not describe the actual positions in space of the addenda to the ring carbon atoms, they indicate correctly their position relative to the mean plane of the ring (*cf.* C. S. Hudson, *Advances in Carbohydrate Chem.* **3**, 1 (1948)). (The representation of a molecule by the planar projection of its three-dimensional structure superimposed on a perspective view of its surface (boundary), as shown in this figure, was first used by Linus Pauling (*Endeavour*, **7**, 43 (1948)), when depicting the structure of the haptenic group of an azoprotein and the combining region of its specific antibody.)

orientated parallel to the enzyme surface with the glucosidic oxygen facing the enzyme, will allow close approximation (1 Å) and juxtaposition of the respective groups of enzyme and substrate. Since in this case the critical demand on the catalyst is to deform the electronic structure of the glucosidic oxygen, closest approach of this atom to the group of the enzyme surface that effects activation seems to be a necessity. Such conditions, taking into account the arrangement in space both of the atoms constituting the receiving region of the protein surface and of the

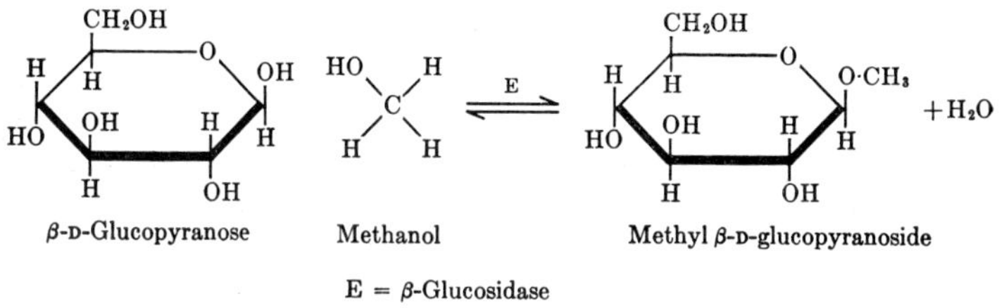

FIG. 3.—Action of β-glucosidase.

glucosidic oxygen and hydroxyl groups representing the substrate's anchoring groups, may provide a basis for the understanding of the highly specific nature of enzyme action (Fig. 2). A concept of this kind seems to be quite satisfactory for all those cases in which the enzyme acts upon a single species of substrate molecules or upon two species which can be accommodated at the enzyme surface side by side, close enough as to allow a chemical bond to be formed between the interacting groups of the two substrates activated by the enzyme, as in the enzymatic synthesis of methyl β-D-glucopyranoside (Fig. 3).[21]

6. *Resolution into Successive Steps of Transfer Reactions between One Enzyme and Two Species of Substrate*

In recent years much attention has been given to reactions in which the enzyme catalyzes the transfer of the glucosidic residue of a disaccharide (or a glucoside) to another partner (acceptor),[22] for example,

α-D-Glucopyranosyl-β-D-fructofuranoside + L-sorbose
$\rightleftarrows$ α-D-glucopyranosyl-α-L-sorbofuranoside + D-fructose.   (3)

As may be seen from Fig. 2, it seems impossible to accommodate at the surface of the enzyme simultaneously a molecule of sucrose and of L-sorbose in such a way as to permit a switch of the fructosidic linkage

---

(21) E. Bourquelot and E. Verdon, *Compt. rend.*, **156**, 1264, 1638 (1913); K. Josephson, *Z. physiol. Chem.*, **147**, 155 (1925).

(22) M. Doudoroff, *Federation Proc.*, **4**, 242 (1945).

O—C2 to C2 of L-sorbose. The length of the C—O bond being equal to the sum of the atomic radii, its formation requires approximate contact between the fructosidic oxygen and carbon atom 2 of L-sorbose. These steric hindrances, however, do not exist if we apply Langenbeck's scheme of carbohydrase action to the transglucosidase reaction. Langenbeck,[23] in his experimental work on enzyme models, found that butyric acid methyl ester, which is only slowly hydrolyzed by hydroxyl ions, on addition of small amounts of benzoylcarbinol forms by transesterification intermediately a much more easily hydrolyzable ester, the carbinol thus acting as a catalyst:

$$C_3H_7 \cdot COO \cdot CH_3 + C_6H_5 \cdot CO \cdot CH_2OH \rightarrow C_3H_7 \cdot COO \cdot CH_2 \cdot CO \cdot C_6H_5 + CH_3OH$$
$$C_3H_7 \cdot COO \cdot CH_2 \cdot CO \cdot C_6H_5 + H_2O \rightarrow C_3H_7 \cdot COOH + C_6H_5 \cdot CO \cdot CH_2OH.$$

This type of reaction was used by Langenbeck[24] to describe the mechanism underlying the enzymatic hydrolysis of a glycoside (Fig. 4). This

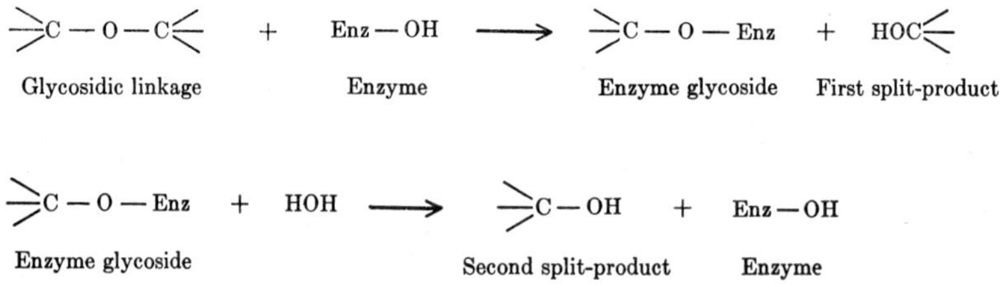

$$>\!\!C-O-C\!\!< \;+\; Enz-OH \;\longrightarrow\; >\!\!C-O-Enz \;+\; HOC\!\!<$$

Glycosidic linkage   Enzyme   Enzyme glycoside   First split-product

$$>\!\!C-O-Enz \;+\; HOH \;\longrightarrow\; >\!\!C-OH \;+\; Enz-OH$$

Enzyme glycoside   Second split-product   Enzyme

Fig. 4.—Langenbeck's formulation of the mechanism underlying the enzymatic hydrolysis of a glycoside.

concept goes far to resolve in transglucosidase reactions the steric difficulties mentioned above. We may assume that a hydroxyl group of an amino acid residue of the enzyme protein interacts with the —C—O—C— grouping of sucrose to form D-fructose and an enzyme—α-D-glucoside which in turn reacts with L-sorbose, producing the sucrose analogue. This sequence of reactions, affording an unhindered interaction between the enzyme and its two substrates as well as preservation of the energy of the fructosidic linkage, is substantiated by recent experimental results. The same enzyme which catalyzes the exchange of ketosugars in disaccharides of the sucrose type is responsible for the following exchange reaction:[22]

Phosphoryl α-D-glucopyranoside + β-D-fructofuranose ⇌ sucrose + phosphate.   (4)

If phosphoryl-α-D-glucopyranoside (glucose-1-phosphate) and radio-

(23) W. Langenbeck and F. Baehren, *Ber.*, **69**, 514 (1936).
(24) See F. F. Nord and R. Weidenhagen, "Handbuch der Enzymologie," Akademische Verlagsgesellschaft, Leipzig, p. 513 (1940).

active inorganic phosphate are added to the enzyme in the absence of a glucose acceptor, a rapid interchange of phosphate occurs between the organic and inorganic fractions without the liberation of any glucose.[25] From this the following reaction was postulated:[25]

D-Glucose-1-phosphate + enzyme ⇌ D-glucose—enzyme + phosphate.

In harmony with the concept of the discoverers of the reaction, Fig. 5 represents the individual steps of this transglucosidase reaction.

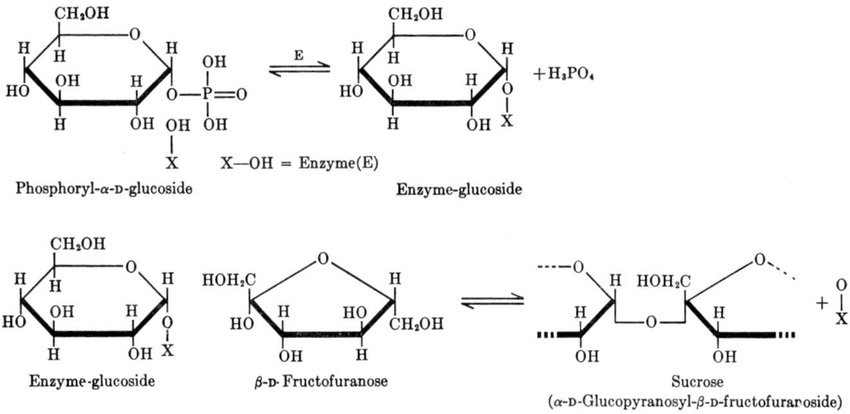

FIG. 5.—Formulation of the mechanism underlying the reversible enzymatic synthesis of sucrose.

It is of interest in this connection that Stearn,[26] in a discussion of enzyme kinetics from the point of view of statistical mechanics and quantum mechanics, regards interaction between a dipole (as part of the enzyme protein) and the reacting groups of the substrate, *e.g.*, C—O, resulting in a redistribution of charge within the C—O bond, as a more rational mechanism of activation than the loosening of the bond by distortion.

7. *Summary*

Summarizing, it would appear justified to assume that the action of glycosidases (carbohydrases) and transglycosidases (phosphorylases) on glycosides is initiated by chemisorption, at the enzyme surface, of the substrate molecule, with the glycosidic oxygen contacting the attacking group of the enzyme, and with hydroxyl groups, *cis*-disposed to the glycosidic oxygen, in juxtaposition to hydrogen-bond-forming groups of the enzyme. If the substrate to be acted upon is a simple hexose, as in

(25) M. Doudoroff, H. A. Barker and W. Z. Hassid, *J. Biol. Chem.*, **168**, 725 (1947).
(26) A. E. Stearn, *Ergeb. Enzymforsch.*, **7**, 1 (1938).

the case of glucose oxidation or glucose phosphorylation by glucose oxidase and hexokinase respectively, the group undergoing the chemical change and at least one additional *cis*-disposed —OH group will be involved in the linkage between substrate and enzyme. By an attachment of this kind proper orientation of the substrate molecule at the enzyme surface is achieved. It is in this precise orientation effected by complementary groups of enzyme and substrate that the specificity of enzyme reactions resides.[27] We may expect, therefore, to obtain some information about the principles controlling enzyme specificity by collecting from the literature data referring to the effect on the rate of enzyme action of changes in the configuration and constitution of the sugar substrates and discussing them on the basis of the two assumptions outlined in this section.

III. The Effect of Configurational and Substitutional Changes in the Sugar Molecule on the Rate of Various Enzyme Reactions and Its Application in Marking the Contacting Groups in the Substrate

1. β-D-*Glucosidase and* β-D-*Glucosides*

Just as the modern concept of enzyme action is based to a large extent on the thorough investigation of the chemical kinetics of a single enzyme, saccharase, the subject of enzyme specificity owes its origin and much of its growth to research into the properties of the enzyme (or enzyme complex) emulsin. Emulsin, extracted more than a hundred years ago from bitter and sweet almonds by Wöhler and Liebig,[27a] was shown by the same authors to decompose the glycoside amygdalin, prepared previously[28] from the kernels of almonds, according to the equation:

$$C_{20}H_{27}O_{11}N + 2\,H_2O = C_7H_6O + HCN + 2\,C_6H_{12}O_6. \quad (5)$$
Amygdalin Benzaldehyde D-Glucose

(27) A. Gottschalk, *Nature*, **160**, 113 (1947); A. Gottschalk, "Report of the 25th Meeting of the Australian and New Zealand Association for the Advancement of Science," Adelaide, p. 292 (1946).

(27a) F. Wöhler and J. Liebig, *Ann.*, **22**, 1 (1837). In a letter from Göttingen, dated October 28th, 1836, Friedrich Wöhler informs Justus Liebig (Giessen) about his fundamental discovery in three simple sentences. "(1) Amygdalin dissolved in water and digested with a crushed sweet almond, starts instantaneously to smell of essence of bitter almonds, which later on can be distilled off in quantity. (2) A strained emulsion from sweet almonds has the same effect. (3) A boiled emulsion from sweet almonds in which the protein is coagulated does not produce a trace of the essence when acting upon amygdalin." From Justus Liebig's and Friedrich Wöhler's "Briefwechsel in den Jahren 1829–1873." Edited by A. W. Hofmann, F. Vieweg, Braunschweig, Vol. 1, p. 89 (1888).

(28) Robiquet and Boutron-Charlard, *Ann. chim. phys.*, [2] **44**, 352 (1830).

Since then it has been established that emulsin is a mixture of enzymes comprising in addition to β-glucosidase[28a] the enzymes α-galactosidase, α-mannosidase, β-glucuronidase and β-(N-acetyl)-glucosaminidase. Concerning the process of purification of the "Rohferment" and the mode of differentiation of the constituent enzymes of emulsin, the most informa-

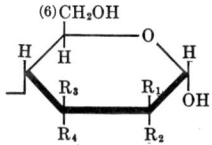

β-D-Glucopyranoside
R = alkyl or aryl group

tive review by Pigman[20] may be consulted. Most of the quantitative work concerning the effect on β-glucosidase activity of structural changes in the substrate has been carried out by Helferich[29] and his collaborators and by Pigman.[20]

β-Glucosidase, prepared from sweet or bitter almonds, acts both on ordinary β-D-glucosides (heterosides) (I) and on disaccharides (holosides),

	R_1	R_2	R_3	R_4
4-D-Glucopyranose residue (in cellobiose)	H	OH	OH	H
4-D-Mannopyranose residue	OH	H	OH	H
4-D-Altropyranose residue (in celtrobiose)	OH	H	H	OH
6-D-Glucopyranose residue (in gentiobiose)	H	OH	OH	H

Fig. 6.—Aldohexopyranose residues of β-D-glucosides (disaccharides)

having an aldohexopyranose residue (Fig. 6) as partner to the glucosidic moiety, the common factor being what may be termed the "glucon," the radical β-D-glucopyranosyl. It is evident that the specificity of β-glucosidase is directed mainly towards the glucon part of its substrate and to a much less degree towards the aglucon which may be an alkyl or aryl group or a residue of an aldohexose. The specificity of the enzyme towards the configuration of the H and OH groups attached to C1, C2,

(28a) The historical names β-glucosidase, β-galactosidase and α-mannosidase, etc. are retained but they are to be understood as short forms for β-D-glucopyranosidase, β-D-galactopyranosidase and α-D-mannopyranosidase, etc.

(29) Cf. B. Helferich, *Ergeb. Enzymforsch.*, **7**, 83 (1938).

C3 and C4 of the glucon is an absolute one. Phenyl α-D-glucoside, phenyl β-D-mannoside (II), methyl β-D-guloside (III), involving reversal of the addenda at C1, at C2, at C3 and C4, respectively, are not appreciably hydrolyzed by almond emulsin.[29] Phenyl β-D-glucoside (I) and phenyl

II   III   IV
Phenyl β-D-mannopyranoside   Methyl β-D-gulopyranoside   Phenyl β-D-galactopyranoside

β-D-galactoside (IV) are split by the "Rohferment" as well as by the silver-purified emulsin in the ratio 10:1.[29] Based mainly on this finding, Helferich assumes identity of β-glucosidase and β-galactosidase in sweet almond emulsin. There may be some doubt regarding the strength of this argument. One could imagine that small amounts of an enzyme

TABLE I[20]

Rates of Enzymatic Hydrolysis of Hexosides and Pentosides with the Same Ring Configuration

Substrate	Ring type	Enzyme value^a (E.V.)	Ratio of enzyme values
Phenyl β-D-glucoside	β-D-Glucopyranose	0.33	100:0.55
Phenyl β-D-xyloside		0.0018	
Phenyl β-D-galactoside	β-D-Galactopyranose	0.032	100:69
Phenyl α-L-arabinoside		0.022	
Phenyl α-D-mannoside	α-D-Mannopyranose	0.10	100:0.94
Phenyl α-D-lyxoside		0.00094	

^{a} Enzyme Value (E.V.), as defined by R. Weidenhagen (*Ergebn. Enzymforschung*, **1**, 168 (1932)) equals $k/(g \times \log 2)$, where k = first order rate constant at 30° (minutes), g = grams of enzyme in 50 ml. of reaction mixture. Substrate concentration in the experiments shown in this table = 0.052 M.

protein closely related in structure to β-glucosidase might survive the purification procedure in about an unchanged proportion to the main enzyme.[30] The following observations seem to favor the existence in almond emulsin of two individual glycosidases rather than that of a single enzyme acting on both β-D-glucosides and β-D-galactosides: (1) Substitution in phenyl β-D-glucoside (I) and phenyl β-D-galactoside (IV) of —CH$_2$OH by H affects in a very different manner the ease of enzymatic

(30) See also E. Hofmann, *Biochem. Z.*, **272**, 133 (1934).

hydrolysis of the substitution products[29] (*cf.* Table I); (2) the ratio k_3 β-glucoside/k_3 β-galactoside[31] varies with pH;[32] (3) the hydrolysis of *o*-cresyl β-D-glucoside is markedly inhibited by D-glucose, but not by D-galactose while the hydrolysis of *o*-cresyl β-D-galactoside is inhibited by D-glucose and by D-galactose to the same extent.[32]

TABLE II
Ratio of Hydrolysis Rates of β-D-Galactoside and β-D-Glucoside with Emulsins from Various Sources

No.	Source of emulsin	Ratio of hydrolysis $\frac{\text{Phenyl β-D-}galactoside}{\text{Phenyl β-D-}glucoside}$
1	Mucor javanicus[33]	0
2	Sacch. fragilis Jørgensen[34]	0
3	Prunus amygdalus[29]	1×10^{-1}
4	Prunus avium[35]	3.3×10^{-1}
5	Aspergillus niger[36]	1.0 (when grown on lactose)
	Aspergillus niger[36]	0.7 (when grown on sucrose)
6	Citrus nobilis[35]	7.2
7	Rosa canina[37]	23.3
8	Glycine soja[35]	∞
9	Lucerne seed[38]	∞
10	Bacillus delbrucki[30]	∞

Furthermore, emulsins from other sources contain β-glucosidases and/or β-galactosidases with markedly different properties.[32a] As may be seen from Table II there is one group of emulsins (Nos. 1–2) hydrolyzing β-D-glucopyranosides, but not β-D-galactopyranosides, another (Nos. 8–10) cleaving β-D-galactopyranosides without effect on β-D-glucopyranosides, and a third group (Nos. 3–7) acting on both stereoisomers but at markedly different rates.[33-38] Very remarkable is the behavior of the

(31) For definition of k_3 see equation 2; substrates used: *o*-cresyl β-D-glucopyranoside and *o*-cresyl β-D-galactopyranoside.

(32) S. Veibel and H. Lillelund, *Enzymologia*, **9**, 161 (1941); S. Veibel, J. Wangel and G. Østrup, *Biochim. Biophys. Acta*, **1**, 126 (1947).

(32a) The first observations in this respect are due to C. Neuberg and E. Hofmann, *Biochem. Z.*, **256**, 450 (1932); E. Hofmann, *ibid.*, **256**, 462 (1932).

(33) E. Hofmann, *Naturwissenschaften*, **22**, 406 (1934).

(34) S. Veibel, C. Møller & J. Wangel, *Kgl. Danske Videnskab. Selskab., Math. fys. Medd.*, **22**, No. 2 (1945).

(35) E. Hofmann, *Biochem. Z.*, **272**, 426 (1934).

(36) E. Hofmann, *Biochem. Z.*, **273**, 198 (1934).

(37) E. Hofmann, *Biochem. Z.*, **267**, 309 (1933).

(38) K. Hill, *Ber. Verhandl. sachs. Akad. Wiss. Leipzig, Math. phys. Klasse*, **86**, 115 (1934); S. Veibel & G. Østrup, *Biochim. Biophys. Acta*, **1**, 1 (1947).

enzyme from *Rosa canina*, which splits phenyl β-D-galactopyranoside and phenyl β-D-glucopyranoside in the ratio 70:3, though the plant belongs to the same family as *Prunus avium* and *Prunus amygdalus*, the enzymes from which show a value for this ratio of less than one. These effects suggest the existence in these emulsins of varying proportions of β-glucosidase and β-galactosidase, an interpretation agreed to by Helferich.[19] It is true that enzymes hydrolyzing the same type of substrates, for example, α-D-glucopyranosides, but derived from different sources (barley maltase and yeast maltase), often vary in their specificity

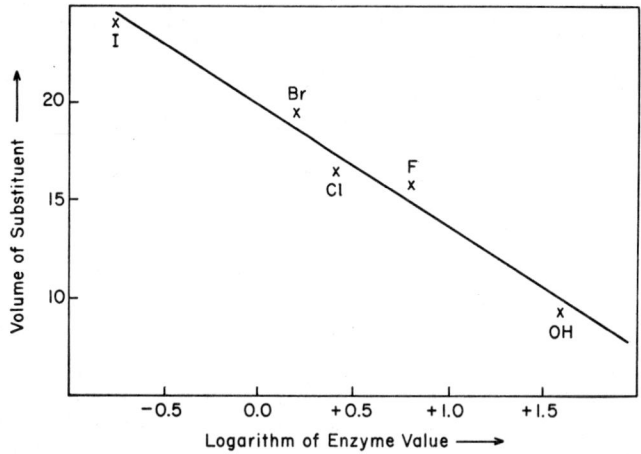

FIG. 7.—Effect of the volume of the substituent at carbon atom six on the hydrolysis rate of vanillin β-D-glucopyranoside (1.04×10^{-3} M conc.) by β-glucosidase.[41]

requirements; but these variations concern pH optimum, affinity constant and the glycosidic partner rather than the configuration or constitution of the glycosyl radical (glycon). There is no other case known where a glycosidase tolerates a change in the configuration at one or more of the carbon atoms 1, 2, 3, 4 of the glycon. It seems preferable, therefore, at the present to assume that β-D-galactopyranosides are not substrates for almond emulsin β-glucosidase, but are acted upon by a specific β-galactosidase.

The effect on β-glucosidase activity of substitution in the glucon has been carefully investigated by the school of Helferich. It has been established that substitution of the hydroxyl groups at C2, C3 and C4 renders the β-D-glucopyranoside unhydrolyzable (i.e., E.V. less than 10^{-5}) by almond emulsin. This refers to phenyl 3-methyl-β-D-glucopyranoside,[29] phenyl 2,4,6-trimethyl-β-D-glucopyranoside[39] and to the

(39) W. W. Pigman and N. K. Richtmyer, *J. Am. Chem. Soc.*, **64**, 374 (1942).

2-tosyl,[40] 3-tosyl and 4-tosyl derivatives of vanillin β-D-glucopyranoside.[29] In contrast to substitution at C1–C4, substitution at C5 and C6 was found to be compatible with enzyme action. The diagram shows that there is a linear relationship between the volumes of the atoms substituting OH at C6 and the logarithms of the enzyme values.[41] As may be predicted from the slope of the line (Fig. 7), substituents having a larger volume than 48, e.g. a tosyl or a glucose residue, render the resulting derivative of β-D-glucopyranoside unattackable by β-glucosidase.[41] Substituting H for CH$_2$OH at C5, thus changing the β-hexoside (phenyl β-D-glucopyranoside (I)) into the corresponding β-pentoside (phenyl β-D-xylopyranoside (V)), has only a quantitative effect on the activity of the enzyme[29] (cf. Table I).

If in β-D-glucopyranoside the linkage between glucon and aglucon is effected by S instead of by O, as is the case in thiophenyl β-D-glucopyranoside, hydrolysis by β-glucosidase does not take place.[42] This

Phenyl β-D-xylopyranoside

result is not surprising since the bond distances O—C (1.42 ± 0.03) and S—C (1.82 ± 0.03) and the electronegativity values for O (3.5) and S (2.5) differ considerably.[43]

As pointed out on page 56, a β-glucosidase will attract the D-glucosyl radical of its substrate by the glucosidic oxygen and one or more OH groups in cis-position to this oxygen. From the above discussion it would appear that of the two OH groups conforming to this requirement (at C3 and C6 of the D-glucopyranosyl moiety) only the OH group at C3 is involved in the bonding with the active center of the enzyme. Concerning the contact between the aglucon and the enzyme, the available information[29] suggests that with alkyl β-D-glucopyranosides as substrate this contact is established by physical adsorption (dispersive forces) rather than by chemisorption (Pigman[20]). When the aglucon is a D-glucose residue (as in cellobiose and gentiobiose) or the residue of an aromatic compound possessing one or more polar groups (as in vanillin β-D-gluco-

(40) Tosyl = p-toluenesulfonyl.
(41) B. Helferich, S. Grünler and A. Gnüchtel, Z. physiol. Chem., **248**, 85 (1937).
(42) W. W. Pigman, J. Research Natl. Bur. Standards, **26**, 197 (1941).
(43) L. Pauling, "The Nature of the Chemical Bond," 2nd edition, p. 60, Cornell University Press, Ithaca, New York (1944).

pyranoside), these groups may add to the forces of attraction between aglucon and enzyme.

2. α-D-*Mannosidase and* α-D-*Mannopyranosides*

α-Mannosidase (*i.e.*, α-D-mannopyranosidase) from almond emulsin splits phenyl α-D-mannopyranoside (VI) and at much lower rates[44] the related pentoside phenyl α-D-lyxopyranoside (VII) and phenyl 2-desoxy-α-D-mannopyranoside (VIII).[29] Phenyl β-D-mannopyranoside[29] and

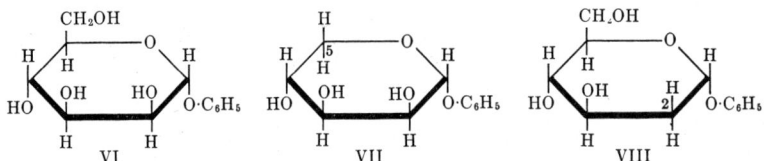

Phenyl α-D-mannopyranoside   Phenyl α-D-lyxopyranoside   Phenyl 2-desoxy-α-D-mannopyranoside

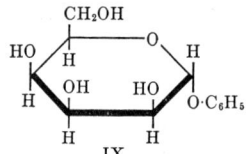

Phenyl α-D-talopyranoside

phenyl α-D-talopyranoside (IX)[42] are not hydrolyzed by almond emulsin. We conclude from these results that in phenyl α-D-mannopyranoside the glycosidic oxygen and the OH group at C4 are the glycon groups establishing contact with the enzyme.

3. α-*Galactosidase and* α-D-*Galactopyranosides*

The activity of α-galactosidase (*i.e.*, α-D-galactopyranosidase) from brewer's yeast towards various α-D-galactopyranosides has been studied by Adams, Richtmyer and Hudson.[45] This enzyme acts on phenyl α-D-galactopyranoside (Xa), melibiose (= 6-α-D-galactopyranosyl-D-glucose (Xb)), methyl α-D-galactopyranoside (Xa) and phenyl β-L-arabinopyranoside (Xc) in the ratio 100:82.1:2.5:0.0015. α-Manninotriose (= 6-α-D-galactopyranosyl-4-α-D-galactopyranosyl-α-D-glucopyranose), *i.e.*, stachyose (*cf.* XIV) minus D-fructofuranosyl, is completely hydrolyzed to its component aldohexoses by the enzyme. However, methyl α-D-manno-D-*gala*-heptopyranoside (Xd), derived from methyl α-D-galactopyranoside by substituting at C6 the group CH_2OH for H, is not a

(44) See Table I. For phenyl 2-desoxy-α-D-mannopyranoside, E.V. = 2.9×10^{-4}.
(45) Mildred Adams, N. K. Richtmyer and C. S. Hudson, *J. Am. Chem. Soc.*, **65**, 1369 (1943).

substrate for yeast α-galactosidase. Emulsin from sweet almond contains an α-galactosidase which is active towards phenyl α-D-galactopyranoside and melibiose in the ratio[46] 100:90; the pentoside phenyl β-L-arabinopyranoside is hydrolyzed at a lower rate, though it would

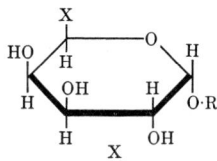

	R	X
a. Methyl or phenyl α-D-galactopyranoside	CH_3 or C_6H_5	CH_2OH
b. Melibiose = 6-α-D-galactopyranosyl D-glucose	glucose residue	CH_2OH
c. Phenyl β-L-arabinopyranoside	C_6H_5	H
d. Methyl α-D-manno-D-*gala*-heptopyranoside	CH_3	$CHOH \cdot CH_2OH$

appear from the published data that the enzyme from almonds is more active towards this substrate than is the yeast enzyme.[29] The corresponding heptoside is not attacked.[42] α-Galactosidase becomes inoperative when the configuration at C2 or at C4 of the α-D-galactopyranoside is changed. These observations indicate that besides the glycosidic oxygen the hydroxyl group at C2 and most probably also that at C6 take a direct part in binding the substrate to the enzyme.

4. β-D-*Fructofuranosidase and Sucrose*

Valuable information regarding the specificity requirements of enzymes acting on carbohydrates may be expected from a discussion of fructofuranosides as substrates for enzyme action. Not only is the hydrolysis of sucrose by β-fructofuranosidase (saccharase, invertase) one

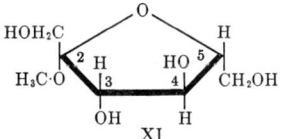

Methyl β-D-fructofuranoside

of the most thoroughly investigated enzyme reactions, but the recent discovery of a phosphorolytic degradation of sucrose affords the opportunity to compare the critical demands of two catalysts acting on the same substrate but directing the chemical change into different channels. It has long been known that yeast saccharase hydrolyzes methyl β-D-fructofuranoside (XI)[47] and the oligosaccharides[47a] sucrose (Fig. 2),

(46) R. Weidenhagen and A. Renner, *Z. Ver. deut. Zucker-Ind.*, **86**, 22 (1936).
(47) H. H. Schlubach and G. Rauchalles, *Ber.*, **58**, 1842 (1925). See also ref. 48.
(47a) See E. Bourquelot and M. Bridel, *Compt. rend.*, **152**, 1060 (1911).

raffinose (XII), gentianose (XIII) and stachyose (XIV). These fructofuranosides possess an unsubstituted D-fructofuranosyl residue linked in β-fashion to an alkyl group or to an α-D-glucopyranosyl residue which may be either unsubstituted (sucrose) or substituted at C6 (raffinose,

Raffinose (6-α-D-Galactopyranosyl-α-D-glucopyranosyl-β-D-fructofuranoside)

gentianose) and C4 (stachyose), respectively. Under comparable conditions yeast saccharase hydrolyzes sucrose 13.5 times as fast as it does methyl β-D-fructofuranoside.[48] The most exact data concerning the relative rates at which sucrose, raffinose and stachyose are acted upon by

Gentianose (6-β-D-Glucopyranosyl-α-D-glucopyranosyl-β-D-fructofuranoside)

highly purified yeast saccharase, were recently provided by Adams, Richtmyer and Hudson.[45] Table III, which gives a summary of their measurements, demonstrates the decrease in reaction rate with increasing bulk of the substituent. Though the ratio a:b varies to some extent with

Stachyose (6-α-D-Galactopyranosyl-4-α-D-galactopyranosyl-α-D-glucopyranosyl-β-D- fructofuranoside)

the source of the enzyme and is not even constant for preparations obtained by different purification methods from the same type of yeast, there can be little doubt that one enzyme is responsible for the hydrolytic cleavage of oligosaccharides characterized by the presence of an unsubstituted β-D-fructofuranosyl residue.[49]

(48) C. B. Purves and C. S. Hudson, *J. Am. Chem. Soc.*, **56**, 702 (1934).
(49) R. Kuhn, *Z. physiol. Chem.*, **125**, 28 (1923).

The indifference of β-fructofuranosidase towards substitution in the "afructon" part of sucrose is contrasted by its extreme sensitivity towards any change in the structure and configuration of the "fructon." Change from the furanose to the pyranose ring structure in methyl β-D-fructoside is incompatible with the action of β-fructofuranosidase.[50]

TABLE III[45]
Hydrolysis Rates of Sucrose, Raffinose and Stachyose by Purified Yeast β-Fructofuranosidase

Type of yeast	Method of purification	Enzyme value		Ratio a/b	Enzyme value		Ratio c/d
		Sucrose^a (a)	Raffinose^b (b)		Sucrose^a (c)	Stachyose^b (d)	
Baker's A	Bentonite, picric acid	791	182	4.35	2460	167	14.7
Baker's B	Acetic acid, acetone	731	156	4.69			
Brewer's A	Bentonite, picric acid	885	103	8.59			
Brewer's B	Acetic acid, ammon. sulfate	544	87.5	6.22			
Brewer's B	Bentonite, ammon. sulfate	884	111	7.96	2749	85	32.3

^{a} Standard conditions. Substrate conc. = 0.1388 M.
^{b} Substrate conc. = 0.02776 M.

The same holds for substitution at C3 and C5 of the fructofuranose moiety of sucrose: neither melezitose (XV)[51] nor α-D-glucopyranosyl-β-D-xyloketofuranoside (XVIc)[22] is hydrolyzed by yeast saccharase. As might have been expected from the experience with glycosidases, any change in the configuration of the "fructon" of sucrose results in stereoisomers unhydrolyzable by β-fructofuranosidase. Thus, methyl and benzyl α-D-fructofuranosides,[52] isosucrose (= β-D-glucopyranosyl-α-D-fructo-

(50) C. S. Hudson and D. H. Brauns, *J. Am. Chem. Soc.*, **38**, 1216 (1916); R. Weidenhagen, *Z. Ver. deut. Zucker-Ind.*, **82**, 912 (1932).
(51) Though there is no direct evidence that the D-fructofuranose unit in melezitose is of β-configuration, the facts that all naturally occurring fructofuranoside sugars known so far are sucrose or sucrose derivatives and that the equilibrium mixture of D-fructose appears to consist almost entirely of β-D-fructopyranose and β-D-fructofuranose, are very suggestive of a β-configuration. For a full discussion of the problem see C. S. Hudson, *Advances in Carbohydrate Chem.*, **II**, 1 (1946).
(52) C. B. Purves and C. S. Hudson, *J. Am. Chem. Soc.*, **56**, 708 (1934); **59**, 49 (1937).

furanoside)[45,53] and α-D-glucopyranosyl-α-L-sorbofuranoside (XVIb),[22] involving reversal of the sucrose configuration at C2 and C5 respectively of the "fructon," are not attacked by yeast saccharase. These results clearly indicate that β-fructofuranosidase requires for action a β-D-fructo-

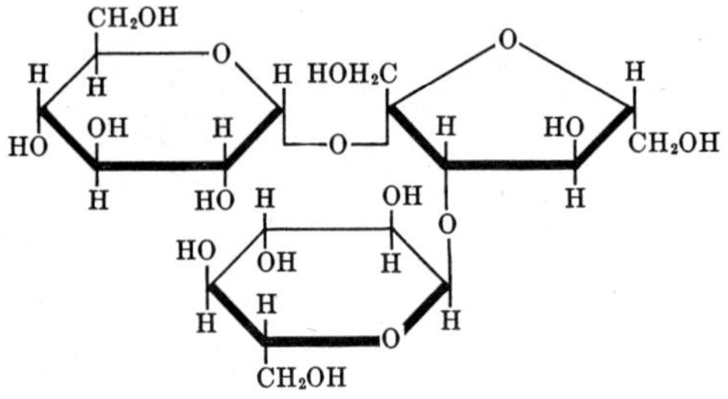

XV

Melezitose (3-α-D-Glucopyranosyl-β(?)-D-fructofuranosyl-α-D-glucopyranoside)

furanosyl radical with OH groups unchanged and not dislocated in space. One may suggest, therefore, that sucrose combines with the β-fructofuranosidase of yeast by the fructosidic oxygen and the hydroxyl groups at C6 and C3 of fructofuranosyl and possibly by the OH group at C2 of glucopyranosyl (see Fig. 2).

5. *Transglucosidase and Sucrose*

Whereas β-fructofuranosidase catalyzes the irreversible[54] hydrolytic separation of β-D-fructofuranosyl from its glucosidic partner in sucrose, transglucosidase (sucrose phosphorylase) accelerates the reversible transfer of the glucose residue of sucrose from fructofuranosyl to another acceptor (*cf.* equations 3 and 4):

α-D-Glucopyranosyl-β-D-fructofuranoside + acceptor
$\rightleftarrows$ α-D-glucopyranosyl–acceptor + β-D-fructofuranose.   (6)

The enzyme has been found in *Leuconostoc mesenteroides*[55] and *Pseudomonas saccharophila*.[56] In a series of outstanding researches Doudoroff and Hassid with associates investigated the specificity requirements of sucrose phosphorylase from *P. saccharophila*. It was found that the enzyme exhibits an absolute specificity for the D-glucose moiety of its

(53) A. Georg, *Helv. Chim. Acta*, **17**, 1566 (1934); C. B. Purves and C. S. Hudson, *J. Am. Chem. Soc.*, **59**, 1170 (1937).
(54) Applying physiological conditions regarding substrate concentration, solvent and temperature, the enzymatic inversion of sucrose is practically irreversible.
(55) B. O. Kagan, S. N. Lyatker and E. M. Tsvasman, *Biokhimiya*, **7**, 93 (1942).
(56) M. Doudoroff, N. Kaplan and W. Z. Hassid, *J. Biol. Chem.*, **148**, 67 (1943).

substrate. Melibiose-1-phosphate[57] (*1*), α-D-xylose-1-phosphate (*2*), α-maltose-1-phosphate (*3*), α-D-galactose-1-phosphate (*4*) and α-D-mannose-1-phosphate (*5*) are unable to replace α-D-glucose-1-phosphate in its reaction with D-fructofuranose.[58a,b] In other words, substitution at C6 (*1*), at C5 (*2*) and C4 (*3*) respectively, or a configurational change at C4

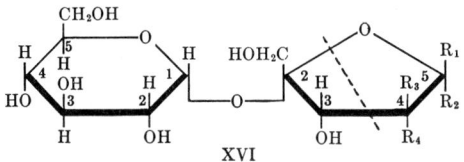

XVI

		R_1	R_2	R_3	R_4
a.	α-D-Glucopyranosyl-β-D-fructofuranoside (*i.e.*, sucrose)	H	CH$_2$OH	OH	H
b.	α-D-Glucopyranosyl-α-L-sorbofuranoside	CH$_2$OH	H	OH	H
c.	α-D-Glucopyranosyl-β-D-xyloketofuranoside	H	H	OH	H
d.	α-D-Glucopyranosyl-β-L-arabinoketofuranoside	H	H	H	OH

(*4*) and C2 (*5*) respectively, in α-D-glucose-1-phosphate render the compound unsuitable for enzyme action. It would appear, therefore, that the glucose donator, in order to combine with the enzyme, must contain an unsubstituted α-D-glucopyranosyl radical.

As glucose acceptor, however, such different substances as inorganic phosphate, keto-monosaccharides and an aldopentose can be used in

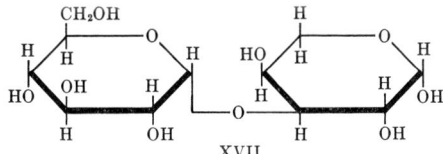

XVII

3-α-D-Glucopyranosyl-L-arabinose

the transfer reaction. If in the presence of the bacterial phosphorylase glucose-1-phosphate is allowed to react with D-fructose or L-sorbose or D-xyloketose or L-araboketose the non-reducing disaccharides XVIa, b, c, d are formed. The addition of the aldopentose L-arabinose to a mixture of glucose-1-phosphate and enzyme results in the production of the reducing disaccharide 3-α-D-glucopyranosyl-L-arabinose[59] which is depicted by Doudoroff, Hassid and coworkers[58a,59] as shown in XVII. On comparing XVI and XVII, Doudoroff, Barker and Hassid[58a] find it

(57) That melibiose-1-phosphate cannot react with fructofuranose is inferred from the inability of the enzyme to phosphorolyze raffinose.

(58) (a) M. Doudoroff, H. A. Barker and W. Z. Hassid, *J. Biol. Chem.*, **168**, 733 (1947); (b) W. Z. Hassid, M. Doudoroff and H. A. Barker, *Arch. Biochem.*, **14**, 29 (1947).

(59) W. Z. Hassid, M. Doudoroff, A. L. Potter and H. A. Barker, *J. Am. Chem. Soc.*, **70**, 306 (1948).

impossible to define exactly the structural feature common to the various glucose acceptors. If, however, the relationship in space between the interacting OH group at C3 of L-arabinose and the glucosidic oxygen of α-D-glucopyranosyl is taken into account, XVIII may be a more appropriate projection of the configuration of 3-α-D-glucopyranosyl-L-arabinose. Visualized in this manner, the close structural relationship of all glucose

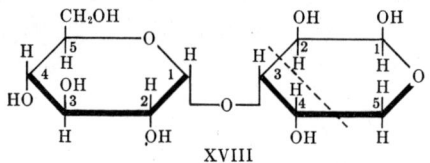

XVIII

3-α-D-Glucopyranosyl-L-arabinopyranose

acceptors, hexose or pentose, ketose or aldose, becomes at once evident (*cf.* dotted lines in XVI and XVIII). In order to satisfy the specificity requirements of transglucosidase, the glucose acceptor must possess adjacent to the glucosidic oxygen an OH group *cis*-disposed and co-directional[60] to the OH group at C2 of the α-D-glucopyranosyl residue. This configuration is not present in D-arabinose, L-xyloketose, or D-xylose,

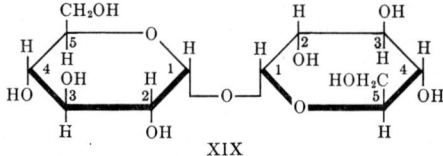

XIX

Trehalose (α-D-Glucopyranosyl-α-D-glucopyranoside)

compounds found to be unsuitable[61] as glucose acceptors in this reaction.[58a] A comparison between XVIII and XIX makes it clear why sucrose phosphorylase is inoperative towards trehalose (XIX).

One might be tempted to bring some structural features of the phosphate ion in line with those of the glucose acceptors represented in XVI and XVIII. Thus, in the tetrahedral ion

$$\overset{\overline{O}}{\underset{\underset{O}{\parallel}}{\overline{O}-P-\overline{O}}}$$

(60) The term *cis*-disposed refers to the mean plane of the disaccharide, co-directional to its longitudinal axis.

(61) The inability of D-tagatose, a stereoisomer of D-fructose with a reverse configuration at C4, to act as glucose acceptor may be due to the accumulation at C3, C4, C5 of three OH groups in *cis*-position to the OH group at C2 of the β-modification or to the existence of D-tagatose in the α-form mainly.

the length of the P—O bonds, all of them being equal due to resonance, was shown by Pauling and Brockway[62] to be 1.55 Å, which approximates the length of the C—O (1.43 Å) and C—C (1.54 Å) bonds in sugars. It would be premature, however, to press this point.

We may conclude from this discussion that in the formation of the phosphorylase–sucrose complex the glucosidic oxygen, the OH groups at C6, C4 and C2 of the glucopyranose moiety and the OH group at C3 of the fructofuranose moiety of the substrate are involved.

When sucrose interacts with β-fructofuranosidase, the hydroxyl groups of the fructose portion of the disaccharide are the main anchoring groups; it is acted upon mainly because of the presence in its molecule of a β-D-fructofuranosyl residue, as proved by the ability of methyl-β-D-fructofuranoside to serve as substrate for this enzyme. In the interaction between sucrose and transglucosidase, on the other hand, sucrose reacts chiefly in the capacity of an α-D-glucopyranoside, as evidenced by the readiness with which phosphoryl-α-D-glucopyranoside is attacked. These differences in the attachment groups of the sucrose molecule when forming intermediate compounds with two different enzymes afford some insight into the mechanism by which enzymes select one out of the several possible reactions.

6. *Hexokinase (Yeast) and* D-*Glucose*

As bacterial transglucosidase is instrumental in the transfer of a D-glucose residue from one acceptor to another, so does yeast hexokinase[63] catalyze a transphosphorylation. The highly specific donator of a labile phosphate group is adenosine triphosphate (XX), the fermentable hexoses D-glucose, D-mannose and D-fructose functioning as acceptors. Hexokinase catalyzes the reaction

Zymohexose + adenosine triphosphate
$$\rightarrow \text{Hexose-6-phosphate} + \text{adenosine diphosphate} \quad (7)$$

Hexokinase is of great biological interest since it would appear that not only in yeast cells but in most, if not all, plant and animal cells phosphorylation at C6 of the common hexoses D-glucose, D-fructose and D-mannose initiates sugar utilization. Since on solution in water the crystalline hexoses quickly undergo mutarotation, resulting in an equilibrium mixture of various tautomeric modifications, the fermentability

(62) L. Pauling and L. O. Brockway, *J. Am. Chem. Soc.*, **59**, 13 (1937).
(63) Hexokinase was first extracted from yeast and partially purified by O. Meyerhof, *Biochem. Z.*, **183**, 176 (1927). It was crystallized by (a) L. Berger, M. W. Slein, S. P. Colowick and C. F. Cori, *J. Gen. Physiol.*, **29**, 379 (1946) and (b) M. Kunitz and M. R. McDonald, *ibid.*, **29**, 393 (1946).

of a sugar does not provide information as to which of the various isomerides present is subject to hexokinase action. In order to obtain more detailed knowledge in this respect Gottschalk carried out experiments at low substrate concentration (0.005 M) under conditions (0°C.; $pH = 4.5$) which reduce the rate of mutarotation to a minimum. It was found that α- and β-D-glucopyranose are fermented by yeast without preliminary interconversion (mutarotation);[64] the same holds for α- and β-D-mannopyranose;[65] with both aldoses the rate of fermentation of the α-modification is about 10% higher than that of the β-form. If β-D-fructopyranose, the only fructose form known to crystallize, is added to a

XX

Adenosine-5′-triphosphate (ATP)
(Adenosine = 9-β-D-ribofuranosyl-adenine)

yeast suspension at 0°C., the rate of fermentation is only a small fraction of that obtained with α-D-glucopyranose as substrate; moreover, in contradistinction to the fermentation rate of α-D-glucopyranose, that of β-D-fructopyranose does not vary with yeast concentration. In fact, the rate almost coincides with that at which β-D-fructopyranose is converted by mutarotation into D-fructofuranose.[64] With β-D-fructofuranose, set free from sucrose by yeast saccharase, as substrate, however, fermentation at 0°C. proceeds readily.[66] Furthermore, in experiments at 25°C. it was possible to make the rate of mutarotation the limiting factor of D-fructose fermentation by allowing a large quantity of yeast to act on a relatively small amount of D-fructose.[67] These findings seem to substantiate the hypothesis, first proposed by Hopkins,[68] that only

(64) A. Gottschalk, *Australian J. Exptl. Biol. Med. Sci.*, **21**, 133 (1943).
(65) A. Gottschalk, *Biochem. J.*, **41**, 276 (1947).
(66) A. Gottschalk, *Australian J. Exptl. Biol. Med. Sci.*, **23**, 261 (1945).
(67) A. Gottschalk, *Biochem. J.*, **41**, 478 (1947).
(68) R. H. Hopkins, *Biochem. J.*, **25**, 245 (1931).

the furanose form of D-fructose is fermentable by yeast, whereas the pyranose modification is unfermentable. By determining the dissociation constants of the hexokinase–glucopyranose and the hexokinase–fructofuranose compounds it was shown that D-fructofuranose has nearly twice the affinity of D-glucopyranose for hexokinase.[69] Using crystalline hexokinase and a substrate concentration of 0.037 M (with excess adenosine triphosphate) Cori and coworkers[63a] found the relative rates of D-glucose, D-fructose and D-mannose phosphorylation to be 1:1.4:0.3. Configurational change in D-glucopyranose at C4 resulting in D-galactose, or substitution at C1 by a bulky group, as is the case in sucrose, renders the compound unsuitable as phosphate acceptor in the reaction catalyzed by hexokinase.[63b] β-D-Fructofuranose[70] (XXI) and the α- and β-modi-

XXI β-D-Fructofuranose

XXII α(β)-D-Glucopyranose

XXIII α(β)-D-Mannopyranose

fications of D-glucopyranose (XXII) and D-mannopyranose (XXIII) have as a common feature the same side-chain at C5 and the same configuration at C5, C4 and C3. Contacting groups of the various substrates will be the OH group at C6 as the group undergoing a chemical change in the enzymatic reaction, and one or more OH groups in *cis*-position to the hydroxyl at C6 and present in that part of the substrate molecule which is common to D-glucose, D-fructose and D-mannose. Since there is free rotation about the C5—C6 single bond, the attachment groups of the zymohexoses may be either the OH groups at C6 and C3 or those at C6 and C4. Though it is not possible to make a final decision between these alternatives, the latter would appear to be the more probable one.[71]

(69) A. Gottschalk, *Australian J. Exptl. Biol. Med. Sci.*, **22**, 291 (1944).

(70) The mutarotation of D-fructose consists for the main part of the reaction β-D-fructopyranose ⇌ β-D-fructofuranose (C. S. Hudson, *J. Am. Chem. Soc.*, **31**, 655 (1909); **52**, 1707 (1930); H. S. Isbell and W. W. Pigman, *J. Research Natl. Bur. Standards*, **20**, 773 (1938)). At 0°C. (pH = 4.3) a solution of D-fructose at equilibrium contains about 12% of the sugar in the furanose form (A. Gottschalk, *Australian J. Exptl. Biol. Med. Sci.*, **21**, 139 (1943)). Whether the furanose fraction contains small amounts of the α-isomer is unknown.

(71) The observation that hexokinase preparations from ox retina react with D-glucose and D-fructose but not with 3-methyl D-glucose (M. Kerly, *Biochem. J.*, **42**, xx (1948)) has no bearing on this question, since animal hexokinase differs from yeast hexokinase (see Gerti T. Cori and M. W. Slein, *Federation Proc.*, **6**, No. 1 (1947)).

IV. Discussion of the Principles Controlling and of the Factors Impairing the Formation of the Intermediate Compound between Carbohydrates and Their Specific Enzymes

In Section II the conclusion was reached that enzyme action involves chemical interaction between enzyme and that group of the substrate undergoing a change in the catalyzed reaction. Enzyme specificity was visualized as resulting from the close approach of atoms or groups of the substrate to atoms or groups of the protein (enzyme) surface, an approach demanding of both partners in the reaction appropriate spacing of the groups within the molecule and suitable electronic structure so as to permit of mutual attraction. The closest fit between a globular protein like an enzyme and a substantial part of the surface area of a sugar molecule is afforded by orientation of the latter with the mean plane of the pyranose or furanose ring parallel to the protein surface, thus bringing short-range forces into operation between atoms of the enzyme protein and of the substrate. The location at the upper or the lower side of the sugar molecule of that atom or group, the electronic structure of which is subject to deformation, will determine which side of the substrate molecule is facing the enzyme when forming the enzyme–substrate compound and which OH groups of the sugar molecule are available as additional attachment groups. By inference from the effect on enzyme activity of structural changes in the substrate, those of the available groups which are essential for binding the carbohydrate to the enzyme can be identified. In Section III this was done for several enzyme reactions. From an analysis of the experimental data it would appear that the number of attachment groups and the area of the surface of the sugar molecule which is in close contact with the active center of the catalyst, vary with different enzymes. There can be little doubt that in the case of β-glucosidase from almonds the combining region of the glucon moiety of the substrate is restricted to the area of the pyranose ring, the distance of the side chain —CH_2OH from the enzyme surface being large enough to allow within certain limits the introduction of substituents at C6. In the formation of the transglucosidase–phosphoryl–α-D-glucopyranoside intermediate compound the glucosidic oxygen and the OH groups at C2, C4 and C6 of the glucopyranosyl radical, *i.e.*, all of the available hydroxyl groups of the sugar, are involved, thus bringing the largest possible surface area of the glucon moiety in close proximity to the surface of the enzyme. Similar conditions seem to obtain with the fructofuranose part of β-D-fructofuranosides when acted upon by yeast β-fructofuranosidase, and with α-D-galactopyranosides in their reaction with yeast α-galactosidase.

In the reactions catalyzed by glycosidases and by transglucosidase the surface *at least* of the pyranose or furanose ring is at the minimum distance of approach to the active center of the enzyme protein; any change, therefore, in the configuration of the ring will either remove from its proper spatial position a contacting group or produce a steric hindrance by displacing a hydrogen atom with the more voluminous hydroxyl group. Both operations will impede the mutual attraction between enzyme and substrate. It is for this reason that the activity of glycosidases is restricted to a single[72] type of the aldopyranose ring.

Another alteration seriously impairing the mutual attraction of enzyme and substrate is any change in the effective amount or the distribution of electric charge within the atoms and groups composing the pyranose ring. This is clearly demonstrated by the observation that substitution by H of the OH group at C2 in phenyl α-D-mannopyranoside reduces the α-mannosidase activity to one-thousandth of that of the non-substituted substrate, though this OH group is spatially arranged in such a manner as not to face the enzyme surface.

With regard to the aglycon, some glycosidases display only a small measure of specificity. Thus, β-glucosidase from almonds splits a large number of alkyl β-D-glucopyranosides, substituted phenyl β-D-glucopyranosides and various disaccharides, the aglucons of the latter differing in constitution as well as in configuration (*cf*. Fig. 6). Other glycosidases are predominantly disaccharases inasmuch as they react preferentially with holosides. In these cases the aglucon specificity is as pronounced as is the glucon specificity; *e.g.*, some α-glucosidases are specific maltases with little effect on heterosides and apparently no effect on disaccharides which differ from maltose only in the constitution of the aglucon without any configurational change (trehalose). It seems likely that enzymes of this kind combine with both moieties of the disaccharide molecule, the target area comprising the two pyranose rings with their side chains; in other words the combining region coincides with the total surface area of that side of the substrate which faces the surface of the enzyme.

Hexokinase tolerates in its substrate, α-D-glucopyranose, a configurational change at C1 and C2 and the structural change from the pyranose 1,5 to the furanose 1,4 ring. The target area in this case includes besides the side-chain only a portion of the ring (*cf*. XXI–XXIII). The small influence on the reaction rate of a configurational change in D-glucopyranose at C1 and the considerable reduction of the rate effected by the same change at C2 indicate that the impairing effect of structural alterations in the substrate molecule outside its combining region with the enzyme is largely dependent on the site of the alteration, increasing with approach to the target area.

(72) W. W. Pigman, *J. Am. Chem. Soc.*, **62**, 1371 (1940).

It would thus appear that it is possible to mark exactly in a great number of carbohydrates the chemical groups which, by establishing contact with complementary groups of the enzyme, orientate the substrate at the enzyme surface. Whereas orientation of the sugar molecule at the surface of the respective catalyst may be regarded as the qualitative factor in enzyme specificity (*conditio sine qua non*), the ratio

$$\frac{\text{Surface area of the substrate molecule effective as combining region}}{\text{Surface area of the substrate molecule available as combining region}}$$

is a quantitative factor determining the degree of specificity, *i.e.*, the sensitivity of the enzyme towards configurational or substitutional changes in the sugar molecule.

ENZYMES ACTING ON PECTIC SUBSTANCES[1]

By Z. I. Kertesz and R. J. McColloch[1a]

New York State Agricultural Experiment Station, Cornell University, Geneva, New York

Contents

I. Introduction.. 79
II. Pectin Chemistry and Nomenclature........................... 80
III. Possible Changes Brought About in Pectic Substances by Enzymes....... 82
IV. Protopectinase... 84
V. Pectin-polygalacturonase (PG)................................ 85
    1. Nomenclature... 85
    2. Occurrence... 85
    3. Methods of Determination................................. 86
    4. Properties... 89
        a. Effect of Substrate................................. 89
        b. Activation and Inhibition........................... 89
        c. Temperature and pH Effects........................ 90
        d. Kinetics of PG Action............................... 91
    5. "Depolymerase"... 92
VI. Pectin-methylesterase (PM).................................. 92
    1. Nomenclature... 92
    2. Occurrence... 93
    3. Methods of Determination................................. 93
    4. Preparation.. 95
    5. Properties... 96
        a. The PM of Higher Plants............................. 96
        b. The PM of Molds..................................... 98
VII. Role and Application of Pectic Enzymes...................... 98
    1. In Undisturbed Plant Tissue.............................. 98
    2. Action of Pectic Enzymes in Plant Tissue Macerates....... 99
    3. Commercial Pectic Enzyme Preparations................... 101

I. Introduction

The general acceptance of the view that the pectic substances are heterogeneous macromolecular polyuronides[2] resulted in rapid advance

(1) Approved for publication as Journal Article No. 721 of the New York State Agricultural Experiment Station.

(1a) Present address: U. S. Fruit and Vegetable Research Laboratory, Pasadena, California.

(2) H. Meyer and H. Mark, "Aufbau der hochmolekularen Naturstoffe"; Akademische Verlagsgesellschaft, Leipzig (1930); J. S. Morell, L. Baur and K. P. Link, *J. Biol. Chem.*, **105**, 1 (1934); L. Baur and K. P. Link, *ibid.*, **109**, 293 (1935); F. A. Henglein and G. Schneider, *Ber.*, **69**, 309 (1936); E. L. Hirst, *J. Chem. Soc.*, **70** (1942).

in the elucidation of their architecture as well as in a somewhat better understanding of the enzymes acting upon them. The main trend in pectin chemistry during the past decade was the application to pectic substances of many of the now standard methods developed for other polysaccharides such as starch and cellulose. As a result, a great deal of useful information has been collected on the macromolecular nature, behavior and types and degrees of heterogeneity of pectinic and pectic acids. Nevertheless, there is still much to be done to arrive at an exact knowledge of the chemical and physical structure of the macromolecules.

In contrast, a comparatively small proportion of recent researches dealing with the enzymes acting on pectic materials proceeded on classical lines involving purification and constitutional studies. With the exception of a few papers, most reports deal with the action and behavior of these enzymes without extensive purification and paying little attention to the chemical composition and constitution of the enzymes themselves. The reason for this might be the general preoccupation with the development of suitable methods for the exact determination of the action of pectic enzymes, which in turn is the result of our lack of knowledge of the exact structure of pectic substances.

The subject of pectic enzymes has been previously reviewed by the senior author[3] and others[4] and therefore this article will be confined to advances made during the past decade. Although the chemistry of pectic materials has been dealt with in a previous volume of this series by Hirst and Jones,[5] we feel that a short review of the general status of pectin chemistry and especially of nomenclature is needed here to facilitate later discussions.

II. Pectin Chemistry and Nomenclature

It is now clear that the basic skeletons of all pectic substances consist essentially of poly-α-D-galacturonic acids the constituent units of which contain the pyranose ring and are linked glycosidically in α configuration through positions 1 and 4.[2,6]

Molecular weight values reported for various pectic substances are obviously averages and cover a wide range[7] from 20,000 to over 200,000.

(3) Z. I. Kertesz, *Ergeb. Enzymforsch.*, **5**, 233 (1936).

(4) F. Ehrlich in E. Abderhalden's "Handbuch der biologischen Arbeitsmethoden"; IV, Part 2, 2405 (1936); H. Bock in E. Bamann and K. Myrbäck's "Die Methoden der Enzymforschung"; George Thieme Verlag, Leipzig, p. 1914 (1941).

(5) E. L. Hirst and J. K. N. Jones, *Advances in Carbohydrate Chem.*, **2**, 235 (1946).

(6) P. A. Levene and L. C. Kreider, *J. Biol. Chem.*, **120**, 591 (1937); P. A. Levene, G. M. Meyer and M. Kuna, *Science*, **89**, 370 (1939).

(7) S. Säverborn, "A Contribution to the Knowledge of the Acid Polyuronides"; Almquist och Wiksells Boktryckery A. B., Uppsala (1945).

The important point is that the qualitative and quantitative heterogeneity of samples of pectic materials is now clearly realized.[8] The main factors of heterogeneity are in the molecular size and weight distribution of the polygalacturonic acid nucleus and the number, distribution and spacing of ester groups along the polygalacturonic acid chains. Attempts to obtain uniform or at least significantly uniform samples have not thus far been successful. The nomenclature recently accepted by the American Chemical Society[9] defines the various pectic substances as follows:

Protopectin.—This is the yet ill-defined water-insoluble parent pectic substance of plants, which upon restricted hydrolysis yields pectinic acids. The structure and indeed even the composition of protopectin is still a debated point. The theory that protopectin is a combination of pectinic acids with cellulose is not entirely discarded, although views that protopectins represent very large pectinic acid molecules[10] involving linkages other than the 1,4-glycosidic connections between the individual galacturonic acid units[11-13] are accepted by an increasing number of workers in this field. The possibility that the insolubility of protopectins is caused by association with proteins[13a] or by the formation of insoluble salts or by a combination of all these various factors should not be overlooked.

Pectinic Acids.—These are colloidal polygalacturonic acids containing more than a negligible proportion of methyl ester groups. Neither the structure of the pectinic acid macromolecule nor the molecular weight limit dividing non-pectic polygalacturonic acids and pectic substances are known. Pectinic acids will form, under suitable conditions, gels with sugar and acid, or, if suitably low in ester content, with certain polyvalent metallic ions.

Pectin.—This general term has been retained to designate pectinic acids which will form the customary pectin–sugar–acid gels.

Pectic Acid.—This term is used for colloidal polygalacturonic acids free of ester groups. The term "colloidal" is used here to define the size or condition in which a polygalacturonic acid becomes a pectic acid.

(8) R. Speiser and C. R. Eddy, *J. Am. Chem. Soc.*, **68**, 287 (1946).

(9) Z. I. Kertesz, G. L. Baker, G. H. Joseph, H. H. Mottern and A. G. Olsen, *Chem. Eng. News*, **22**, 105 (1944).

(10) Z. I. Kertesz, "Note on the nature of protopectin." Paper given at the Memphis, Tennessee, meeting of the Am. Chem. Soc., April, 1942.

(11) Z. I. Kertesz, *J. Am. Chem. Soc.*, **61**, 2544 (1939).

(12) F. A. Henglein, *J. makromol. Chem.*, **1**, 121 (1943).

(13) H. Bock, "Theorie und Praxis der Pektingewinnung"; Dissertation, Karlsruhe, 1943.

(13a) H. Colin and A. Chaudun, *Compt. rend.*, **198**, 2116 (1934); A. Dauphine, *Compt. rend.*, **196**, 1738 (1933).

It will be clear that in this nomenclature arabans and galactans are not included as components of pectinic or pectic acids. They have been found in pectic preparations, but it is not known whether they represent chemically bound constituents such as side chains or are to be regarded as separate substances closely associated physically with the predominating pectinic or pectic acids. The nomenclature is thus based upon an ideal conception and should be regarded as provisional and subject to more accurate specification as knowledge increases.

As will be seen below, the nomenclature of enzymes acting on pectic substances is even less satisfactory.

III. Possible Changes Brought About in Pectic Substances by Enzymes

At the present time three pectic enzymes are assumed to exist. *Protopectinase* is alleged to catalyze the hydrolysis of protopectin. *Pectin-polygalacturonase*[3] (pectinase, pectolase,[4] hereafter designated as PG) catalyzes the hydrolysis of the 1,4-glycosidic linkages between adjacent galacturonic acid units in the polygalacturonic acid chains. *Pectin-methylesterase*[14] (pectinesterase,[15] pectase, which will be called PM) catalyzes the hydrolysis of the methyl ester groups on the terminal (6) carbon atom of the galacturonic acid units. A further possible pectic enzyme is the recently reported *depolymerase*, the action of which seems to be the reduction of the molecular size of pectic acids without eventual hydrolysis to the monomer D-galacturonic acid.[16]

It would indeed be a fallacy to assume at this time that these four enzymes are the only ones capable of acting on pectic substances. The structure of protopectin and of the pectinic acid macromolecule are not entirely clear and if it can be eventually shown that linkages other than the 1,4-glycosidic bondages occur in them, the likelihood of the occurrence of further pectic enzymes will be greatly increased. It would be a serious omission, furthermore, to overlook the possibility that different hydrolytic enzymes (PG) exist for the hydrolyses of the various sizes of polygalacturonic acids. There is a considerable, although yet uncrystallized, feeling that the existence of such different polygalacturonases may explain some of the irregularities observed in PG action, a point to which we shall return later. Other enzymes acting on galacturonic acid or on polygalacturonic acids might also be assumed to exist. One of the problems which fascinated many workers in this field is the genesis and

(14) R. J. McColloch and Z. I. Kertesz, *J. Biol. Chem.*, **160**, 149 (1945).
(15) H. Lineweaver and G. A. Ballou, *Arch. Biochem.*, **6**, 373 (1945).
(16) R. J. McColloch and Z. I. Kertesz, *Arch. Biochem.*, **17**, 197 (1948).

fate of pectic materials in plants. It has been stated often that in nature the polygalacturonides are derived by the oxidation of the primary hydroxyl group on carbon atom 6 of the D-galactose units in galactans to carboxyl groups. The polygalacturonic acids, in turn, are supposed to be enzymatically decarboxylated to give arabans. If such reactions occur in plants, there are indications that their mechanism may be quite complex. In any case, there is no evidence concerning the occurrence in plants of enzymes or enzyme systems which may bring about these transformations. During growth and storage the total content of polygalacturonides and galacturonic acid of plant tissues often decreases (not only by "dilution" through the formation of other components) and it is clear that pectic polygalacturonides disappear in some manner. A preliminary search in the writers' laboratory to find in plants an enzyme system capable of decarboxylating pectic substances or galacturonic acid gave negative results.

Perhaps we shall have to revise our views that transformations in natural substances must occur as direct results of enzymatic catalysis. A noteworthy development in these lines is the effect of ascorbic acid and peroxides on pectic constituents and other polysaccharides. Ascorbic acid, when itself oxidized, may cause the degradation of both insoluble and soluble pectic substances.[17] Peroxides have a similar action, and a combination of ascorbic acid and peroxide is especially active in bringing about the above changes, the chemistry of which is far from clear. It is difficult to escape the suggestion that perhaps these agents, known to occur in plants, might perform in plants some of the changes for which we have always assumed the existence of specific enzymes. This view gains much in weight by the frequent failure to find in various plants the pectic enzymes alleged to catalyze the changes that can be demonstrated by chemical analyses.

Needless to say, there is a definite possibility that, if reactions as just described participate in the transformations of polyuronide polysaccharides, enzymatic control of these may exist by systems leading to the oxidation of ascorbic acid or producing peroxide-type intermediaries.

There is the further, although remote, possibility that the only substantial variation in the composition of pectinic acids obtained from fruits and beets, and from some other sources, might be the result of differences in the occurrence in these of natural enzymes acting upon pectic substances. Pectinic acids isolated from beets have been shown to contain acetyl groups,[3] the mode of attachment of which to the poly-

(17) W. R. Robertson, M. W. Ropes and W. Bauer, *Biochem. J.*, **35**, 903 (1941); Z. I. Kertesz, *Plant Physiol.*, **18**, 308 (1943); H. Deuel, *Helv. Chim. Acta*, **26**, 2002 (1943); Jean H. Griffin and Z. I. Kertesz, *Botan. Gaz.*, **108**, 279 (1946).

galacturonic acid skeleton is not known. Perhaps acetyl groups occur more frequently in the pectinic acids as they exist in plant tissues and are only removed by very active enzymes which naturally occur in some plants but not in others.

Leaving the realm of speculations, it must be admitted that we yet know only the three afore-mentioned enzymes acting on pectic substances, namely, protopectinase, pectin-polygalacturonase (PG) and pectin-methylesterase (PM). Two of these might be identical. In addition, there is now some certainty that the enzyme provisionally designed as depolymerase is different in nature from true PG, even if the former might commonly occur with PG. Developments in our knowledge of these enzymes during the past decade will be discussed below.

IV. Protopectinase

There is no major advance which can be reported in our knowledge of protopectinases. The possibility of the existence of a protopectinase different from PG seems now somewhat more remote than it was ten years ago. As stated above, the trend in thinking about protopectin is definitely in the direction that protopectin is not a specific cellulose-pectinic acid compound but pectinic acid which is insoluble either on account of large molecular size or due to the effect of polyvalent cations or both. In this sense there is really no need to assume the existence of a separate protopectinase. There is little doubt that this uncertainty is in a large degree attributable to the lack of precise knowledge of protopectin.

There are a great number of reports in the literature dealing with the transformation of insoluble pectic constituents of plants into soluble ones and the presence of protopectinase-like enzymes has been often assumed in this connection. The technique of measuring protopectinase action is still where it was ten years ago[3] and increased solubility and the disintegration of plant tissues are mostly used for this purpose. Ehrlich[4] dealt with protopectinase in great detail and measured a number of properties, using a great variety of source material including higher plants, molds, commercial pectinases, and the digestive juices of snails. There is not the slightest indication in Ehrlich's results that he was dealing with protopectinase and not pectin-polygalacturonase (PG). His findings were, furthermore, entirely in line with observations reported previously;[3] therefore these results will not be discussed here.

An interesting sideline of the protopectinase problem is the possible action on this purported enzyme of agents which speed up the softening of fruits like apples, pears and peaches upon treatment with ethylene.

Hansen[18] suggested that inasmuch as such quicker ripening is closely associated with the more rapid transformation of the insoluble pectic constituents into soluble ones, the effect might be, partly at least, the result of an activation of protopectinase by the ethylene. The activating effect of ethylene on isolated protopectinase (or on PG) has never been demonstrated.

V. Pectin-polygalacturonase (PG)

1. *Nomenclature*

This enzyme is now generally accepted as responsible for the catalysis of the hydrolysis of the 1,4-glycosidic linkages between adjoining units of galacturonic acids in the basic polygalacturonic acid skeletons of all pectic substances. PG is identical with the main hydrolytic enzyme of pectinases as well as with Ehrlich's pectolase[19] and several other enzymes which lost their right of existence (in name) when it became clear that Ehrlich's pectic compounds were principally pectic and pectinic acids of various degrees of degradation. The name pectin-polygalacturonase (PG) has been suggested in the previous review[3] and is now widely accepted. The retention of "pectin" in "pectin-polygalacturonase" is desirable especially since PG is unable to hydrolyze methyl α-D-galacturonic acid and pneumococcus polysaccharide Type I which contains some 60% anhydro-D-galacturonic acid of apparently the same configuration and ring structure as exists in pectic substances.[19a]

2. *Occurrence*

There seems to be considerable delay in establishing the extent of occurrence of PG in higher plants. In spite of extensive efforts to demonstrate the presence of this enzyme in apples, there is no conclusive evidence as yet that PG normally occurs in this fruit. On the other hand, while some doubt was expressed earlier concerning the occurrence of PG in ripe tomatoes[20] it is clear now that pectolytic action can be demonstrated in this fruit.[21,22] As will be seen below, the nature of action of this enzyme seems to be different from that of mold PG but it might be a PG capable of hydrolyzing only large units of polygalacturonic acids.[16]

(18) E. Hansen, *Science*, **86**, 272 (1937); also *Proc. Am. Soc. Hort. Sci.*, **36**, 427 (1939).
(19) F. Ehrlich, *Biochem. Z.*, **250**, 525 (1932), **251**, 204 (1932).
(19a) H. Lineweaver, Rosie Jang and E. F. Jansen, *Arch. Biochem.*, **20**, 137 (1949).
(20) Z. I. Kertesz, *Food Research*, **3**, 481 (1938).
(21) L. R. MacDonnell, E. F. Jansen and H. Lineweaver, *Arch. Biochem.*, **6**, 389 (1945).
(22) H. H. Mottern and C. H. Hills, *Ind. Eng. Chem.*, **38**, 1153 (1946).

Molds are very good sources of PG. Pectinases[3,23,24] showing high PG activity in addition to some PM action are manufactured now in the United States as well as in other countries,[25] and are offered for sale in the form of a standardized dry preparation for the clarification of fruit juices, and for other purposes. The PM can be removed from some such commercial pectinase preparations by ion exchange resins[14] or selective inactivation with heat and acid.[26,27] One of these commercial pectinases (Pectinol) has been shown to contain saccharase, amylase, maltase[28] and some other enzymes which act on nitrogenous compounds[29] and on hemicellulose isolated from apple juice.[30] A purified sample of PG[19a] however, failed to hydrolyze a long list of somewhat related compounds like alginic acid, hyaluronic acid, starch, inulin, mesquite gum, gum ghatti, gum arabic, gum tragacanth, etc.

3. Methods of Determination

A great variety of interrelated changes occur during the enzymatic hydrolysis of polygalacturonic acids by PG and many of these changes have been used at times for the determination of PG action. They often seem to give dependable means of following this reaction, but the interpretation of the results is in most cases very difficult. The discrepancies observed in the interpretation of results obtained by different methods in following the hydrolysis of a sample of pectinic or pectic acid will bring to the fore most forcefully our lack of knowledge of the structure of pectic substances.

Measurement of the increase in reducing power (free aldehyde groups) by the Willstätter-Schudel hypoiodite method and its modifications is still the most popular method for following PG action. This method permits a high degree of accuracy in the latter stages of hydrolysis but it

(23) N. J. Proskuriakov and F. M. Ossipov; *Biokhimiya*, **4**, 50 (1939), quoted from Phaff (ref. 24).

(24) H. J. Phaff, *Arch. Biochem.*, **13**, 67 (1947).

(25) "Pectinol," J. J. Willaman and Z. I. Kertesz, U. S. Pat. 1,932,833 (1933); "Filtragol," A. Mehlitz, German Pat. 680,602 (1932); a Danish preparation is mentioned by A. Hansen, *J. Chem. Education*, **24**, 223 (1947), and a Swiss preparation by H. Pallmann, J. Matus, H. Deuel and F. Weber, *Rec. trav. chim.*, **65**, 633 (1946).

(26) Z. I. Kertesz, *Food Research*, **4**, 113 (1939).

(27) E. F. Jansen and L. R. MacDonnell, *Arch. Biochem.*, **8**, 97 (1945).

(28) V. B. Fish and R. B. Dustman, *J. Am. Chem. Soc.*, **67**, 1155 (1945).

(29) D. C. Carpenter and W. F. Walsh, *N. Y. State Agr. Expt. Sta., Tech. Bull.* No. 202 (1932).

(30) R. L. Messier, "An Investigation of the Action of Clarifying Enzymes on Apple Juice"; Dissertation, Cornell University, Geneva, N. Y., 1945.

is unreliable when used on the original high molecular weight pectinic or pectic acid. The different common methods of sugar determinations[31] when used on a pectinic acid solution will give results in a considerable range, depending on the conditions used but this difficulty disappears after the colloidal properties of the pectinic acid are eliminated early in the hydrolysis by PG.

Perhaps the main advantage of the reducing-power method for following PG action is the fact that a chemical reaction is measured and that therefore the meaning of the results is clear. Strictly speaking, this method measures not only the hydrolysis of the pectic compounds but also the continued hydrolysis of nonpectic polygalacturonic acids below the threshold of size where they will not qualify any more as "pectic substances." Another commonly used method, the determination of the calcium pectate value[32] (weight of calcium pectate obtainable) is free from this objection but does not follow the course of hydrolysis once the magnitude of all molecules has been reduced below the above threshold. Therefore the changes of precipitability as calcium pectate will give a true picture of the disappearance of pectinic or pectic acids from a solution. In turn, the objection to the calcium pectate method is that it will distinguish only between the pectic substances above a certain threshold size without expressing in any manner the extent of colloidality of the pectic material governed by the weight–average molecular weight values. Furthermore, a considerable extent of degradation by certain means can be attained in pectinic and pectic acids (as for instance by heating or by the action of ascorbic acid or peroxides, or both) without essentially reducing the weight of calcium pectate which may be obtained from the sample.

Part of this objection to the calcium pectate as a means of following the hydrolysis of pectic materials can be met by using the simple procedure developed by Fellers and Rice[33] for the estimation of pectic substances as pectic acid. This approximate method measures the volume of the pectic acid which can be produced from a sample of soluble pectic material and will therefore show the loss of colloidality by the rapidly decreasing volume even if the weight of the precipitate remains the same. Unfortunately the Fellers-Rice method is not sufficiently accurate for exact kinetic studies.

Many physical methods of measurement have been applied in follow-

(31) C. A. Browne and F. W. Zerban "Physical and chemical methods of sugar analysis"; Third Ed., John Wiley and Sons, Inc., New York, 1941.

(32) M. H. Carré and Dorothy Haynes, *Biochem. J.*, **16**, 60 (1922); H. D. Poore, *Ind. Eng. Chem.*, **26**, 637 (1934).

(33) C. R. Fellers and C. C. Rice, *Ind. Eng. Chem. (Anal. Ed.)*, **4**, 268 (1932).

ing PG action. Changes in the optical rotation[34] may be used, but the method is of little use in the early stages of the hydrolysis. Stating this fact in a different manner, above a certain molecular weight the optical rotation of pectic substances is affected so little by differences in molecular weight that during the early part of hydrolysis (which is usually of the most interest from the kinetic standpoint) the observed change as a rule is not suitable for accurate measurements.

Viscosity measurements have been used most extensively for following PG action. There are a number of excellent reports on the factors which determine the viscosity of solutions of pectinic and pectic acids,[7,35] and it is comparatively easy to adjust the conditions of measurements (concentration, pH and salt concentration) in such a manner that the results are regular and reproducible. Changes in the viscosity are eminently suitable for following the PG action in its very early stages[24] although the viscosity will be quickly reduced to a constant value approaching that of the solvent and this will actually occur before the calcium pectate values will start to diminish significantly. There are still major difficulties in the interpretation of viscosity measurements during hydrolysis because the mode of action of PG (especially on polygalacturonic acids of different sizes) is not entirely clear.

It may be added that viscosity measurements have often been applied for the evaluation of the PG in commercial pectinases.[36] This method, as will be seen below, might give dependable results under certain conditions but, due to the many complicating factors, it should be carefully adjusted before too much trust is placed in the results.

The method most commonly used until a few years ago, namely precipitation of the pectinic acid with ethanol, gives fairly good results in the case of pure pectinic acids and has a larger range than does that of the determination of the calcium pectate value. In other words, polygalacturonic acids still will be precipitated by ethanol after they have ceased to give precipitates as calcium pectate. Jirak and Niederle[37] used repeated precipitation with acetone but, although the method gave good results in their study of the PG in various microorganisms isolated from fruits, the general applicability of the procedure is questionable.

(34) F. Ehrlich, *Enzymologia*, **3**, 185 (1937).

(35) H. S. Owens, H. Lotzkar, R. C. Merrill and M. Peterson, *J. Am. Chem. Soc.*, **66**, 1178 (1944); H. Deuel and F. Weber, *Helv. Chim. Acta*, **28**, 69 (1945).

(36) A. Mehlitz and M. Scheuer, *Biochem. Z.*, **268**, 355 (1934); H. Pallmann, J. Matus, H. Deuel and F. Weber, *Rec. trav. chim.*, **65**, 633 (1946); G. Weitnauer, *Helv. Chim. Acta*, **29**, 1382 (1946); F. Weber and H. Deuel, *Mitt. Lebensm. Hyg.*, **36**, 368 (1946).

(37) L. Jirak and M. Niederle, *Vorratspflege u. Lebensmittelforsch.*, **4**, 513 (1941).

4. Properties

a. Effect of Substrate.—In 1933 Waksman and Allen[38] found that PG acted more rapidly on pectic acid than on pectinic acid. Recent work by Jansen and MacDonnell[27] and Jansen, MacDonnell and Jang[39] confirmed the finding that pectic acids (polygalacturonic acids) are the true substrates of PG. The rate of hydrolysis (with PG essentially free of PM) seems to be inversely proportional to the methyl ester content or at a given methoxyl content is proportional to the amount of PM added to the PG preparation. The unfortunate result of this relationship is that it greatly reduces the value of previous determinations in which pectinic acids served as substrates for PG and in which PM could be assumed to have been present. It is clear that in kinetic studies only pectic acids should be used as a substrate for PG, or a rapid deesterification should be assured by the addition of ample PM. Such complementary use of PM is not easy and might not be free of objections on account of the differences in the pH optima of PG and PM and the possible susceptibility of these enzymes to activation by various ions, points which will be discussed below.

Recently Matus[39a] found that PG was entirely without action on the (neutral) glycol ester of pectic acid. As the extent of esterification was reduced, PG showed increasing activity.

b. Activation and Inhibition.—Both tomato PG and that found in a Swiss commercial pectinase[25] (presumably of fungal origin) are activated by the addition of sodium and other chlorides. The extent of activation of the tomato PG (in the absence of PM) seems to be much less than that reported for the fungal PG. Such activation has been measured on the tomato PG both on pectinic acid after complete removal of PM and on pectic acid in the presence of PM. We shall return later to the extent of activation of mixtures of PG and PM. Bodansky[40] recently found that glycine inhibits phosphatase; Otto and Winkler[41] claimed that the PG in mold cultures, when treated with this amino acid, is completely inactivated. Tests made in the writers' laboratory to establish the inactivation of PG by glycine gave disappointing results. The concentration of glycine which had to be used to obtain any great extent of inactivation was such that the claim of a specific inactivating effect of

(38) S. A. Waksman and M. C. Allen, *J. Am. Chem. Soc.*, **55**, 3408 (1933).
(39) E. F. Jansen, L. R. MacDonnell and Rosie Jang, *Arch. Biochem.*, **8**, 113 (1945).
(39a) J. Matus, *Ber. Schweitz. botan. Ges.*, **58**, 319 (1948).
(40) O. Bodansky, *J. Biol. Chem.*, **165**, 605 (1946).
(41) R. Otto and G. Winkler, German Pat. 729,667 (1942).

glycine on PG seems to be unjustified. The possibility exists, nevertheless, that the glycine might act more in accordance with the claim of Otto and Winkler in the case of mold cultures. It is of interest to note in this connection that gelatin, a common protein containing very high proportions of glycine, has been successfully used for the stabilization of enzyme solutions.[42]

Matus[39a] claims that the activation of PG by various ions increases with the valence of the cation with an especially great increase in activity from the monovalent sodium to the bivalent calcium ions:

$$Na^+ < Ca^{++} < Al^{+++} < Th^{++++}$$

Detergents are known to inhibit many enzymes.[43] The action of Nacconol NRSF (an alkyl aryl sulfonate) is equally pronounced on tomato PM[44] and tomato PG. Depending on various conditions, both enzymes are completely inactive in the presence of 10–20 mg. % of the active ingredient. Nacconol was also quite effective in the inactivation of fungal PG but further work is needed to establish the quantitative relationships.

c. Temperature and pH Effects.—The pH optimum of fungal (Pectinol) PG has been repeatedly determined in the writers' laboratory and the previous value[3] of 3.5 confirmed. Here is one case in which viscosity measurements, especially in the presence of PM and salts, will give erroneous results. As a rule the pH optimum of the fungal PG appears to be higher when determined in the presence of PM using pectinic acid substrates. The pH optimum of tomato PG has been found to be at 4.5 when measured by various methods.[21]

In a recent electrophoretic study the isoelectric point of tomato PG has been found[45] to be at pH 6.75 to 6.80.

There is little new information available on the heat inactivation of fungal PG. Weitnauer,[46] using a questionable technique[47] of viscosity measurements and entirely disregarding the possible effect of PM, found that the PG solutions were inactivated by holding for a few hours at 40° and that there is a noticeable drop in activity even at 30°. This is in good harmony with the common experience that the PG activity in solu-

(42) M. H. Adams and J. M. Nelson, *J. Am. Chem. Soc.*, **60**, 2474 (1938).
(43) S. Freeman, M. W. Burrill, T. W. Li and A. C. Ivy, *Gastroenterology*, **4**, 332 (1945).
(44) R. J. McColloch and Z. I. Kertesz, *Arch. Biochem.*, **13**, 217 (1947).
(45) R. J. McColloch, "An Electrophoretic Investigation of Some Pectic Enzymes"; Dissertation, Kansas State College, Manhattan, Kansas, 1948.
(46) G. Weitnauer, *Helv. Chim. Acta*, **29**, 1382 (1946).
(47) H. Deuel and F. Weber, *Helv. Chim. Acta*, **29**, 1872 (1946).

tions of mold pectinases will diminish rapidly at room temperatures. Weitnauer did not use temperatures above 40°. At pH 4.0, holding for 3 hours at 50° caused in one case complete loss of PG activity.[39a] Purified PG preparations are even more sensitive to heat. On the other hand, dry mold PG preparations will retain some of their activities even after heating at temperatures over 80° for several hours.[39a]

d. *Kinetics of PG Action.*—The present status of the kinetics of PG action might be cynically cited as a good illustration that formulas can be found to fit almost any set of conditions, even if such will shed little light on the problem. It is clear now that the extent of esterification and the presence or absence of PM will have a governing action on the hydrolysis of the polygalacturonic acids by PG. But even when PM-free PG or pectic acid substrate is used the kinetics of hydrolysis are far from clear. Weitnauer[46] developed a formula to obtain fairly constant k values for certain phases of the hydrolysis. Jansen and MacDonnell[27] state that the hydrolysis of pectic acid by PG goes through two stages: an initial rapid stage which proceeds to the point at which approximately 45–50% of the glycosidic bonds are fissured and a second stage at which the residual bonds are hydrolyzed. In a way these authors assume the existence of "limit polygalacturonic acids" similar to the limit dextrins ("Grenzdextrine") studied by Myrbäck and collaborators.[48] Jansen and MacDonnell assume that this slowing down of the hydrolysis is the result of either the changing affinities of various polygalacturonic acids to PG or because different PG enzymes act on the lower polymers. We have noted above that a feeling that polygalacturonides of different sizes are hydrolyzed by different enzymes seems to be shared by some other workers in the field. It would be foolish indeed to exclude the possible existence of either of the above suppositions. However, we feel that it is at least equally possible that yet unknown structural components occur in polygalacturonic acids, resulting in masking effects on reducing power, viscosity and other properties which are used for the following of the hydrolysis. This brings us inevitably to the theory of occurrence of different types of structures (linkages) in pectic polygalacturonides, a suggestion which was made some time ago by one of us[11] and which precipitated the wrath of many workers in the field. Clearly, we will have to know a great deal more about the structure of polygalacturonides and must have on hand defined and homogeneous polygalacturonic acids of various sizes before a conclusive explanation can be given to the kinetic data.

(48) K. Myrbäck, *Current Sci.*, **6**, 47 (1937); K. Myrbäck, G. Stenlid and G. Nycander, *Biochem. Z.*, **316**, 433 (1944).

5. *"Depolymerase"*

Difficulties were encountered for some time in explaining the mode of action of the pectolytic factor occurring in tomatoes.[48a] Recent studies[16,48b,48c] indicate that this enzyme, formerly often designated as a PG, has some unique properties. The enzyme attacks only low-ester pectinic acids and pectic acid, but ceases to act after about half of the theoretically present glycoside bonds have been hydrolyzed. Some of the end points of the action appear to be polygalacturonic acids containing an average of five anhydrogalacturonic acid units. In the course of enzyme action the typical colloidal properties of the substrate are very rapidly destroyed, apparently with a smaller number of fissures (as measured by the liberation of the reducing groups) than in the case of mold PG. Further addition of this enzyme does not effect any increase in reducing power but mold PG will continue the hydrolysis to the monomer. Thus the enzyme appears to cause a depolymerization in the sense allowed by the suggested structure for pectic substances[11] and is called provisionally "depolymerase" in order to distinguish it from mold PG. It might be regarded as a specific PG.

This tomato enzyme can be obtained from tomato tissues by extraction with sodium chloride solutions of 10% and above and in such solutions shows considerable resistance to heating.[16] This pectolytic factor also has been shown to be present in some commercially canned tomato juice.[48c]

Apparently, yeasts are also capable of forming a pectolytic factor[48d] similar to the depolymerase of tomatoes.

VI. Pectin-methylesterase (PM)

1. *Nomenclature*

Pectin methylesterase (PM) is the enzyme which catalyzes the hydrolysis of the methyl ester groups in pectinic acids. In contrast to PG and protopectinase, a great deal of progress has been made during the past decade in our knowledge of PM. The older term "pectase,"[3] invented before the chemical action of the enzyme was understood, is now used by a decreasing number of workers while the name pectin-

(48a) Z. I. Kertesz and J. D. Loconti, *New York State Agr. Expt. Sta., Techn. Bull. No. 274* (1944).

(48b) R. J. McColloch, *Fruit Products J.*, **27**, 319 (1948).

(48c) R. J. McColloch and Z. I. Kertesz, *Food Technol.*, **3**, 84 (1949).

(48d) B. S. Luh and H. J. Phaff. Paper given at the A.C.S. meetings, Spring 1948.

esterase[49] is regarded by some as superior to the term PM used in this laboratory. Both pectin-methylesterase and pectinesterase are better than the "pectin-methoxylase," suggested by one of us some time ago.[50] However, the term "methyl" in pectin-methylesterase is regarded as necessary on the grounds that it more completely defines the substrate specificity and esterase character of the enzyme. Unlike many esterases, PM exhibits a high degree of substrate specificity having practically no action on any but the methyl ester of pectic acid. The glycol and glycerol esters of pectic acid are not hydrolyzed by PM,[50a] which is a further indication of its specificity and an additional reason for retaining the "methylesterase" in PM.

2. Occurrence

PM seems to occur in small amounts in the vegetative tissues of all higher and lower plants investigated. It is very abundant in the tomato fruit,[50] orange flavedo and albedo,[21] the tobacco plant,[51] eggplant[52] and alfalfa; it usually occurs in the commercial fungal pectinase preparations that are manufactured for the clarification of fruit juices.[25] In the natural products in which the enzyme is found, the major portion of it is usually strongly adsorbed on the water-insoluble cellular components of the tissue macerates.[50,52,53]

3. Methods of Determination

A great variety of different methods has been used for the determination of PM action. The old test requiring gel formation[3] is now little used and methods involving the chemical measurement of changes are preferred. Some workers still use a determination of the methanol liberated,[54] but by far the simplest and most accurate is the method originally suggested by one of us[50] in which the enzyme action is followed by continuous titration. Since the original description of the procedure in 1937, several modifications were developed by Fish and Dustman,[28] Lineweaver and Ballou[49] and McColloch and Kertesz.[44] The following procedure, now used in the writers' laboratory, incorporates the advantages of most of these modifications:

(49) H. Lineweaver and G. A. Ballou, *Arch. Biochem.*, **6**, 373 (1945).
(50) Z. I. Kertesz, *J. Biol. Chem.*, **121**, 589 (1937).
(50a) H. Deuel, *Helv. Chim. Acta*, **30**, 1523 (1947).
(51) C. Neuberg and M. Kobel, *Biochem. Z.*, **190**, 232 (1927); **197**, 1492 (1928); **229**, 455 (1930).
(52) J. J. Willaman, H. H. Mottern, C. H. Hills and G. L. Baker, U. S. Pat. 2,358,430 (1944).
(53) J. J. Willaman and C. H. Hills, U. S. Pat. 2,358,429 (1944).
(54) Margaret Holden, *Biochem. J.*, **40**, 103 (1946).

Fifty to 100 ml. of a solution of good-grade pectin containing at least 8% methoxyl and made 0.1 M with respect to sodium chloride is placed in a 250-ml. beaker and a suitable stirring arrangement is provided. The extension electrodes of a good pH meter are inserted and the reaction mixture is titrated with 0.1 N sodium hydroxide to the pH desired for the experiment. The amount of base used in this adjustment is noted. The solution containing PM is then added and a timer started. As the enzyme acts the pH of the solution will begin to fall due to the liberation of free carboxyl groups and must be kept at the experimental value by continuous titration with the base. At the end of the reaction period the amount of base consumed is noted; the mixture is then rapidly titrated to pH 7.5 (the equivalent point of pectinic acids) and the amount of base used for this titration also noted.[28] A blank is obtained by titrating an identical mixture of substrate and inactive enzyme to pH 7.5. The total base consumed during the reaction minus the base consumed by the blank gives the base consumption due to PM activity.

Inactivation of the enzyme may be accomplished by placing in a boiling water bath for five minutes. When working with tomato PM the enzyme is inactivated before titrating the reaction mixture to pH 7.5 and for the blank titration by adding two ml. of 1% Nacconol solution.[44] The addition of Nacconol does not affect the titration and is of considerable advantage in that it prevents action of the enzyme during the final titration. Otherwise the final titration must be made in considerable haste to prevent introducing this activity in the final result and there is considerable danger of overshooting the very sharp end point of the pectinic acid. This is not an important consideration in the case of mold PM since the activity there falls off rapidly above pH 5.

The activity is expressed in terms of the number of mg. methoxyl split off in thirty minutes[50] (total ml. of 0.1 N base $\times$ 3.1). The pectinmethylesterase units (PMU) may then be given as PMU per ml. enzyme solution, per g. dry matter in the solution, or per mg. protein nitrogen under the experimental conditions. The experimental conditions of pH, salt concentration and source of enzyme should always be specified. The amount of enzyme should not be greater than will give a base consumption of 1 ml. in five minutes if a pH meter is used, or 0.3 ml. in five minutes if methyl red indicator[50] is used. The reaction rate follows a straight-line relationship only during the first 50% of demethylation. Higher than ordinary activities may be determined by diluting the sample or by using shorter reaction periods.

Lineweaver and Ballou[49] have proposed a pectinesterase unit ("PE. u.") for expressing PM activity. One such unit is equivalent to 1/930 PMU under the same experimental conditions or the quantity of enzyme that, at 30° and optimum pH, will catalyze the hydrolysis of pectin at an initial rate of one milliequivalent ester bonds per minute in a standard substrate (0.5% citrus pectin containing 8–11% methoxyl) and 0.15 M sodium chloride. The use of the latter unit is unfortunate since the values obtained for the activity in ordinary plant materials are obtained in the third decimal place and because the experimental conditions are so rigidly defined that they do not accommodate the variations in properties of the enzyme as obtained from different sources and of different stages of purity. Mold pectin methylesterase for example, undergoes heat inactivation even at 30° and the optimum salt concentration for tomato PM varies considerably with the purity of the preparation.

Holden[54] used a method of measurement in which the methanol liberated by PM in a buffered solution of pectinic acid in ten minutes has been estimated by the Zeisel method. This procedure is much more complicated than that described above and must be regarded as unreliable on account of the rapid change in the pH of such reaction mixtures, even if buffered.

4. *Preparation*

As stated above, the PM occurring in higher plants is usually adsorbed in the water-insoluble cellular tissues. However, the enzyme may be easily desorbed and obtained in solution by extraction with fairly strong (10%) salt solutions,[55] by raising the pH of the tissue macerate above 6[53] or by a combination of these two conditions. Several methods have been suggested for the preparation of the enzyme by precipitation with ethanol of the insoluble materials on which the PM is adsorbed, or by the extraction of the tissues at an elevated pH. The former yields a very crude and weak enzyme while the latter does not allow for obtaining the dry enzyme from the solution. It is now realized that the PM of tomatoes, tobacco and probably of oranges once resorbed from the plant tissue, cannot be precipitated by ethanol without extensive inactivation. However, the enzyme can be precipitated from its solutions by dialysis and is somewhat selectively redissolved (together with PG) from the precipitate by 10% salt solution. A method for obtaining highly purified and concentrated dry preparations of tomato PM has been based on these observations.[55] The PM of oranges and tobacco has also been adsorbed under the proper conditions on materials such as Celite and can be then eluted by salt solutions or by raising the pH above 6. The mechanism of the latter reaction is apparently similar to precipitation by dialysis.

A recent patent issued to Leo and Taylor[56] describes a method for the preparation of a colorless solution of PM from alfalfa. The ground leaves are pressed out and the juice is clarified with aluminum hydroxide. The latter may be produced in the press juice by the addition of calcium carbonate, and aluminum chloride hexahydrate. The precipitate carries

(55) R. J. McColloch, J. C. Moyer and Z. I. Kertesz, *Arch. Biochem.*, **10**, 479 (1946).

(56) H. T. Leo and C. C. Taylor, U. S. Pat. 2,406,840 (1946).

down the chlorophyll and other colored bodies, giving a clear solution after filtration or centrifuging. The calcium chloride formed in the solution is claimed not to interfere with the PM in solution. As will be clear from the discussion below, the salt even might be beneficial in effecting a certain degree of activation of the PM.

Solutions of PM when kept with salt are fairly stable. No loss in activity could be observed when PM solutions were held frozen for several months. Highly active PM solutions from tomatoes have also been dried in a vacuum[55] without much loss in activity.

5. *Properties*

a. The PM of Higher Plants.—Since there seems to be a qualitative difference between the PM of molds and that of higher plants their properties will be dealt with separately. Varations also occur in the properties of the PM of higher plants but they are small and of a quantitative rather than qualitative nature.

Rather thorough studies have now been made of the properties of the PM of tomatoes,[20,21,44,52,53,55] orange peel,[21] tobacco[3,51,54] and alfalfa.[49] The activity of PM is profoundly affected by the pH of the reaction mixture, the salt concentration and the cation component of the salt. In salt-free solutions, the activity of the PM of higher plants is nearly zero at pH 4.25 and increases linearly with a steep slope as the pH is increased to 8. Above 8 the pectinic acids are also demethylated by the action of alkali and the enzyme activity measurements therefore become unreliable. The relationship between activity and pH is also dependent on the salt content of the reaction mixture.

An extensive study of the effect of salts on the pH-activity curves of PM was made on alfalfa[49] and on orange[21] and tomatoes.[44] In general, the effect of salts is to lower the pH at which maximum activity is attained and to extend the activity into lower pH regions. At the higher pH values (7–8), salts have practically no activating effect. The main usefulness of salt activation of PM seems to lie in counteracting adverse pH conditions.

Salt activation falls into two classes with respect to the valency of the cation component. With divalent cations and at pH 6, maximum activation is produced at about 0.03 M concentration. At higher concentrations with the same pH, suppression of activity occurs. Monovalent cations in general produce maximum activation at pH 6 in 0.10 M concentration and do not suppress activity below molarities of 1.0. Maximum activity is obtained at a lower pH and lower concentration of divalent cations than monovalent. The maximum activity obtained at the optimum salt concentration for a given pH value is, within experimental error, the same as the maximum obtained at any other pH value at the optimum salt concentration, at least between the pH values 4.5 to 8.0. In 0.05 M calcium chloride solution the pH–activity curve of orange PM forms a nearly level plateau from pH 5.0 to 8.0.

Lineweaver and Ballou[49] proposed a mechanism to explain the salt activation of PM based upon the assumption that salts prevent the binding of the enzymes to the demethylated carboxyl groups of the pectinic acids. This proposal is regarded with reservations since the concentration of salts required for a given activation can be greatly reduced by purification of the enzymes. It seems likely, therefore, that the salts also act by peptizing or desorbing the enzyme from insoluble high molecular weight compounds other than pectinic acids which occur as impurities in its preparations. It can be easily shown that the enzyme is reversibly absorbed and desorbed in its crude preparations either by lowering and raising the pH or by lowering and raising the salt content of the mixture at the lower pH values.

Recent electrophoretic studies[45] indicated that the isoelectric point of tomato PM is at pH 6.80 to 6.85.

The PM of higher plants is very resistant to inactivation by chemical agents.[44] Such classical enzyme inhibitors as hydrogen cyanide, iodoacetic acid, iodine, formaldehyde, and mercuric chloride do not easily inactivate tomato PM in the presence of its substrate. When dilute solutions of pectinic acids are titrated to pH 7.5 with pyridine or quinoline and then tomato PM is added, the activity is inhibited. The inhibition may be removed by adjusting to a lower pH or by the addition of salts, and therefore seems to be of the competitive type, the competition being for the carboxyl groups of pectin. Small quantities of synthetic detergents of the sodium lauryl sulfate and alkyl aryl sulfonate types (as Nacconol NRSF, Swerl) will inactivate tomato PM completely and immediately. The inactivation is not reversed by changing the pH or by the addition of salts. This fact is useful for stopping the activity of the enzyme at any desired point in the demethylation of pectinic acids without resorting to extreme conditions of temperature or acidity.

As stated above, the PM of tomatoes is inactivated easily and completely by precipitation with ethanol. This method can be used for a quantitative separation of the PM and "PG" of tomatoes.

Solutions of tomato and tobacco PM in the pH range 4–6 will resist heat inactivation at temperatures up to 60° for at least an hour.[26] Above 60° the enzyme is rapidly inactivated. Orange PM is reported[21] to be inactivated at 45° at pH 7.5 and to suffer slow inactivation even at 5° especially at pH values higher or lower than 7.5. This is in contrast to tomato PM, which retains its activity practically undiminished for many months when stored even at room temperature in dilute salt solutions between pH 4 and 6.

The Q_{10} (the ratio of the rate at Centigrade temperature $t + 10$ to the rate at t) of tobacco[54] and tomato[44] PM between 20° and 60° has been variously reported as about 1.45. Tomato PM has a Q_{10} of 1.80 between 0° and 20°. Values for the energy of activation of tomato PM are reported between 5500 and 7500 cal. per mole by various workers,[44,57] and depending upon the assumed order of the reaction. The value is about 9,500 cal. per mole between 0° and 20°.

  b. *The PM of Molds.*—PM is widespread among the mold genera and it always occurs in commercial pectinase preparations of fungal origin. The PM of molds apparently differs sharply from that of higher plants.[44] Mold PM has a well-defined[28,44] optimum at pH 4.5 to 5.0, the activity falling off considerably on both sides of this range. It is active in salt-free solutions below pH 4.25 although its activity may be increased by sodium chloride at its optimum pH. In contrast to the enzyme in higher plants, the maximum salt activation occurs at the optimum pH and decreases on either side of it. The mold enzyme is more resistant to chemical inactivation than tomato PM and is not inactivated by similar amounts of detergents. However, mold PM is much more sensitive to temperature and its inactivation by heat begins to occur at 30°. Below this temperature it has about the same Q_{10} value and energy of activation[44] as tomato PM. The mold PM is not inactivated by precipitation from its solutions with ethanol and it is not precipitated from its solutions by dialysis. When mold and tomato PM are allowed to act simultaneously on pectinic acids, their actions can be shown to be independent of each other.[44]

Apparently the electrophoretic characteristics of mold PM and tomato PM are also different.[45]

VII. Role and Application of Pectic Enzymes

1. *In Undisturbed Plant Tissue*

The dynamics of plant life are still essentially unknown. It is clear now, however, that the occurrence of an enzyme in a tissue does not necessarily mean that it will act on its substrate which is also present,[58] nor does our inability to demonstrate the presence of certain enzymes exclude the occurrence of changes usually attributed to enzyme action. In other words, our knowledge of *in vitro* and *in vivo* relationships of plant enzymes is as yet very much in its infancy.

  (57) R. Speiser, C. R. Eddy and C. H. Hills, *J. Phys. Chem.*, **49**, 563 (1945).
  (58) Z. I. Kertesz, *Plant Physiol.*, **12**, 845 (1937).

Because of the well-demonstrated relationship in the case of apples between changes in pectic constituents, texture changes and commercial value, one may take this fruit as an example for discussion. Ever since the occurrence of pectic transformations in apples became clear,[59] the existence of specific enzymes catalyzing these changes has been assumed.[60] Yet, contrary to the reports of Joslyn and Sedky[61] and Zimmermann, Malsch and Weber,[62] and in spite of vigorous efforts mentioned before,[17] we have never succeeded in demonstrating the presence of protopectinase or PG in apple juice or apple tissue macerates. Attempts to remove possible inhibitors in the sense demonstrated by Harley, Fisher and Masure[63] for the amylase of apples were also of no avail. One type of nonenzymatic mechanism through which such changes might occur has been discussed[16] on page 83, but it is only fair to state that we have no exact information at hand as to just how the pectic changes in apples (and other plant tissues) are actually brought about.

Without extending this discussion unduly, one may state that there is not a single case in which the function of any of the known enzymes acting on pectic substances demonstrated in a plant tissue could be definitely associated with the pectic changes which occur in the tissue. Such information would be of great value since it might provide better means for controlling pectic transformations related to ripening and storage behavior. Little comfort can be derived from the fact that the situation concerning enzymes acting on other important natural components of plants is similarly unsatisfactory.

2. *Action of Pectic Enzymes in Plant Tissue Macerates*

The fate of pectin in plant macerates and juices has been extensively studied.[64] Perhaps the case that has received the most attention is that of tomato juice, and especially of the various concentrates made from it, since here again pectin has a dominant effect on consistency,[48a,65] and this

(59) M. H. Branfoot (neé Carré), "A Critical and Historical Study of the Pectic Substances of Plants." Food Investigation Special Report No. 33, H. M. Stationery Office, London (1929).
(60) R. W. Thatcher, *J. Agr. Research*, **5**, 103 (1915).
(61) M. A. Joslyn and A. Sedky, *Plant Physiol.*, **15**, 675 (1940).
(62) W. Zimmermann, L. Malsch and R. Weber, *Vorratspffege u. Lebensmittelforsch.*, **2**, 271 (1939).
(63) C. P. Harley, S. F. Fisher and M. P. Masure, *Proc. Am. Soc. Hort. Sci.*, **28**, 561 (1931).
(64) D. K. Tressler, M. A. Joslyn and G. Marsh, "Fruit and Vegetable Juices"; Avi Pub. Co., New York, N. Y. (1940).
(65) Z. I. Kertesz, *Canner*, **88**, (1), 11 (1938).

property, in turn, on commercial value. This interest led to the discovery of PM in tomatoes[20] and to the extensive investigation of possible means of preventing the enzymatic degradation of pectic constituents in tomato macerate (juice). Thus far, thermal inactivation is the only practical means at our disposal for preventing such detrimental changes, but it is hoped that continued investigations will lead to other and less drastic methods.[44] The pectic changes which occur in tomato macerates in the absence of adequate heating might be summarized as follows.[48c] The pectolytic factor of tomatoes (depolymerase) does not break down highly methylated pectinic acids as they naturally occur in the tomato fruit. However, if the PM is not inactivated, deesterification will commence immediately upon maceration of the tomato tissue and this will enable the pectolytic factor (whether a PG or not) to proceed with the depolymerization of the resulting low ester pectinic acids. The addition of salt (sodium chloride) to the macerate will enable the PM to exert higher activity and will pave the way for quicker depolymerization by the depolymerase. Thus concerted action of the two enzymes will be necessary to destroy the useful colloidal properties of the pectinic acids. This will result in poor, watery consistency in the tomato juice and will require excessive concentration in order to produce tomato puree, catsup, and other similar products of desirably viscous properties. On the other hand, if the tomatoes are sufficiently heated immediately upon or even before the destruction of the natural cell structure, the full benefits from the colloidal properties of the pectic constituents are realized.

Similarly, there are present in citrus juices active enzyme systems that destroy the important pectic constituents. These enzymes have to be rapidly inactivated to protect the desirable "cloud" (turbidity) which is apparently stabilized by pectic substances.[66]

With the exception of tomato and perhaps citrus fruits, there is no instance in which the presence of a PG in higher plants or macerates has been conclusively demonstrated. Often the types of changes reported suggest[61] that infection by microorganisms rather than naturally occurring enzymes was the cause of the observed chemical transformations. This is especially true with fruit juices, where the rapidly increasing bacterial and mold flora may eventually cause reactions which do not occur when a sterile macerate is incubated.

This review would be incomplete without mention of an important and rapidly developing new field in pectin chemistry and technology, namely that of the low-ester pectinic acids. In contrast to pectinic acids (pectins) of over 7% methoxyl content, pectinic acids containing 3–6% methoxyl are capable, in the presence of the required proportion of

(66) J. W. Stevens, U. S. Pat. 2,217,261 (1940).

bivalent ions, of forming solid gels with reduced proportions of sugars.[67] The point which concerns us here is that PM has been extensively used for the deesterification of pectinic acids. Owens, McCready and Maclay[68] described a method in which the deesterification is performed *in situ* in the macerate by the PM naturally present in the tissue prior to or during the extraction of the pectinic acid. This method seems to be especially suitable for the preparation of low-ester pectinic acids from citrus peel, inasmuch as the latter contains a very active PM.[51] The main points in the process are the adjustment and control of the pH of the macerate followed by efficient thermal inactivation of the PM after the desirable extent of saponification has been reached.

3. *Commercial Pectic Enzyme Preparations*

A further important field of practical application that developed during the early thirties is the use of pectinases for the clarification of fruit juices. There are apparently no recent major developments in the use of clarifying enzymes although their action is now much better understood. Such enzyme preparations, which are apparently of fungal origin exclusively, are now used extensively in many countries.[25] It is known that their PM content will have much to do with the velocity of the PG action and that they are, as a rule, comparatively low in PM activity. Several of the above studies of PM from higher plants have been conducted with a view of the possible use of such PM in conjunction with these mold pectinases,[68a] but this matter still seems to be in the speculative stage. These enzymes have been successfully used in the removal of pectin from fruit juices and extracts before concentration. Such hydrolysis of the pectic substances by enzymatic means is especially desirable when flavoring constituents are apt to be destroyed by the use of the cheaper but more drastic methods.

There are important developments to be expected in the commercial production of PM. As noted above, this enzyme proved to be a useful demethylating agent in the production of low-ester pectins. The use of enzymes has certain advantages over acid or alkali, and this fact might eventually create considerable market for PM preparations that are free or essentially free of PG.

(67) C. H. Hills, J. W. White and G. L. Baker, *Proc. Inst. Food Technol.*, **47** (1942); R. Speiser, C. R. Eddy and C. H. Hills, *J. Phys. Chem.*, **49**, 563 (1945); R. M. McCready, H. S. Owens, A. D. Shephard and W. D. Maclay, *Ind. Eng. Chem.*, **38**, 1254 (1946).

(68) H. S. Owens, R. M. McCready and W. D. Maclay, *Ind. Eng. Chem.*, **36**, 936 (1944).

(68a) E. F. Jansen and H. Lineweaver, U. S. Pat. 2,457,560 (1949).

Commercial pectic enzyme preparations[25] have been used for some time[19] in the production of D-galacturonic acid from pectic substances. This matter attained increased importance with the recently discovered double salts of galacturonic acid[69,70] which make more economical production of galacturonic acid possible. Galacturonic acid can serve as a starting material in a comparatively simple synthesis of ascorbic acid.[71]

Pectic enzymes have been used in this laboratory and elsewhere for many years in the analysis of plant materials. Isbell and Frush[72] described recently a method using Pectinol for the determination of "pectic enzyme soluble substances," meaning the substances that become soluble through the enzyme actions. A gravely objectionable feature of the method lies in the fact that pectic enzyme preparations contain many other enzymes that act on polysaccharides[28,30] and other complex compounds contained in plant tissues.[29] Therefore the digestion with the pectic enzyme will cause solubilization of the latter constituents in addition to that of the pectic substances. Furthermore, the presence of various ions such as those of magnesium and calcium, for instance, might also seriously alter the solubilization of the pectic materials.[39,73]

Pectic enzymes have also been employed for the detection of pectin when it is used as a stabilizing or thickening agent in various food products.[74]

(69) H. S. Isbell and Harriet L. Frush, *J. Research Natl. Bur. Standards*, **32**, 77 (1944).
(70) R. Pasternack and P. P. Regna, U. S. Pat. 2,338,534 (1944).
(71) H. S. Isbell, *J. Research Natl. Bur. Standards*, **33**, 45 (1944).
(72) H. S. Isbell and Harriet L. Frush, *J. Research Natl. Bur. Standards*, **33**, 389 (1944).
(73) J. D. Loconti and Z. I. Kertesz, *Food Research*, **6**, 499 (1941).
(74) E. Letzig, *Z. Untersuch. Lebensm.*, **72**, 312 (1936), **84**, 289 (1942).

THE RELATIVE CRYSTALLINITY OF CELLULOSES

By R. F. Nickerson

Textile Research Laboratory, Monsanto Chemical Company, Everett, Massachusetts

Contents

I. Introduction.. 103
II. Purpose... 104
III. Historical.. 105
IV. X-ray Diffraction Studies..................................... 106
    1. Crystalline Cellulose...................................... 106
    2. Non-crystalline Cellulose.................................. 107
V. Chemical Studies.. 109
    1. Acid Hydrolysis Methods.................................... 109
    2. Swelling and Density Investigations........................ 120
    3. Oxidation and Deuterium Exchange........................... 121
    4. Esterification and Etherification.......................... 122
VI. Discussion.. 124
VII. Implications... 125

I. Introduction

Just over a century ago Payen[1(a)] conducted a series of careful purifications and analyses of plant tissues and reached the conclusion that the fibrous skeleton of all young plant cells is a uniform chemical entity. This skeletal substance appeared to be a polysaccharide derived from glucose residues and to be isomeric with starch. It differed from starch, however, in that it had to be swollen with sulfuric acid before it gave the blue iodine color which starch gives almost instantly. Payen correctly decided that the purified fibrous material must be more compact or more highly aggregated than its isomer, starch, and he applied to it the name *cellulose* which had been proposed by Dumas' committee[1(b)] in their report to the French Academy on Payen's researches.

These conclusions were not universally accepted. Payen's cellulose was regarded by many workers in the field as an artifact produced by chemical changes in the materials during purification and isolation. A common concept was that native plant materials were homogeneous complexes characteristic of each species of plant. Even Cross and

---

(1) (a) A. Payen, *Compt. rend.*, **7**, 1052 (1838); **8**, 169 (1839); **9**, 149 (1839). (b) See also the report of a committee (J. B. Dumas, chairman), *Compt. rend.*, **8**, 51 (1839).

Bevan, the great research team who originated the viscose process, subscribed to this contrary idea well into the present century. They believed that lignin, cellulose, pectin, fatty matter and other constituents of plant tissue blended homogeneously into one another as "ligno-", "pecto-", and "adipo-" cellulose. Thus, jute fiber was regarded as a homogeneous aggregate of "bastose," $C_{76}H_{80}O_{87}$, rather than as a heterogeneous mixture of cellulose, lignin, hemicellulose and other constituents.

There were a few investigators, principally botanists, who stoutly supported the Payen hypothesis during this period. Their position was not a particularly easy one to maintain because the lignocellulose theory had to be disproved without resort to chemicals or other "modifying" influences. Microscopic studies and physical investigations of birefringence and swelling of cellulosic fibers, however, yielded strong evidence to support the Payen views.

The turning point came about 1920. Studies of cellulosic fibers by X-ray diffraction methods showed that lignified structures like jute and wood yielded crystalline diffraction patterns identical with those given by cotton, ramie and flax, the celluloses of which occur naturally in relatively high purity. Thereafter, the lignocellulose theory lost ground rapidly and the modern concept of cellulose as a crystalline network structure connected by regions of disordered cellulose began to emerge. Lignin and other non-cellulosic fiber constituents are now generally believed to occur as incrustants shot through this network structure, although it still remains a possibility that lignin may be joined to the cellulose by true chemical bonds which are readily broken by customary purification treatments. In any event, it is possible to remove the bulk of the non-cellulosic constituents without drastically modifying the underlying skeleton, and in this sense the view of Payen is now generally believed to be correct.

II. Purpose

It has grown increasingly apparent that the non-crystalline portions of cellulose structures may play as important a role in the properties and behavior of cellulosic materials as the crystalline parts. X-ray diffraction studies have greatly extended knowledge of crystalline cellulose but in the case of the amorphous or disordered fraction the methods of study have necessarily been indirect and not completely reliable.

The need of fully dependable information on the relative proportions of crystalline and amorphous cellulose in different cellulosic materials has given rise to a variety of approaches to the problem. These alternative methods in many instances fail to support X-ray findings and often show large discrepancies among themselves. It is the purpose of this review to summarize some of the available information on relative crystallinity and to indicate some of its implications.

III. Historical

It is fairly certain that pure cellulose is a condensation product of β-D-glucose. In the natural formation of a single cellulose molecule, n moles of D-glucose apparently condense and lose $n - 1$ moles of water. However, since n is relatively large and differs but little from $n - 1$, the two values are regarded as substantially equal and the cellulose molecule is considered to be composed entirely of anhydro glucose units ($C_6H_{10}O_5$). A careful review by Purves[2] leaves little doubt that D-glucose is the sole constituent of pure cellulose. The Armstrongs[3] and Haworth[4] have adequately described the structure of β-D-glucose while Purves[2] and Compton[5] have shown in excellent reviews that the nature of the glucose-to-glucose linkage in cellulose is well established. Anhydroglucopyranose residues apparently are condensed head-to-tail through β-1,4 linkages to form long, unbranched chains.

Sisson[6] has traced the evolution of current concepts of the crystalline part of cellulose structures. The fiber diagram obtained by X-ray diffraction is now known to be produced by a series of elementary crystals, called crystallites, which have a definite arrangement with respect to the fiber axis. It is also known that the crystallites in regenerated cellulose may be oriented to varying degrees with respect to the fiber axis and that the crystallites in regenerated cellulose and mercerized cotton differ from those in native fibers. These "hydrate" type crystallites appear to be more reactive chemically than the native type.

Bear[7] has pointed out that there are small but important structural differences between starch and cellulose. The cellulose molecule may be regarded as a long ribbon because the constituent β-anhydroglucopyranose rings lie largely in one plane while the starch molecule is somewhat kinked because adjacent α-anhydroglucopyranose units are prevented by their oxygen bridges from extending in the same plane. In

---

(2) C. B. Purves in "Cellulose and Cellulose Derivatives," (E. Ott, editor), Interscience, New York, p. 54 *et seq.* (1943).

(3) E. F. Armstrong and K. F. Armstrong, "The Carbohydrates," 5th edition, Longmans, Green & Co., London (1934).

(4) W. N. Haworth, "The Constitution of Sugars," Edward Arnold and Co,, London (1929).

(5) J. Compton, *Advances in Carbohydrate Chem.*, **3**, 185 (1948).

(6) W. A. Sisson, in "Cellulose and Cellulose Derivatives" (E. Ott, editor), Interscience, New York, p. 203 *et seq.* (1943).

(7) R. S. Bear, *The Technology Review* (Mass. Inst. Tech.), **45**, (No. 7) p. 363, (May, 1943).

addition, some starch molecules (amylopectin) are of the branched-chain type. These differences may explain in part why cellulose is much more dense and crystalline than starch. Kratky[8] has suggested a free-rotation mechanism to account for the marked tendency of cellulose to crystallize as it precipitates from solutions.

The average number of anhydroglucose units in the chain length or degree of polymerization (DP) of celluloses varies widely and depends upon a number of factors in the history of the sample. Thorough discussions of this phase of the subject are available. A few typical weight average DP data from the literature are given in Table I.

TABLE I
Estimated Weight Average DP of Various Cellulosic Materials

Material	DP	Reference
	(Number of glucose units)	
Raw cotton	2520, 10,800	9
Unbleached American linters	2700, 9,300	9
Bleached American linters	920–1540, 1000–3000	9, 10
Wood pulp	949–1100, 600–1000	9, 10
Commercial regenerated celluloses	200–600	10
Commercial cellulose acetates	170, 175–360	9, 10
Plastics (Nitrocellulose)	500–600	10

The values shown represent probable orders of magnitude only and a wide range undoubtedly exists for each type of material.

IV. X-ray Diffraction Studies

1. *Crystalline Cellulose*

Cellulosic materials are now generally regarded as polycrystalline structures in which the cellulose molecules may traverse several crystallites and the intervening amorphous or disordered areas. This concept has grown out of the work of Meyer and Mark,[11] Gerngross, Herrmann and Abitz,[12] Frey-Wyssling[13] and a number of other investigators. A

(8) O. Kratky, *Silk and Rayon*, **13**, 480, 571, 634 (1939).
(9) W. Badgley, V. J. Frilette, and H. Mark, *Ind. Eng. Chem.*, **37**, 227 (1945), from data of N. Gralén.
(10) E. O. Kraemer, *Ind. Eng. Chem.*, **30**, 1200 (1938).
(11) K. H. Meyer and H. Mark, *Ber.*, **61**, 593 (1928).
(12) O. Gerngross, K. Herrmann and W. Abitz, *Biochem. Z.*, **228**, 409 (1930).
(13) A. Frey-Wyssling, "Submikroskopische Morphologie," Gebrüder Borntraeger, Berlin (1938).

schematic representation of such a structure is shown in Fig. 1. Mark[14] has summarized and discussed some of the X-ray diffraction studies designed to elucidate crystallite size. A fair estimate from this type of investigation seems to be that the average crystallite in native cellulose is about 50 Å in width and somewhat more than 600 Å in length.

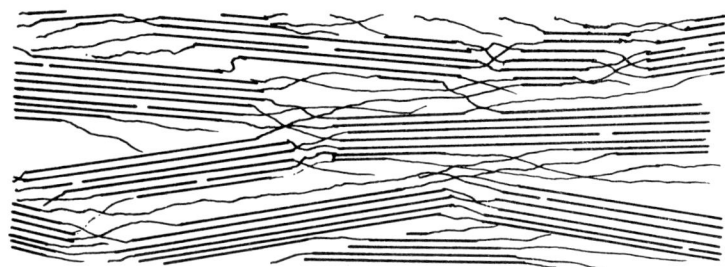

Fig. 1.—Possible microstructure of cellulose, showing crystallites (darkened regions) and noncrystalline network.

The cellulose molecule is estimated to be about 5 Å in length. This means that the average chain in cotton of DP 2000 may be about 10,000 Å in length. In such a case a single cellulose molecule may participate in several crystallites and intercrystalline regions. As average DP decreases, for example in regenerated celluloses, the chains must extend through fewer crystallites and non-crystalline regions.

2. Non-crystalline Cellulose

Mark[14] has also discussed the problem of determining the amount and characteristics of intercrystalline cellulose by the X-ray diffraction method. This approach is based on the fact that while crystals give spot or sharp-ring diagrams, liquids give diffuse rings which indicate an inferior degree of molecular ordering. Since practically all natural and synthetic fiber diagrams show considerable blackening of background, it seemed possible to estimate the amount of disordered cellulose from the diffuse radiation under carefully controlled conditions.

Hemp, ramie, and highly oriented viscose rayon fibers produce different amounts of diffuse radiation. A similar experiment with viscose rayon filaments swollen and increasingly stretched yielded diagrams which indicate that crystallite orientation increases and diffuse background decreases with extent of stretching. It was also found that strong swelling agents for cellulose, such as phosphoric acid, calcium thiocyanate solutions and quaternary bases, convert the normal fiber diagram of dry or water-swollen cellulose into a diffuse liquid diagram.

(14) H. Mark, *J. Phys. Chem.*, **44**, 764 (1940).

As a result of these studies it became evident that mechanical and chemical treatments may alter the relative amounts of crystallized and amorphous cellulose in a sample. For example, it was estimated that a normally coagulated viscose filament was about 40% crystalline and 60% amorphous while filaments of the same material, after being stretched, appeared to be 70% crystalline and 30% amorphous. This effect was observed to be more pronounced in cellulose derivatives than in unsubstituted cellulose where apparently the blocking of hydroxyl groups reduces the lateral forces of cohesion between chains, facilitates slipping and consequently promotes parallelization of chains.

Further analysis of X-ray fiber diagrams indicated that the diffuse radiation simulated that of a liquid consisting of long-chain molecules. Thus, the amorphous portions may not be completely disordered but may rather behave like bundles of nearly parallel chains which are not well enough oriented to give sharp interference spots. As Mark[14] observed, this picture suggests that the amorphous portions of a cellulosic fiber should have a higher energy content than the crystalline regions. That is, the intercrystalline areas may be regarded as highly distorted crystal lattices in which single atoms are shifted considerably from their normal equilibrium positions. In other words, the crystallites represent a minimum potential energy condition and the less ordered regions, a somewhat higher level.

This concept suggests that the crystallites form the firm reenforcing part of the structure while the intercrystalline regions are vulnerable and in many cases constitute actual weak spots. When a fiber or film is soaked in water, the X-ray fiber diagram apparently is not changed from that of the dry fiber.[12] It is probable, therefore, that water penetrates only the intercrystalline regions or that the lateral molecular forces in crystallites are too strong to permit distortion by water alone. The effects that water has on fiber properties must consequently be closely allied with swelling of, and other influences on, the intercrystalline cellulose. Furthermore, the chemical reactivity of the disordered cellulose is probably much greater than that of the crystallites.

X-ray diffraction methods have been indispensable in the development of the present concept of cellulose structure, but Mark[15] has offered the opinion that such methods alone are ill-suited to the quantitative determination of crystalline and non-crystalline fractions. Chemical methods which depend upon the greater reactivity or accessibility of the incompletely ordered regions have offered an alternative approach to the problem.

(15) H. Mark, in "Wood Chemistry" (L. E. Wise, editor), Reinhold Publishing Co., New York, p. 129 (1944).

V. Chemical Studies

1. *Acid Hydrolysis Methods*

Determination of the relative amounts of crystalline and non-crystalline cellulose was undertaken by hydrolysis and oxidation. The method[16] was based on the observation that a boiling mixture of ferric chloride and hydrochloric acid evolves carbon dioxide relatively rapidly from glucose. Apparently ferric chloride and hydrochloric acid concentrations can be so chosen as to yield a practically linear evolution of carbon dioxide from glucose with time. However, there is a time lag between the start of the reaction and the appearance of carbon dioxide in appreciable quantities, but when allowance is made for this lag period the rate of carbon dioxide evolution is roughly proportional to glucose concentration for a period of about 6 hours. A mixture containing 2.45 moles of hydrochloric acid and 0.6 moles of ferric chloride per liter was found to be a suitable one for the cellulose investigations at sample-to-liquid ratios of 1:40. Carbon dioxide was periodically determined by both a volumetric method similar to that of Dickson, Otterson and Link[17] and a gravimetric method similar to that of Whistler, Martin and Harris.[18] The latter was found to be somewhat easier to use.[19]

The cellulose specimen under examination is refluxed in the acid-oxidant mixture and the gases formed are swept continuously into an absorption train by a carrier stream of air free of carbon dioxide. Conrad and Scroggie[20] have added a number of important improvements which apparently increase the reproducibility of results. One of their modifications is a stirrer in the reaction chamber which reduces the danger of bumping caused by superheating. The latter is undesirable since the reaction is apparently quite sensitive to the temperature.

The specimen and reaction mixture are placed in the reaction vessel, which is attached to a reflux condenser. Heat is then applied and the onset of boiling in the reaction mixture is taken as the start of the process. Periodically thereafter, carbon dioxide absorption tubes are removed from the train and replaced by tubes of known weight. Some experiments have been conducted[19] in which the reflux mixture was brought to a boil and the test specimen thereafter introduced through a side tube in the reaction flask. This procedure insures that the reaction period

(16) R. F. Nickerson, *Ind. Eng. Chem., Anal. Ed.*, **13**, 423 (1941).
(17) A. D. Dickson, H. Otterson and K. P. Link, *J. Am. Chem. Soc.*, **52**, 775 (1930).
(18) R. L. Whistler, A. R. Martin and M. Harris, *J. Research Natl. Bur. Standards*, **24**, 13 (1940).
(19) R F. Nickerson and J. A. Habrle, *Ind. Eng. Chem.*, **37**, 1115 (1945).
(20) C. C. Conrad and A. G. Scroggie, *Ind. Eng. Chem.*, **37**, 592 (1945).

is not preceded by any indefinite effect prior to zero time. Finally, Conrad and Scroggie[20] reached the conclusion that the initial lag period between onset of boiling and rapid carbon dioxide evolution from glucose actually constituted a period of slow but steadily increasing carbon dioxide output. This finding, however, does not grossly alter the interpretation of data.

The acid–oxidant method is based on the idea that the hydrolysis of cellulose might be continuously determined from the rate of carbon dioxide evolution. Since, under controlled conditions, the rate of evolution of carbon dioxide is proportional to glucose concentration, it should be possible to follow the course of cellulose hydrolysis by means of the rate of carbon dioxide evolution provided that the sole final product of hydrolysis of cellulose is glucose. The latter assumption appears to be justified where the sample is reasonably pure.

It was shown[16] that dry corn starch yields carbon dioxide in approximately linear fashion with time, just as glucose does. However, the rate of carbon dioxide evolution, determined graphically as the slope of the best straight line for a carbon dioxide–time plot, was 1.10 times as fast for starch as for an equal weight of glucose. Such would be the case if the starch consisted only of anhydroglucose since unit weight of starch should yield 111% of glucose on hydrolysis. The absence of any lag period other than that already mentioned for glucose indicated that hydrolysis was extremely rapid.

Under similar conditions, unit weight of mercerized cotton yields carbon dioxide at a slow initial rate which appears to increase rapidly with time but which is always low compared to that of starch. Differences between c. p. glucose and glucose derived from cellulose were apparently ruled out in the following way.[21] A dry sample of purified cotton of accurately known weight was placed in 25 ml. of ice-cold 40% hydrochloric acid and kept at 2°C. for 48 hours. The liquid was then carefully removed by aspiration, first at low temperature and finally at higher temperatures. The concentration was completed by aspiration at 60°C. with successive additions of absolute ethanol. A light-brown sirup so obtained was dried at reduced pressure over phosphoric anhydride, a dish of soda-lime being placed in the desiccator to absorb any hydrogen chloride.

The final product was subjected to ferric chloride–hydrochloric acid treatment in the same manner as commercial glucose and starch. A linear carbon dioxide–time relationship was observed which was practically identical with that for starch. In other words purified cotton cellulose, on relatively complete hydrolysis, appeared to give glucose in a

(21) R. F. Nickerson, *Ind. Eng. Chem.*, **33**, 1022 (1941).

yield of about 99% of the theoretical. This result suggested that there is little difference in oxidative behavior between c. p. glucose and the glucose from cellulose, and that the observed effects in the case of mercerized cotton require some other explanation.

The character of the carbon dioxide–time curves for unsubstituted celluloses made impossible the use of simple graphical methods. It was found,[22] however, that the data gave reasonably straight lines on logarithm–logarithm paper and the following scheme was employed to obtain estimates of the apparent hydrolysis. Glucose was assumed to be represented by the linear relationship

$$CO_2/W = S(t - x), \qquad (1)$$

in which CO_2 = cumulative carbon dioxide yield, W = weight of glucose taken, S = slope to be determined, t = observed time and x = induction period derived by projecting the best straight line through the CO_2-time data and taking the intercept on the time axis. The value so found was assumed to hold for all similar experimental conditions. The derived value of x was then deducted from all observed times and the differences so obtained were used as abscissas. For unit weight of glucose, equation 1 then reduces to

$$CO_2/T = S, \qquad (2)$$

in which $T = t - x$. Plotted on logarithm paper against T, cumulative carbon dioxide data for cellulose yielded reasonably straight lines except for the early stages of the process. Being of the type

$$CO_2 = A \cdot T^B, \qquad (3)$$

in which A and B could be determined graphically with fair accuracy, the exponential equation was differentiated to give a relationship for instantaneous rates of carbon dioxide evolution, namely,

$$\frac{d(CO_2)}{dT} = A \cdot B \cdot T^{B-1} \qquad (4)$$

The ratio of the instantaneous rate at a given time for unit weight cellulose to that for unit weight of glucose should represent the relative glucose concentration. By incorporation of the factor 100 to convert to a percentage basis and the factor 1.11 to correct glucose to cellulose, an equation for estimating percentages of cellulose hydrolyzed is finally obtained, namely,

$$\text{Per cent cellulose hydrolyzed} = 100 \cdot A \cdot B \cdot T^{B-1}/1.11S \qquad (5)$$

Since S is a predetermined numerical value and A and B are constants, percentages of cellulose hydrolyzed can be calculated for a series of T

(22) R. F. Nickerson, *Ind. Eng. Chem.*, **34**, 1480 (1942).

values. Conrad and Scroggie[20] have suggested two alternative methods of converting observed carbon dioxide–time data to percentages of cellulose hydrolyzed.

Typical hydrolysis curves for a variety of celluloses are shown in Fig. 2. As has been noted by various investigators,[9,20,21] the curves indicate rapid initial hydrolysis followed by a slower hydrolysis and are

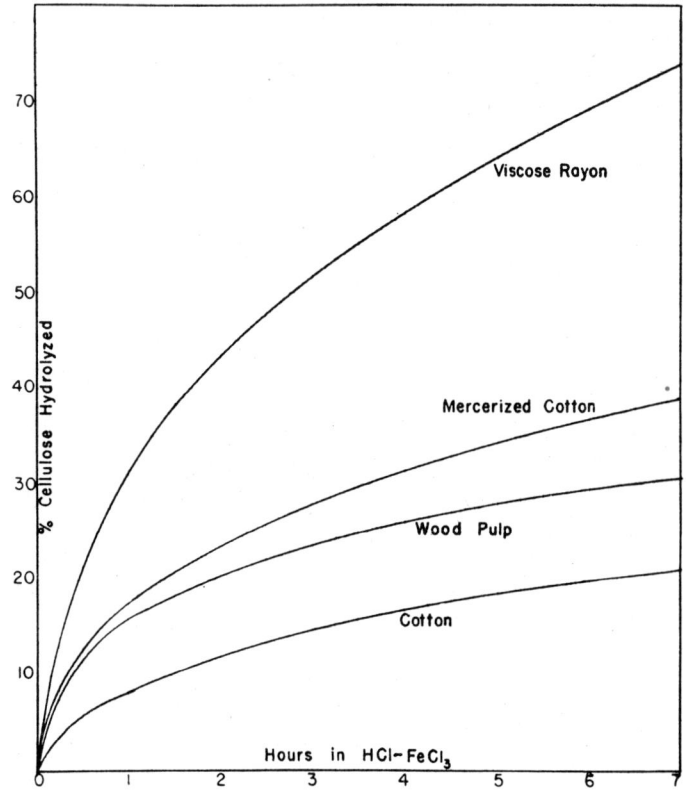

FIG. 2.—Hydrolysis of various cellulosic materials with time in acid-oxidant.

taken to mean that a part of the cellulose structure hydrolyzes readily because it is more reactive or more accessible than the remainder of the structure. The latter is presumed to consist of the dense, less reactive crystallites which are not penetrated or distorted by the hydrolyzing medium. In other words, hydrochloric acid–ferric chloride solution as used in the test is not an "intramicellar" swelling agent. Furthermore, the resistant flock which remains at the end of a seven hour run gives a sharp crystalline diffraction pattern.[21]

A series of estimates of non-crystalline cellulose in various types of cellulosic material is presented in Table II. These estimates were obtained by extrapolating the relatively flat portions of curves such as those given in Figure 2 to zero time and are admittedly first approxima-

TABLE II
Content of Non-crystalline Cellulose and Reactivity of Crystalline Cellulose in Various Materials[22]

Material	Non-crystalline cellulose, percent	Reactivity of crystalline cellulose[a]
Unmercerized cotton	5	7.4
Mercerized cotton	11	13.0
High grade linters	6	8.6
Wood pulp	10	8.5
Viscose rayon from linters	21	26

[a] In arbitrary units.

tions only.[22] The crystallite reactivity values given in the table are also estimates based on the assumption that the flattened part of the hydrolysis curve represents primarily the hydrolysis of crystallites. If this is true, then the slopes of the flattened parts indicate the rates of hydrolysis

TABLE III
Accessibilities of Cellulose in Original and Modified Wood Pulps and Cotton Linters[20]

Description	Percent alpha cellulose	Percent accessibility
Wood pulp from beech	88.5–89.0	11.5
Wood pulp from Southern pine	93	10
Wood pulp from Western hemlock	91.5	9
High-alpha pulp from Southern pine	94.5–95.0	7.5
Cotton linters (DP ca. 1000)	98.5	5.3
Cotton linters alkali cellulose after shredding, DP 850		14.5
Cotton linters alkali cellulose after shredding and aging, DP 610		11.0
Cotton linters, dried by solvent exchange		7.5

of the crystallites. The crystallite reactivity data so derived suggest that mercerization causes a substantial increase in crystallite accessibility and an even more pronounced effect is shown in the case of viscose rayon.

Data on accessibility that were obtained in a manner similar to that described above are given in Table III. These values were reported by

Conrad and Scroggie[20] and, in so far as direct comparison is possible, they agree fairly well with the data in Table II. It is apparent in this case also that treatments with strong alkali increase the apparent accessibility.

In a series of further experiments Conrad and Scroggie[20] obtained accessibility measurements on several rayons of known strength and elongation. Their data are shown in Table IV where it can be seen that

TABLE IV
Accessibility of Regenerated Cellulose Yarns[20]

Type of rayon	Dry yarn strength g. per denier	Elongation, percent	Accessibility, percent
High tenacity, low elongation viscose	4.9	7.4	15.0
High tenacity, deacetylated cellulose acetate	4.2	7.2	16.5
Textile, delustered viscose	1.8	17.8	27.0
Textile, bright viscose	1.8	18.6	27.0
High tenacity, tire cord yarn viscose	3.7	9.7	27.0
High tenacity, experimental viscose	3.9	10.8	31.5

the accessibility varies considerably from yarn to yarn and seems to depend on the method of manufacture. Through X-ray determinations of "crystallinity numbers," they noted some inverse correlation between accessibilities and this measure of crystallinity. They concluded that this evidence favored the idea that accessibility is a measure of disordered cellulose.

The method of converting raw carbon dioxide data to percentages of cellulose hydrolyzed, described above, apparently is reasonably accurate after the first stages of the process. For example, on washing, drying and weighing the resistant residues present in the reaction vessel after regular seven-hour runs, it was found[22] that practically the entire calculated amount of unhydrolyzed material was recovered. In other words, the sum of the residue recovered and of the estimated amount hydrolyzed accounted for the original sample within experimental error. This observation suggested a more direct approach to the problem than that of determining carbon dioxide, and two independent investigations were undertaken in which resistant residues were used as a means of following the course of hydrolysis.

The first of these was made by Lovell and Goldschmid.[23] These workers determined weight losses produced by the action of ferric

(23) E. L. Lovell and O. Goldschmid, *Ind. Eng. Chem.*, **38**, 811 (1946).

chloride–hydrochloric acid for varying times, plotted the observed losses against time of treatment, and obtained curves similar to those in Figure 2. Extrapolation of the flattened parts of curves to zero time yielded estimates of accessibility, while the slope of the flat part of a curve was used as an estimate of crystallite reactivity. Their data are given in part in Table V. By comparison with data previously cited,

TABLE V

Accessible Cellulose in Wood Pulp, Linters and Regenerated Cellulose. Crystalline Reactivity and Yarn Properties[23]

Material tested	Accessibility percent	Reactivity[a]	Yarn strength g. per denier, dry	Yarn elongation percent, dry
Wood pulp from Southern pine	5.9	1.6		
Rayon, high stretch	25	5.2	3.4	10.0
Rayon, normal	27	4.5	2.1	23.5
Spinner's fluff	21	4.3	0.06	28.7
Cotton linters	4.5	1.2		
Rayon, high stretch, from linters	23	5.7	3.4	11.8

[a] Arbitrary units, percent per hour for crystalline portion.

these values indicate that agreement is reasonably good for similar samples. Spinner's fluff is filamentous cellulose fiber regenerated without the application of tension. The similarities in accessibilities and reactivities of the fluff compared to those of stretch-spun yarns led the investigators to conclude, as did Conrad and Scroggie, that the crystallinity of rayon depends principally on the processes of coagulation and regeneration and not on the crystallinity of the starting material or on the spinning stretch.

The second investigation by the method of weighing residues is that of Philipp, Nelson and Ziifle.[24] These workers prepared an excellent review of the prior literature on cellulose structure and acid treatments and suggested that the degradation may represent two first-order reaction processes. One first-order reaction involving the hydrolysis of amorphous cellulose predominates initially but soon becomes negligible; the other involving the crystallites ultimately becomes the principal process. In this case a plot of weight loss against time on semi-logarithm paper should yield a straight line except for the initial stages of the process and the straight line should extrapolate to a reliable zero time value. A

(24) H. J. Philipp, Mary L. Nelson and Hilda M. Ziifle, *Textile Research J.*, **17**, 585 (1947).

variety of materials were accordingly treated, the oxidant (ferric chloride) being omitted.

The omission of the oxidant did not, however, prove to be a simplification. With increasing time of hydrolysis at or near the boiling point the breakdown products of cellulose darken and apparently undergo a polymerization to insoluble substances that adhere to the resistant solid residues. It was necessary, therefore, to introduce a correction for these so-called "humic" substances before the semi-logarithm plots gave linear relationships that could be extrapolated. This being done, the assumption was made that the intercept at zero time gave the relative amounts of crystalline and amorphous cellulose in the intact samples. The ratio of the crystalline cellulose so determined to the total cellulose found by complete oxidation was taken as the degree of crystallinity.

Estimates reached by Philipp and coworkers for a variety of materials are shown in Table VI. These data indicate somewhat lower percentages

TABLE VI

Degree of Crystallinity of Various Fibers and Some Rayon Yarn Properties[24]

Material	Degree of crystallinity	Tenacity, g. per denier	Elongation, percent
Ramie	.95	—	—
Raw cottons	.84–.88	—	—
Purified linters	.88	—	—
Mercerized cottons	.68–.78	—	—
High tenacity saponified acetate	.83	7.6	5.9
High tenacity viscose rayon	.81	5.0	7.2
Special viscose rayon	.69	0.3	8.2
Textile, viscose rayon	.68	2.5	15.8
Tire cord viscose rayon	.62	3.9	9.4

of crystalline cellulose, and hence relatively higher percentages of amorphous material, than are shown for corresponding substances in the preceding tables. In general, however, orders of magnitude do not differ appreciably, and the data for regenerated cellulose, excepting those for saponified cellulose acetate and high tenacity viscose, tend to confirm the conclusion of Lovell and Goldschmid, and Conrad and Scroggie, that crystallinity may depend upon coagulation conditions rather than upon subsequent processing.

Further insight into the decomposition of sugars which precedes formation of humic substances has been supplied by Saeman.[25] Investigating the hydrolysis of wood in the temperature range 170–190°C. with

(25) J. F. Saeman, *Ind. Eng. Chem.* **37**, 43 (1945).

dilute sulfuric acid, he found the decomposition of sugars to follow a first-order reaction with an activation energy of 32,800 calories.

A number of different cellulosic materials examined by Saeman[25] showed the presence of an easily hydrolyzed portion in every instance. The hydrolysis of the resistant portion was noted in all cases to be a reaction of the first order. By extrapolation of hydrolysis–time data Saeman found 44% of resistant cellulose in Douglas fir wood having a total of 66.6% potential reducing sugar. The difference, 22.6%, represents easily hydrolyzed polysaccharides of which only ⅔ were fermentable sugars, presumably D-glucose. If the assumption is made that the fermentable sugars come only from α-cellulose, a maximum of about 20% easily hydrolyzed cellulose is obtained for pure wood cellulose. Any fermentable sugars arising from hemicellulose would reduce this estimate in proportion.

A recent modification of the hydrolysis–oxidation technique[19] has produced strong support for the acid hydrolysis results already mentioned. In this instance samples of purified cotton linters were treated for varying times with boiling 2.5 N hydrochloric acid or with 2.5 N hydrochloric acid– 0.6 M ferric chloride. The latter was necessary where the hydrolysis extended over periods longer than 12 minutes and acted to hinder the formation of "humic" materials. By filtration and washing, a series of hydrocelluloses was obtained which corresponded to times of hydrolysis varying from 0 to 7 hours.

The moisture-regain of these hydrocelluloses at 70°F. and 65% relative humidity was determined and is shown in Table VII. It is evident

TABLE VII
Effect of Hydrolysis on Moisture-Regain (at 70°F., 65% rel. humidity) of Cotton Linters[19]

Time Hydrolyzed, Hours	Moisture-Regain of Residue, Percent
0	7.66
0.07	6.26
0.2	6.43
0.8	6.77
2.0	7.14
4.0	7.71
7.0	8.17

that the first 3 or 4 minutes of hydrolysis causes a short, sharp drop in regain after which a slow rise in regain occurs. Davidson[26] has reported a similar moisture-regain behavior as a result of acid treatment.

The same series of hydrocelluloses was subsequently subjected to the carbon dioxide evolution test with the results shown in Fig. 3. These curves indicate that 4 minutes (.07 hr.) of prior hydrolysis caused an

(26) G. F. Davidson, J. Textile Inst., **34**, T87 (1943).

appreciable decrease in available or accessible cellulose. Further prior hydrolysis for times up to about an hour apparently caused additional slight reductions in accessibility but not thereafter. In other words, prior hydrolysis for about an hour lowered accessibility to an irreducible minimum.

By calculations based on differences in carbon dioxide evolution, it was found that about 3% of the intact linters structure was removed in .07 hours of prior hydrolysis. Partial confirmation of these calculations was obtained experimentally when it was found that a mixture containing 97% of .07-hour hydrocellulose and 3.3% of glucose simulated

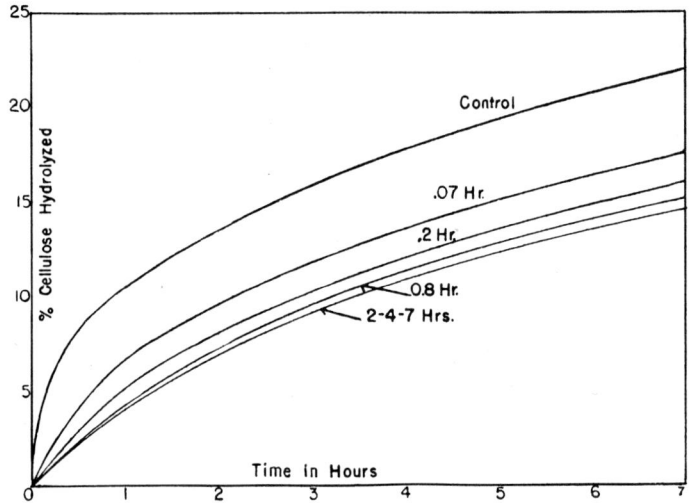

Fig. 3.—Effects of varying amounts of hydrolytic pretreatment on hydrolysis of cotton linters.

the evolution behavior of intact linters almost perfectly. With this 3% of the intact structure there apparently is associated an appreciable hygroscopicity. The ready availability and high regain capacity of this small fraction of the structure suggested it to be extremely open.

An additional 3% of the linters was also estimated to be accessible. This additional material, which hydrolyzed between .07 and 1 hour, was regarded as fairly well ordered since its removal was effected less readily than the first 3% and since it exhibited a hygroscopic behavior similar to highly acid-resistant cellulose. By difference then, the crystalline material should constitute 94% of the intact linters. A value of 92% crystalline or highly resistant cellulose was derived from the weights of residues recovered. This result confirmed, at least qualitatively, the amounts found by the differential method.

Some additional experiments have been conducted in which a direct analysis was employed.[27] Cellulosic materials were boiled in 2.5 N sulfuric acid for varying times and the hydrolyzates were quantitatively examined for glucose. A modification of the glucose method of Hassid[28] was employed. This approach was designed to give more accurate information on the early stages of the hydrolysis than could be obtained by other methods.

The investigation showed two significant things. First, hydrolysis with sulfuric acid also caused a sharp reduction in moisture-regain but the minimum is reached only after 30 minutes of boiling. This indicated that sulfuric acid is considerably less active as a hydrolyzing agent than

TABLE VIII

Moisture-Regain Reduction, Percentage of Cellulose Hydrolyzed and Estimates of Crystallite Length[27]

Material	Cellulose hydrolyzed, percent	Moisture-Regain reduction,[a] percent	Crystallite length, DP
Viscose rayon	5.70	3.28	110
Mercerized cotton	3.25	1.91	179
Unmercerized cotton	1.80	1.07	253
Cotton linters	2.00	1.14	244

[a] Actual difference caused by hydrolysis.

hydrochloric acid of equivalent concentration. The second significant observation was that the cuprammonium viscosity of hydrocelluloses fell rapidly during the first 30 minutes of boiling with sulfuric acid and thereafter remained constant for several hours at least and probably much longer. Simultaneous glucose determinations indicated, moreover, that hydrolysis was initially rapid and was still appreciable after chain shortening had apparently ceased. These observations are in excellent agreement with those of Davidson[26] whose experimental conditions were, however, quite different.

The findings suggest that shortening of the cellulose molecule ceases in 30 minutes but hydrolysis does not. A possible explanation is that disordered hygroscopic cellulose between crystallites is first attacked and that thereafter attack occurs mainly on the lateral surfaces of crystallites. If this is indeed the case, then the effective length of crystallites is indicated by the minimum viscosity attained.

Data for a few representative materials are presented in Table VIII

(27) R. F. Nickerson and J. A. Habrle, *Ind. Eng. Chem.*, **39**, 1507 (1947).
(28) W. Z. Hassid, *Ind. Eng. Chem., Anal. Ed.*, **8**, 138 (1936).

and illustrate the simultaneous changes in regain, and percentages of cellulose hydrolyzed, and also possible crystallite lengths in DP. It should be understood that the given percentages of cellulose hydrolyzed probably correspond to only a part of the total accessible cellulose, perhaps one half or less. In any event, it can be seen that both mercerization and regeneration processes increase the readily available fraction of cellulose, with corresponding increases in hygroscopicity. These changes also appear to coincide with reduction in length of crystallites.

Nelson and Conrad[29] have recently confirmed the viscosity behavior observed by Davidson[26] and Nickerson and Habrle[27] and have drawn a similar conclusion, namely, that after the rapid destruction of about 2% of the intercrystalline network, hydrolysis occurs mainly on lateral crystallite surfaces. They also show that the apparent degree of crystallinity is reduced by fine grinding of cotton fibers.

It is obvious that acid hydrolysis methods leave a number of unsolved problems and many minor disagreements to be ironed out. In general, however, the available results suggest that the natural celluloses consist chiefly of crystalline material which is only slowly eroded by acids. The non-crystalline fraction appears to be relatively more susceptible to hydrolysis than the crystalline fraction and to have a greater capacity to absorb moisture. In other words, the non-crystalline fraction is probably more reactive than the crystalline material, as Mark[14] has suggested. In this connection the fact should not be overlooked that the surface layer of the crystallites is probably amorphous and hence relatively more reactive than the underlying layers.

2. *Swelling and Density Investigations*

A study conducted under the direction of Elöd at Badenweiler, Germany, and reported by Schiefer and Kropf[30] clearly shows that the amount of fiber swelling may depend on fiber density. Dry fiber buoyancies were determined by means of a micro-balance contained in a carbon tetrachloride system, and were calculated to densities. Fiber swelling was estimated by spinning water-swollen fibers in a screen basket centrifuge for one minute at an acceleration of 750,000 cm./sec.^{2} The samples were then weighed, dried and reweighed. Swelling values so obtained were found to decrease as fiber densities increased. For example, a rayon which yielded a swelling value of 160% had a density of 1.505, while cotton with a swelling value of about 50% had a density of 1.534. If the reasonable assumption is made that fiber density decreases

(29) Mary L. Nelson and C. M. Conrad, *Textile Research J.*, **18**, 149, 155 (1948).
(30) H. F. Schiefer and R. T. Kropf, *Textile Research J.*, **16**, 432 (1946).

as non-crystalline cellulose increases, then swelling may occur chiefly in the non-crystalline regions.

Hermans[31] reported a somewhat similar investigation. Densities of natural fibers were determined by immersing them in carbon tetrachloride and varying the temperature. The density of crystalline cellulose as derived from X-ray data was taken as 1.59. Hermans concluded that, in general, low density cellulosic fibers are those known to have less crystalline substance and higher moisture-regain than higher density fibers. By assuming amorphous cellulose to have a 6% lower density than crystalline material, he arrived at 40%, 50%, and 75% as estimates of the amorphous cellulose in natural fibers, wood pulp, and rayon, respectively. This approach is interesting, but the final data are so far out of line that some refinement of technique may ultimately be required.

3. *Oxidation and Deuterium Exchange*

Goldfinger, Mark, and Siggia[32] oxidized cellulose with sodium periodate in aqueous solution. The observed rates of oxidation were explained on the assumption that the materials consisted of easily accessible and difficulty accessible regions. By extrapolation of the ultimate slow rates,

TABLE IX

Amorphous Cellulose in Cotton and Cuprammonium Rayon by Periodate Oxidation[11]

Material	Amorphous, percent
Purified cotton linters	6.0
Cotton (average of three)	1.4
Cuprammonium rayon, dried from water	7.4
Cuprammonium rayon, dried from benzene	19.5

which appeared to correspond to oxidation of the inaccessible fraction in each case, estimates were reached for the amounts of amorphous component. The periodate ion is known to attack the 2 and 3 positions of the glucose units of cellulose, to split the glycol configuration, and to convert it to two carbonyl groups.[33]

Goldfinger and coworkers obtained the data shown in Table IX by this method. The difference in amorphous cellulose between linters and cotton is appreciable but attention is directed particularly to the behavior of cuprammonium rayon after being dried from water and after being

(31) P. H. Hermans, *J. Textile Inst.*, **38**, P63 (1947).
(32) G. Goldfinger, H. Mark and S. Siggia, *Ind. Eng. Chem.*, **35**, 1083 (1943).
(33) E. L. Jackson and C. S. Hudson, *J. Am. Chem. Soc.*, **59**, 2049 (1937); **60**, 989 (1938).

dried by solvent exchange. Conrad and Scroggie[20] found no difference in hydrolysis–oxidation behavior to be caused by solvent exchange dehydration. Since the periodate oxidation is conducted in aqueous medium, no differences should be expected inasmuch as fiber swelling should eliminate the solvent exchange effect. The fact that a difference is reported suggests that further investigation is needed to clarify the issue and possibly also to explain the linters–cotton discrepancy.

Badgley, Frilette, and Mark[9] devised a method of digesting pulp in water of high deuterium content. Exchange was permitted to occur and the rate of absorption of D_2O was determined as a function of time. Variations in conditions produced in all cases curves with a steep initial portion followed by a flattened linear branch. The investigators emphasized that a multistep exchange reaction of D_2O with cellulose hydroxyls apparently occurred but the work was too immature to permit a strict interpretation. It seemed possible, however, that the results might mean that exchange takes place rapidly with all primary hydroxyls whether inside or outside the crystalline regions because the initial rapid phase of the process appears to involve about one-third of all the cellulose hydroxyls.

In a note on the decomposition of cellulose by a mixed culture of *Vibrio perimastrix*, Perlin and Michaelis[34] reported about 5.6% of filter paper was fermented in 8 days. The agreement between this percentage and the percentages of accessible cellulose in native cellulose as found by hydrolysis may be fortuitous. However, there is enough similarity in acid and enzymatic hydrolysis to warrant further work along this line.

As Spurlin[35] has pointed out, the effects of water on amorphous cellulose are akin to a solvent action. In effect, water converts the intermicellar regions to a condition analogous to agar or gelatin gels. The diffusion of moderately large molecules through such gels is almost as rapid as through water. Penetration of reagents into the intermicellar regions must also be rapid as the fiber imbibes liquids and swells.

4. *Esterification and Etherification*

Elöd and Schmid-Bielenberg[36] observed that the speed of acetylation of dry native fibers increases with decreasing degree of micellar (crystallite) orientation. Arranged in order of increasing reactivity the dry native fibers were flax, hemp, ramie and cotton. On being pretreated with water or acetic acid, however, the fibers were alike in rates of reac-

(34) A. S. Perlin and M. Michaelis, *Science*, **103**, 673 (1946).
(35) H. M. Spurlin in "Cellulose and Cellulose Derivatives" (E. Ott, Editor), p. 615 (1943), Interscience, New York.
(36) E. Elöd and H. Schmid-Bielenberg, *Z. physik. Chem.*, **B25**, 27 (1934).

tion and reached approximately the same ultimate acetyl contents. They concluded that the reaction rates of the dry fibers are determined by rates of diffusion of reactants into the fibers and that the more parallel the micelles are, the slower are the diffusion rates.

Acetylation rates have also been studied by Centola[37] who treated natural and mercerized ramie fibers for varying times with acetic anhydride and sodium acetate and examined the reaction products chemically and by X-ray diffraction. The reagent was considered to penetrate into the interior of fibers. A heterogeneous micellar reaction was believed to occur that converted a semi-permeable elastic membrane around the micelles into the triacetate. The rate of acetylation of mercerized ramie was observed to be faster than that of unmercerized fiber. Centola concluded that about 40% of the cellulose in native ramie is amorphous and acetylates rapidly.

The effects of swelling on estimates of accessible cellulose have been clearly illustrated by Assaf, Haas and Purves.[38] These investigators prepared cellulose ethers by way of thallous ethylate which is strongly basic and reacts with cellulose according to the following equation:

$$H-\overset{|}{\underset{|}{C}}-OH + Tl-O-C_2H_5 \rightarrow H\overset{|}{\underset{|}{C}}-O-Tl + C_2H_5OH.$$

The thallous derivative so formed was found to undergo metathesis with methyl iodide and to yield thallous iodide and methyl cellulose. Analyses of the latter product for methoxyl content after suitable purification yielded a measure of the accessible hydroxyl groups. The reactions were all conducted in non-aqueous systems.

Evidence was presented that thallous ethylate did not penetrate or alter the crystalline parts of the fiber. Moreover, it was possible to conduct the thallation with different solvents for thallous ethylate. When this was done with normal ethers, the extent of methylation was observed to decrease as the molecular volume of the thallous ethylate solvent increased. These results suggested that accessibility is dependent upon the penetrating power of the ether solvent. Amorphous cellulose was, therefore, defined as the percentage of cellulose wetted by an ether of zero molecular volume and was estimated by determining methylation–molecular volume values for three or more straight-chain ethers, plotting the data and extrapolating to obtain methoxyl content for an ether of zero molecular volume. The amount of cellulose corre-

(37) G. Centola, *Gazz. chim. ital.*, **65**, 1021 (1935); *Chem Abstracts*, **30**, 5027(1936); *Atti X congr. intern. chim.*, **4**, 129 (1939); *Chem Abstracts*, **34**, 2169 (1940).

(38) A. G. Assaf, R. H. Haas and C. B. Purves, *J. Am. Chem. Soc.*, **66**, 59 (1944).

sponding to this theoretical condition was calculated from the derived methoxyl value.

Amorphous cellulose, so defined, was reported for two simple but noteworthy modifications of cotton linters. First, linters which had been swollen with cold 10% sodium hydroxide, washed, and dried by solvent exchange prior to thallation and methylation, showed an amorphous content as high as 27%. Secondly, unswollen linters appeared to contain only 0.25 to 0.50% of amorphous cellulose. Similarly, swollen ramie appeared to contain 18% of amorphous cellulose; unswollen ramie, 0.25%.

The investigators also presented a comparison of data on crystallinity drawn from the literature. This tabulation clearly shows the wide variations which have been reported. At first sight the values just presented, particularly for linters, may seem to be inconsistent with acid hydrolysis estimates. However, it was recognized that swelling could increase the amorphous fraction to 50 or even 100 times the amount occurring in dry unswollen material and that determinations based on aqueous solutions such as acids involve considerably swollen material. Hence, values obtained by the latter methods may be reasonably correct as implicitly defined by test conditions.

VI. Discussion

The evidence presented fails to suggest the causes for the large variations in crystallinity estimates which have been reported for similar cellulosic materials. There is a possibility that the different methods may not measure precisely the same characteristic of the material. It also may be that relative crystallinity is not a fixed quantity in any case but depends on circumstances involved in the measurement, such as the amount of swelling. The estimates reached by different methods need to be reconciled. At present, crystallinity estimates which depend wholly or in part on X-ray diffraction seem to be much higher than those obtained by chemical methods. The fact is that X-ray diffraction methods are ideal for studies of the crystalline fraction but are necessarily indirect in application to the non-crystalline fraction. The converse is true for the chemical approach. Apparently a combination of diffraction and chemical methods may adjust the existing differences.

The conditions of swelling during the measurement undoubtedly have an important bearing on crystallinity estimates. Area determinations by Assaf, Haas and Purves[39] indicate that up to 98% of the surface of moist cellulose is obliterated by direct drying while about 75% remains

(39) A. G. Assaf, R. H. Haas and C. B. Purves, *J. Am. Chem. Soc.*, **66**, 66 (1944).

after solvent exchange dehydration. Other investigators have found similar large changes in internal surface with swelling.

Swelling may influence crystallinity estimates in either of two ways or in both ways simultaneously. First, swelling may convert some crystallized cellulose to amorphous, that is, it may distort the internal fiber structure and increase chain disorder. Second, the laying open of the fiber by swelling may make accessible some non-crystalline areas which are screened off in the unswollen, dry material. In addition, there exists in the case of aqueous swelling a hysteresis effect, since the condition of the sample depends upon the direction of approach to equilibrium that is, whether from the drier or wetter condition. Also, the normal shrinkage can be almost completely prevented by solvent exchange dehydration. Taken together, these factors make the specification of a reference standard swelling state for crystallinity purposes a difficult undertaking.

It is probable that varying degrees of ordering of chains exist in a cellulosic material and that a sharp differentiation of crystalline and non-crystalline celluloses may not be feasible or even possible. Theoretically, the lateral surfaces of crystallites are amorphous but may have far less importance in determining such properties as strength, flexibility and extensibility than the non-crystalline cellulose which supplies continuity of structure in the direction of crystallite orientation. Yet properties like moisture absorption and swelling may be more dependent upon the amount of cellulose which exceeds a certain degree of disorder (permeability) than upon location. The definition of crystallinity may, therefore, be made ultimately in terms of practical objectives.

VII. Implications

The regions in a cellulosic structure that are arbitrarily designated as amorphous, accessible, or disordered are unquestionably more reactive than the crystalline substance. The augmented reactivity of this intercrystalline material is manifest in its hygroscopic nature, in its low density and ease of swelling, and in the relatively high velocity at which it undergoes hydrolysis, oxidation, esterification and etherification. In practice this behavior is highly important because processing, or the effects of processing, may tend to localize in the more reactive areas. The intercrystalline regions may, therefore, be vulnerable parts of the structure which are relatively susceptible to alteration or degradation.

A few examples may serve to illustrate the implications of relative crystallinity and especially of intercrystalline reactivity. Two independent investigations[27,29] have shown that cotton fibers are reduced to a powder by the hydrolytic removal of 2% or less of the fiber weight. Fiber continuity is, accordingly, almost completely depend on this small, relatively susceptible fraction. It is not unlikely that this fraction is also first attacked by oxidizing agents and microorganisms. Apparently, modification of this small fraction so as to confer resistance to hydrolytic, oxidative and enzymatic attack would impart a large measure of immunity to the fiber. Such modifications of specific small regions in the fiber would be facilitated by their inherently greater reactivity.

The case of regenerated cellulosic fibers is interesting in this connection. It is well known that ordinary unsubstituted rayons swell appreciably in water and lose considerable strength, which is recovered as the materials dry and shrink. The swelling presumably occurs chiefly in the intercrystalline regions where chains are pushed apart laterally, forces of cohesion are reduced, and slippage can then take place. In an excellent review Heuser[40] has cited data which indicate that one per cent or less of formaldehyde combined with viscose rayon sharply decreases both swelling and strength losses. Formaldehyde presumably forms methylene ether cross linkages between chains in the intercrystalline regions and thus achieves these effects.

Mercerization apparently causes an appreciable increase in the amount of accessible cellulose in cotton while unsubstituted rayons appear to contain a greater proportion of intercrystalline cellulose than either type of cotton. The dye affinities and moisture-regain capacities of these fibers generally seem to be in the order of increasing accessibilities.

Relative crystallinity undoubtedly influences such properties of cellulosic materials as rigidity, flexibility, plasticity and extensibility. Likewise the amount and reactivity of intercrystalline cellulose are major factors in common processing treatments such as bleaching, dyeing, pulping and wet finishing. Further refinement of measuring methods and the development of further correlations between crystallinity and fiber properties would contribute much to this important field.

(40) E. Heuser, *Paper Trade J.*, **122**, (No. 3) 43 (1946).

THE COMMERCIAL PRODUCTION OF CRYSTALLINE DEXTROSE

By G. R. Dean and J. B. Gottfried

Corn Products Refining Company, Argo, Illinois

Contents

I. Introduction	127
II. History	127
III. Manufacture	131
1. Alpha Dextrose Monohydrate	131
2. Anhydrous Alpha Dextrose	135
3. Beta Dextrose	136
IV. Application of Ion Exchange Refining to the Commercial Manufacture of Crystalline Dextrose	137
1. Introduction	137
2. Factors Governing Application of Ion Exchange	139
3. Methods of Application	140
4. Advantages	140
5. Description of Operation	142

I. Introduction

The carbohydrate chemist is familiar with the general aspects of the process for the manufacture of crystalline dextrose because of the important position this compound occupies in the whole carbohydrate field. The special features of the process, however, are less familiar and it is hoped that this discussion will afford a better idea of modern methods.

More than a century elapsed between the discovery of the hydrolysis of starch and the successful large scale manufacture of crystalline dextrose of high purity. Examination of some less familiar properties of dextrose will indicate the difficulties which have been encountered in developing the processes used today.

II. History

Dextrose was doubtless known to the ancients because of its occurrence in granulated honey and evaporated wine musts. Efforts to prepare the sugar as an article of commerce were not made until the beginning of the nineteenth century.

The shortage of sugar in Europe occasioned during the Napoleonic Wars by the Continental System stimulated searches for substitutes. Numerous attempts were made during this period to develop continental sources of sugar. One of the first of these was the preparation of dextrose from grapes by Proust in 1801.[1] A prize was awarded him by Napoleon Bonaparte in order to stimulate work to develop his process further. The older name, grape sugar, owes its origin to this process. This work was of short duration because of Kirchoff's preparation of dextrose, in 1811, from potato starch by hydrolysis with sulfuric acid.[2] Although, as early as 1781, it was known that starch reacted in the presence of acids to produce a soluble sweet flavored substance, Kirchoff is generally credited with independently discovering the method of obtaining dextrose in this manner. Moreover, he recognized the value of the reaction and proceeded energetically to develop his discovery. Kirchoff succeeded in preparing a sirup which crystallized upon standing and from which a solid product was obtained by pressing the mass in cloth sacks. The product contained large proportions of mother liquor and was far from the pure dextrose produced today. Serious efforts were made to manufacture the product on a considerable scale. In 1813 the continental blockade was lifted, imported sugar was again available, and on economic grounds the work on dextrose was abandoned.

Interest in the manufacture of dextrose from starch was revived from time to time during the early part of the nineteenth century. Although small quantities of dextrose in a fairly pure form were prepared by repeated crystallization and the use of non-aqueous solvents, large scale efforts were confined to preparing the sugar either in sirup form or as a solid mass of crystals containing the mother liquor. To a limited extent, products of the latter type are manufactured today and marketed under the names of chip sugar, brewer's sugar, "70 sugar" and "80 sugar."

The problem of crystallizing dextrose from starch conversion liquor in a form which would allow separation from the viscous mother liquor was attacked in the latter part of the nineteenth century by various workers. The most successful efforts were made by Behr in 1881.[3] Corn starch was hydrolyzed at low concentration, the refined product was concentrated, seeded with a very small proportion of pure anhydrous dextrose crystals, and it was then allowed to stand without agitation in heated rooms. Care was taken to exclude any traces of dextrose hydrate crystals. After several days the crystalline magma was separated and washed in centrifugal filters. By this means individual crystals of anhydrous dextrose of suitable size were obtained and the product

(1) H. Wichelhaus, "Der Stärkezucker," Akademische Verlagsgesellschaft, Leipzig, (1913).

(2) G. S. C. Kirchoff, *Academie imperiale des sciences de St. Petersbourg, Memoires,* **4**, 27 (1811).

(3) A. Behr, U. S. Pat. 250,333 (Dec. 6, 1881); 256,622 (April 18, 1882).

proved to be about 96% pure. Further purification was effected by dissolving the crystals and repeating the process. A process quite similar to this was described later by Soxhlet.[4]

Wagner in 1906[5] investigated the use of slow agitation similar to that practiced in the cane sugar industry. The conditions of temperature, concentration, and type and proportion of seed crystals used were the same as those described previously by Behr. Simultaneous growth of crystals of the anhydrous and monohydrate types gave products which were difficult to separate.

During the first World War, the Corn Products Refining Company attempted to improve the methods of manufacturing dextrose to obtain a pure product at low cost. A process was developed in 1918 by Porst[6] which produced chemically pure dextrose but cost of production was still high.

In 1920 W. B. Newkirk joined that company and began an intensive investigation of the variables affecting the crystallization of dextrose from starch conversion liquors. The previous methods were reviewed, particularly those of Behr and Wagner, and it was concluded that in order to obtain satisfactory yields by crystallization of anhydrous dextrose, successive crystalline crops were required. It was found, however, that the influence of the high proportion of impurities in the mother liquor was too great for successful operation. On the other hand, it was observed that crystallization of dextrose in the monohydrate rather than anhydrous form proceeded satisfactorily despite the presence of impurities.[7] Another recognized advantage of crystallizing dextrose in this form was the fairly large change in solubility with change in temperature as contrasted with that for the anhydrous form (Fig. 1). Higher crystalline yield per crop resulted.

Efforts to crystallize dextrose in the anhydrous form were abandoned and work was concentrated on finding the best conditions for controlling the crystallization of the monohydrate. The proper concentration and temperature of the sirup before crystallization commenced were determined as well as the effect of agitation. Great care was needed in cooling the mass in order that the hydrate crystals might grow properly without excessive deposition of minute crystal nuclei, known to the industry as "false grain." A great many highly skilled operators were required, which added seriously to the cost of production.

(4) F. Soxhlet, U. S. Pat. 335,044 (Jan. 26, 1886).
(5) T. Wagner, U. S. Pat. 835,145 (Nov. 6, 1906).
(6) C. E. J. Porst and N. V. S. Mumford, *Ind. Eng. Chem.*, **14**, 217 (1922).
(7) W. B. Newkirk, *Ind. Eng. Chem.*, **16**, 1173 (1924); U. S. Pat. 1,471,347 (Oct. 23, 1923); 1,508,569 (Sept. 16, 1924); 1,571,212 (Feb. 2, 1926).

All experiments up to this time employed only minute quantities of seed crystals. In investigating the variables affecting the growth of the dextrose crystals, Newkirk found that the operation could be controlled by using much greater proportions of seed crystals than had hitherto been employed.[8] The excessive formation of crystal nuclei too small and numerous to be able to grow to satisfactory size could be avoided by this means. The operation was most economically carried out by leaving in the crystallizer 25 to 30% of a finished batch to act as seed for the

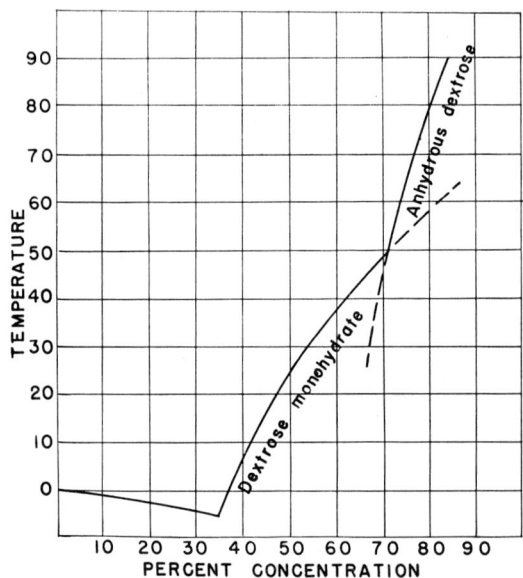

FIG. 1.—Solubility of dextrose in water.[8a]

sirup of the succeeding batch. As the mixture was slowly agitated a large area of available crystal surface was constantly present throughout the crystallizing liquor. Growth of the seed crystals reduced the local degree of supersaturation, and the tendency for nucleation. The net effect was that the final crystal mixture consisted of seed crystals which had undergone further growth, together with crystals which were originally spontaneously formed nuclei but which had opportunity to grow to a satisfactory size. The resulting product was suitable for separation of mother liquor by centrifuging and washing and required far less super-

(8) W. B. Newkirk, U. S. Pat. 1,521,830 (Jan. 6, 1925).

(8a) R. F. Jackson and Clara G. Silsbee, *Bur. Standards Sci. Paper No.* **437**, 715 (1922).

vision by skilled operators. In fact, with proper control of cooling the operation could be made almost automatic. Essentially, this is the method used today for crystallizing dextrose from starch conversion liquors. Some recent improvements in the manufacture of dextrose will be described in a later section but these comprise the steps of refining the liquor prior to crystallization.

III. Manufacture

1. *Alpha Dextrose Monohydrate*

Commercial dextrose is marketed in three forms, the alpha monohydrate, alpha anhydrous, and beta anhydrous modifications. The principal proportion is the monohydrate; the other forms are marketed for special consumers.

A pure grade of starch is used as raw material for the process. In the corn wet-milling industry this appears as a suspension in water subsequent to the separation steps by which the gluten, hull, and germ of the original grain are removed. The starch suspension is adjusted to the required concentration, usually 15 to 20%, hydrochloric acid is added to a concentration of about $0.03\ N$, and the suspension is fed to an autoclave for conversion. The autoclave is heated to about 150°C. for a period of about thirty minutes and the contents are then discharged into a neutralizing vessel. Soda ash solution is added until the pH is between 4 and 5. At this point a large proportion of the non-carbohydrate impurities such as insoluble protein, fats, and colloidal matter coagulate. The optimum pH of neutralization is determined by trial. The liquor is then run through skimming tanks where those impurities which float are removed. The underflow is filtered and the clear, dark-colored liquor passes through filters of bone char. The partially decolorized liquor is evaporated under reduced pressure to a concentration of about 60% and then given a second decolorizing treatment with bone char, followed by additional treatment with activated carbon. The liquor is finally evaporated under reduced pressure to a sirup containing 75 to 78% dry substance. The sirup proceeds to coolers, where its temperature is adjusted to about 45°C., and then into crystallizers. These are large cylindrical tanks mounted horizontally and fitted with slowly moving agitators, cooling jackets, and coils. A substantial bed of seed crystals consisting of about 25% of a previous batch is mingled with the liquor which is now at a temperature of about 41°C. Crystallization is carried out with slow cooling over a period of several days until about 60% of the mass has crystallized. The resulting magma of alpha dextrose monohydrate crystals known as the massecuite is centrifuged. A spray of water is used to wash out the last traces of mother liquor. The wet sugar is passed through rotary kilns where it is dried in a stream of filtered air to remove all but the water of crystallization. Actually the final moisture content is slightly less than the theoretical 9.1%; by this means freedom from caking is ensured.[9] The mother liquor and washings from the centrifuge are processed for recovery of a second crop of dextrose. Originally this was carried out with little further treatment other than concentration to the required degree. Today the mother liquor is subjected to "reconversion" at high temperature in the presence of acid.[10] By this means the polysaccharides which result from reversion of dextrose and which limit the yield of dextrose in the original starch hydrolyzate are hydrolyzed to dextrose. After neutralization, refining, and concentrating, the liquor is crystallized in much the same manner as the first crop. Longer crystallizing times are required because of the presence of more impurities. The final magma is separated in centrifugal filters and the crystals washed with a spray of water. This second crop is not marketed as such but is redissolved in water and added to the starch hydrolyzate before the first crystallization. The total yield of crystalline dextrose from two crops is about 80%. The mother liquor called "hydrol" yields further crops of dextrose with difficulty and consequently is not processed further. Hydrol is sold to various consumers such as the tanning, rayon, and caramel industries.

One process for recovering further amounts of dextrose from hydrol is based upon the decidedly greater ease of crystallization of the sodium chloride addition compound, $(C_6H_{12}O_6)_2 \cdot NaCl \cdot H_2O$.[11] This compound forms large rhombohedral crystals which separate rapidly even from impure solutions and in a form quite ideal for separation from the mother liquor. Hydrol contains considerable sodium chloride due to the repeated acidification with hydrochloric acid and neutralization with soda ash. In actual practice sodium chloride is added to the hydrol to give close to the theoretical 13.4% for the addition compound. After separation from the hydrol the crystalline product is dissolved in some liquor at an earlier stage of the process. The ratio of sodium chloride to dextrose is thereby reduced below the point where the addition compound separates; instead, free dextrose is recovered. Although sodium chloride is added to the process, the net effect is the recovery of further quantities of dextrose. This occurs because the total proportion of

(9) W. B. Newkirk, U. S. Pat. 1,559,176 (Oct. 27, 1925).

(10) C. Ebert, W. B. Newkirk and M. Moscowitz, U. S. Pat. 1,673,187 (June 12, 1928); 1,704,037 (March 5, 1929).

(11) T. H. Barnard and P. L. Stern, U. S. Pat. 2,150,146 (March 14, 1939).

impurities of carbohydrate type which retard dextrose crystallization is reduced.

The dextrose monohydrate obtained today by the foregoing method is a high grade product practically pure chemically. Where higher purity is occasionally desired, this product is recrystallized by essentially the same methods as described earlier.[12] The absence of appreciable amounts of impurities renders crystallization somewhat easier. Nevertheless, the tendency for uncontrolled spontaneous crystallization is still encountered so that care is necessary in cooling the mass at the

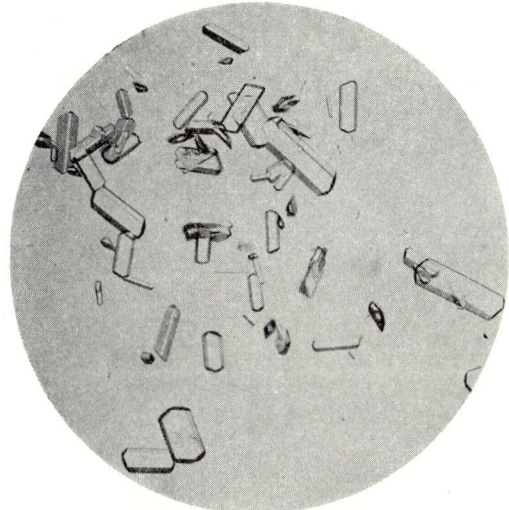

Fig. 2.—Alpha dextrose monohydrate from freshly prepared alpha dextrose solution. Magnification, 50.

proper rate. The crystallization of alpha dextrose, even from chemically pure solution, is strongly influenced in two ways by the presence of the beta isomer. First, the transformation in solution from the more soluble beta form to the less soluble alpha form is slow at room temperature and imposes a limit on the rate of crystallization. Secondly, in a dextrose solution containing the equilibrium proportions of the two forms, the beta dextrose acts as an impurity or foreign adjunct retarding the rate of crystallization and modifying the crystal habit of the alpha dextrose (Figs. 2, 3). The crystals from a chemically pure solution differ considerably in size and shape from those obtained directly from starch conversion liquors (Figs. 3, 4).

(12) W. B. Newkirk, *Ind. Eng. Chem.*, **31**, 18 (1939).

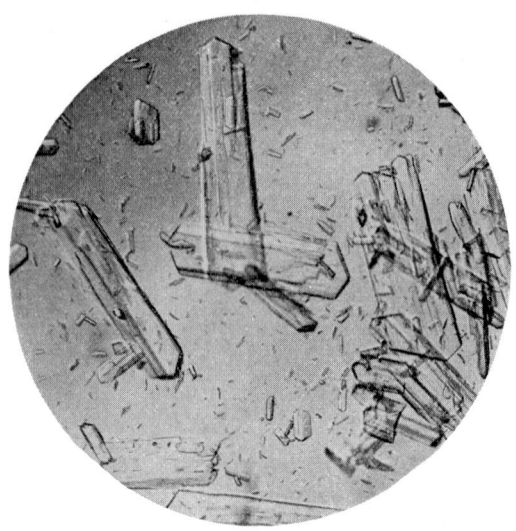

Fig. 3.—Alpha dextrose monohydrate from chemically pure (equilibrium) solution. Magnification, 50.

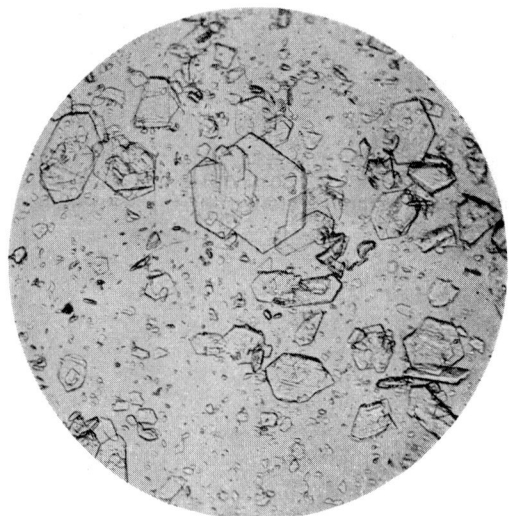

Fig. 4.—Alpha dextrose monohydrate from refined starch conversion liquor. Magnification, 50.

2. *Anhydrous Alpha Dextrose*

Anhydrous alpha dextrose may be obtained by drying the hydrate in a stream of warm air.[6] For a very pure grade of the anhydrous product a better method is used which consists of crystallizing from solution in a vacuum pan at a temperature above 50°C.[13,14] This temperature is the hydrate–anhydrous transition point. Dextrose from the first crystallization of the starch conversion liquor is dissolved in water to give about a 60% solution which is treated with a small quantity of activated carbon

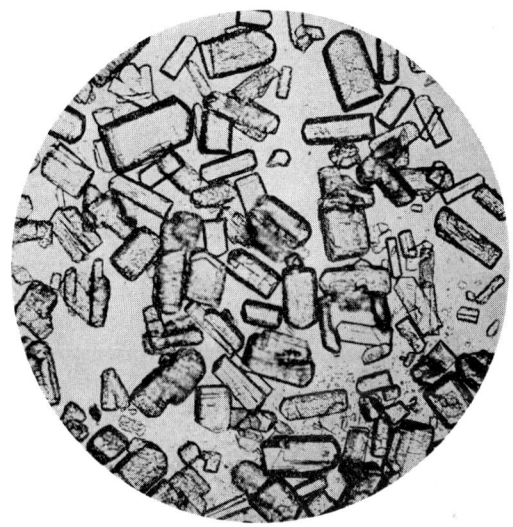

FIG. 5.—Anhydrous alpha dextrose. Magnification, 50.

to remove traces of coloring matter. A limited quantity of the solution is fed to a vacuum pan where it is carefully evaporated at about 65°C. to a concentrated sirup until crystal nuclei of anhydrous dextrose form. This is continued until, according to the judgment of the operator, sufficient crystal nuclei are present. Sufficient of the dilute dextrose solution is then fed into the evaporating pan to give a dextrose content lower than that required for spontaneous nucleus formation but above the saturation point. The crystal nuclei are then allowed to grow. At intervals, further increments of dextrose solution are added and evaporation maintains the required concentration. The temperature is held at about

(13) W. B. Newkirk, U. S. Pat. 1,521,829 (Jan. 6, 1925); 1,693,118 (Nov. 27, 1928); 1,722,761 (July 30, 1929); 1,783,626 (Dec. 2, 1930); 1,976,361 (Oct. 10, 1934); 2,065,724 (Dec. 29, 1936).

(14) W. B. Newkirk, *Ind. Eng. Chem.*, **28**, 760 (1936).

65°C. throughout this period. This higher temperature accelerates the beta-alpha transformation in the solution. Finally, after 6 to 8 hours, the batch has filled the evaporating pan and at the same time the original crystal nuclei have grown to a desirable size. The mass of crystals is then fed to a centrifugal filter, and the product is dried to a moisture content of less than 0.1%. The product is a very pure, free flowing granular material which quickly hydrates or "sets" in the presence of a limited amount of water.[15] This taking up of water of hydration is utilized to some extent in certain applications. The sugar in this form is particularly useful as a raw material in chemical synthesis, especially in non-aqueous systems. Figure 5 shows the form of the crystals.

3. *Beta Dextrose*

Where concentrated solutions are desired, as they are in special applications, the rather slow rate of solution of dextrose is encountered. This is due to the fact that a saturated solution of dextrose consists of an equilibrium mixture of the alpha and beta isomers and the attainment of this equilibrium at room temperature is slow if crystalline alpha dextrose is the starting material. If, on the other hand, beta dextrose or a mixture rich in this isomer is used, the rate of solution is very rapid.[16]

Beta dextrose has long been known as a laboratory product obtained by crystallizing dextrose at high temperature in the presence of non-aqueous solvents, particularly glacial acetic acid.[17] On a commercial scale, however, these methods are impractical and the operation is performed directly from aqueous solution.

One method consists of crystallizing dextrose solution in a vacuum pan in a manner similar to that for anhydrous alpha dextrose except that much higher temperatures are used.[18] The solution is concentrated to a sirup containing higher than 90% solids and at a temperature slightly above 100°C. A small proportion of seed crystals of beta dextrose is added and the mass is allowed to crystallize. As the operation proceeds the temperature is gradually lowered. Beta dextrose is metastable at these lower temperatures but continues to crystallize if care is taken to exclude seed crystals of alpha dextrose. The injurious effect of prolonged high temperature with incipient decomposition of the sugar is thus avoided. Figure 6 shows the form of the beta dextrose crystals.

(15) T. A. Bruce, U. S. Pat. 2,058,852 (Oct. 27, 1936).
(16) C. S. Hudson and E. Yanovsky, *J. Am. Chem. Soc.*, **39**, 1013 (1917).
(17) C. S. Hudson and J. K. Dale, *J. Am. Chem. Soc.*, **39**, 320 (1917).
(18) W. B. Newkirk, U. S. Pat. 1,693,118 (Nov. 27, 1928); 1,722,761 (July 30 1929).

Another commercial method of preparing a beta dextrose-containing product consists of the spray drying of a hot concentrated dextrose solution. To aid in removing last traces of water the solution is sprayed onto a moving bed of anhydrous solid product, conveniently obtained from a previous batch. The product is finally kiln dried and appears as pellets or "pearls" containing a mixture of beta and alpha dextrose.[19]

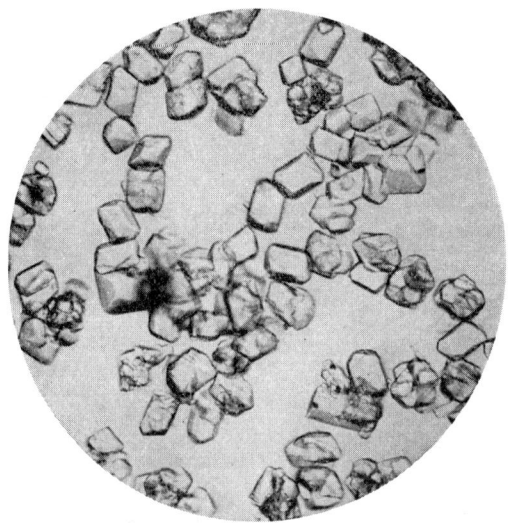

Fig. 6.—Beta dextrose. Magnification, 50.

The spray drying process can be thought of as a method of preserving in the solid state the equilibrium between the two forms of dextrose which prevails in solution. The composition of the solid does not agree strictly with that of the solution, undoubtedly because some fractional crystallization takes place before all of the water has been removed from the solid.

IV. Application of Ion Exchange Refining to the Commercial Manufacture of Crystalline Dextrose

1. *Introduction*

Application of ion exchange to the manufacture of dextrose has been in the process of development for several years. Numerous problems, not present in the application of ion exchange to water purification, were met which required solution. Although some problems remain,

(19) A. T. Harding, U. S. Pat. 2,369,231 (Feb. 13, 1945).

many have been solved, and sufficient progress has been made to permit practical application of ion exchange to dextrose production.

The use of acid and base exchangers was introduced by Adams and Holmes in 1935.[20] Progress in both application and improvement of these materials has been rapid since that time. Ion exchangers have the property of removing specific ion types from solution by a process of combined adsorption and chemical combination, and may be regenerated by appropriate treatment with the opposite type of ion. The mechanism of "exchange," though not completely defined, is supposed to proceed either by elimination of the ion in question, or by the exchange of a hydrogen or hydroxyl ion, as the case may be, for the ion removed from solution. Ordinarily, cation exchange materials operate by the actual exchange of a hydrogen ion for another cation in solution, while the so-called anion exchange resins may more properly be described as "acid-sorbing" resins. Some of the latter, particularly the newer resins, possess salt-splitting properties by which a hydroxyl ion is actually exchanged for another anion in solution. The cation exchange materials need not be limited to the hydrogen cycle, but may also be operated on the sodium cycle. In this case the resin is regenerated with sodium chloride and sodium ions are exchanged for other cations. The reactions involved may be illustrated as follows:

CATION EXCHANGER

Hydrogen Cycle,

$$R-SO_3H + NaCl \underset{\text{Regeneration}}{\overset{\text{Operation}}{\rightleftarrows}} R-SO_3Na + HCl \text{ or}$$

Sodium Cycle,

$$2\ R-SO_3Na + CaCl_2 \underset{\text{Regeneration}}{\overset{\text{Operation}}{\rightleftarrows}} (R-SO_3)_2Ca + 2\ NaCl$$

ANION EXCHANGER

$$R_3-N + HCl \xrightarrow{\text{Operation}} R_3-N-HCl$$

$$R_3-N-HCl + NaOH \xrightarrow{\text{Regeneration}} R_3-N + NaCl + H_2O$$

In general, the acid-sorbing resins may be classified as high molecular weight polyamines or polyimines. Thus, the original Adams and Holmes material was a polymer of *m*-phenylenediamine. Cation exchange materials include synthetic resins, such as sulfonated phenol–formaldehyde or polystyrene types, and sulfonated coal. Some manufacturers have a variety of sub-types which are considered superior for particular applications.

(20) B. A. Adams and E. L. Holmes, *J. Soc. Chem. Ind.*, **54**, 1T (1935).

2. Factors Governing Application of Ion Exchange

The application of ion exchangers to dextrose process liquors involved considerable experimental work because of a number of factors which do not enter into their application to water purification. The accumulation of fats and proteins on the resin surfaces must be guarded against by proper clarification of the liquors to be treated. Such accumulation may result from precipitation as the neutralization progresses, and may soon destroy the effective acid-removing capacity of the anion exchange resin. This difficulty can effectively be eliminated by prior precipitation of the refinery residue from the acid liquor by bentonite, a colloidal clay of opposite electrical charge to the colloids,[21] followed by filtration.

Despite proper clarification of the liquor to be treated, the anion exchange resin may show progressively decreased capacity on continued use. One factor contributing to such capacity loss was found to be caused by adsorption on the resin surface of trace metal constituents, specifically iron and copper. Such adsorption of metallic ions causes a loss of capacity of the resin. This problem was solved, to a large extent, when it was found possible to remove iron and copper by passing the acid liquor through a cation exchange bed placed ahead of the anion exchange bed. This was an unexpected use of cation exchange material, since it is generally employed in the treatment of neutral or alkaline solutions. The pre-treatment under acidic conditions, though effective in keeping the iron and copper figures at harmless levels, introduces another difficulty when cation exchange materials of comparatively low capacity are used. The acid liquor passing through the resin near the end of the cycle acts as regenerant, thereby increasing the ash content of the liquor. This phenomenon of "ash-throwing" does not occur if cation exchange resin of high capacity, now available, is used, or if no attempt is made to utilize the normally full capacity of the cation exchange resin. A further capacity lowering effect was observed from the continued action of dextrose on the polyamine resin. This was thought to be due to the aldehydic character of dextrose and the ability of aldehydes to react with amines, especially under neutral or slightly basic conditions, to give irreversible products. This difficulty is largely eliminated by use of resins of other chemical constitution, for example, those which contain both imine and tertiary amino groups. These groups do not react with dextrose.

Finally, the effect of temperature on the life of anion exchange resins must be considered. Many of these materials are sensitive to high temperatures, and degradation of the resin may result if process tem-

(21) M. S. Badollet and H. S. Paine, *Ind. Eng. Chem.*, **19**, 1245 (1927).

peratures are not held to relatively low levels. Some of the newer anion exchange resins, however, possess a greater degree of thermostability and may be used at higher, though not at excessively high, temperatures.

In general, the foregoing factors do not apply to cation exchangers, particularly of the polystyrene or sulfonated coal types, since these are remarkably stable materials. Even so, their life is prolonged by limiting operating temperatures to moderate levels and by thorough preclarification of the liquors passed through them.

3. *Methods of Application*

In early attempts to apply ion exchange refining to the process for the production of dextrose the single pass system of liquor treatment was used. This simply involved passing the liquor through two single cells—one containing cation exchanger, and the other containing anion exchanger. A multiple pass system, introduced several years ago, is now more commonly employed. Such a system consists of a series of four sets of cation-anion exchange cells, each set of which consists of one cation exchange unit and one anion exchange unit. The liquor flow is countercurrent so that the untreated liquor first passes through the oldest unit and progresses in series to the least exhausted unit. One set is thus out for regeneration at all times. The multiple pass system makes possible the utilization of the full capacity of each unit. Another advantage is that more nearly complete refining is obtained, since the freshly regenerated cells are used almost completely for refining and later in a cycle for demineralization.

4. *Advantages*

The application of ion exchange refining to the dextrose manufacturing process affords a number of advantages, some of which are briefly discussed below.[22]

(1) The refining effect of ion exchange treatment is of sufficient magnitude to permit the elimination of a substantial portion of activated carbon, or bone char requirement. At least a portion of the cost of operation of the ion exchange unit is thus paid for by the savings in requirement of color adsorbent.

(2) Sugar quality is enhanced by removal of colored impurities and by the extent and type of refining which is obtained. It is believed that this type of refining is the most valuable not only because of its extent but also because it is applied early in the process. This reduction of the impurity load near the forepart of the process avoids potential trouble

---

(22) S. M. Cantor, U. S. Pat. 2,328,191 (Aug. 31, 1943); 2,389,119 (Nov. 20, 1945).

from continued decomposition of impurities, as well as the catalytic effect of those impurities initially present, on the production of more decomposition products.

In addition to removing colored impurities and other colloids, ion exchange resins effectively remove the uncolored precursor of these colored substances. It has long been recognized that a small amount of 5-hydroxymethylfurfural (HMF), formed from dextrose by the action of acid catalysts, decomposes to form organic acids as well as pigments which polymerize to form highly colored bodies.[23] Normal bone char

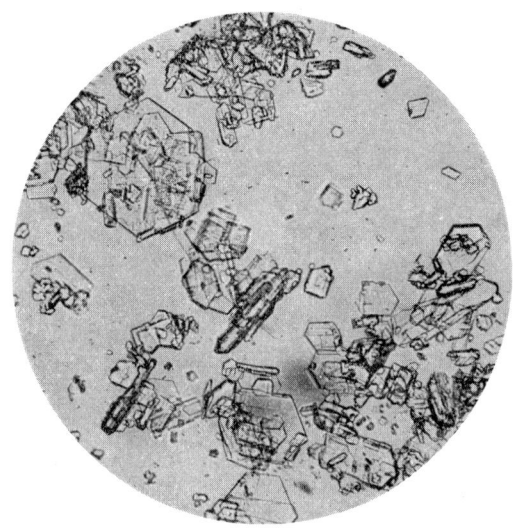

Fig. 7.—Alpha dextrose monohydrate from ion exchange refined starch conversion liquor. Magnification, 50.

or carbon treatment removes the colored bodies but only a small portion of the colorless HMF. This material continues to react slowly, and it is this slow reaction which produces browning in sugar under storage conditions. Ion exchangers remove a larger portion of HMF.

(3) Crystallization yield is increased by removal of colloidal inhibitors and also by removal of ash, which, when present, causes an increase in dextrose solubility. Growth of more uniform crystals, which permits better separation from the mother liquor, takes place. Figure 7 shows the type of crystals usually obtained from liquors refined by ion exchange treatment.

(4) One of the principal effects of ash constituents in liquors undergoing conversion is their tendency to catalyze the rate and extent of the

(23) B. Singh, G. R. Dean, and S. M. Cantor, *J. Am. Chem. Soc.*, **70**, 517 (1948).

reversion reaction. The elimination of salts from such liquors by ion exchange refining, therefore, minimizes this effect during subsequent reconversion. Thus, no limiting number is placed on the reconversion cycles that can be employed in the process, since there is no appreciable buildup of ash.

5. *Description of Operation*

Application of ion exchange refining to the process for the manufacture of dextrose may be understood from the following description of such a process. Triple-washed starch is diluted with ion-free water to the desired concentration and is acidified with a definite quantity of mineral acid such as hydrochloric or sulfuric. It is desirable that the acidified starch slurry be held for at least thirty minutes in order to permit an effective equilibrium acidity to be reached. The starch slurry may contain a quantity of soft water salts which consume acid, and since this consumption is variable, the acidity is checked and adjusted to the desired level following the holding period. The conversion is then carried out at elevated temperature and under pressure for a period sufficient to yield maximum dextrose. The conversion may be carried out batchwise in an autoclave, or continuously.

The fats and proteins are coagulated by treatment of the hydrolyzate with a small quantity of bentonite. The coagulated residue is filtered from the hydrolyzate with the addition, if desired, of a small quantity of filtercel to improve the filtration rate. The hydrolyzate must be properly clarified in order to reduce to a minimum the potential contamination of the ion exchange resin with extraneous material.

The clarified hydrolyzate is now refined by ion exchange treatment. For most efficient use of the exchange resins and for satisfactory refining a triple-pass system is utilized. This means a requirement of four cells containing cation exchange material and four cells containing anion exchange resin. Three sets are operated in series so that the liquor passes through alternate cation and anion exchange beds, while one set is always out for regeneration. The size of ion exchange beds to be chosen for a particular installation depends upon the capacity of the resin, the size of the plant, and the operating cycle desired.

The degree of exhaustion of a cation exchange bed can be determined by testing the effluent for acidity or ash content. The acidity is easily and quickly determined by titration. In the case of an anion exchange bed the point of exhaustion is considered as having been reached when the pH value of the effluent drops to a predetermined level such as 4.0. When these tests show that the first anion exchange bed in the series has been exhausted, it is removed from the operating cycle along with the companion cation exchange bed. It is desirable that the cation exchange bed have a somewhat greater capacity than the corresponding anion exchange cell in order that adequate de-ashing be accomplished at all times. The second unit in series is then moved into first position, the third into second position, and a freshly regenerated pair of cells is placed on stream in the third position. In this way countercurrent operation is maintained so that the most nearly exhausted units receive the raw liquors.

When a unit is removed from service it is "sweetened-off," that is, the remaining sugar liquor is recovered by rinsing the resin with water, and such "sweetwater" is returned to the stream ahead of the ion exchange station. Here again the countercurrent principle may be applied by returning the most concentrated portion to the stream and saving the more dilute portion for the initial part of the succeeding sweetening-off operation. When all sugar substance has been rinsed from the bed, the resin is further washed with water passing upflow through the bed. This operation removes foreign material deposited on the surface of the resin and also reclassifies the bed so that any air or gas pockets are eliminated. The beds are then regenerated, the cation exchange resin with hydrochloric or sulfuric acid and the anion exchange resin with a solution of base such as sodium or ammonium hydroxide or sodium carbonate. The excess regenerant is rinsed from the resin beds, preferably with ion-free water. The beds are then ready for the next cycle of operation.

The effluent from the ion exchange station is adjusted, if necessary, to the desired pH level (4.5 to 4.8) before it is further decolorized with activated carbon. This step is usually carried out in two stages and the adsorbent is utilized countercurrently. Thus, the virgin carbon is applied to the liquor concentrated to about 50 to 55% solids and this once used adsorbent is re-used on the more dilute effluent from the ion exchange station. The liquor must be filtered thoroughly to remove all traces of carbon prior to concentration, *in vacuo*, to about 74% solids.

Crystallization is carried out in motion over a period of several days, during which the temperature is gradually reduced. Because ion exchange refining removes a greater proportion of impurities than is possible by conventional methods, the crystallization rate may be accelerated and the yield increased. Furthermore, it is possible to obtain a substantial second crop of crystals from the mother liquor without reconversion, thereby affording an over-all reduction of processing required per unit weight of dextrose hydrate produced.

THE METHYL ETHERS OF D-GLUCOSE

By E. J. Bourne and Stanley Peat

The University, The University College of
Birmingham, England North Wales, Bangor,
Caernarvonshire, Wales

Contents

I. Introduction	145
II. Monomethyl Ethers	148
1. 2-Methyl-D-Glucose	148
2. 3-Methyl-D-Glucose	151
3. 4-Methyl-D-Glucose	154
4. 5-Methyl-D-Glucose	157
5. 6-Methyl-D-Glucose	158
III. Dimethyl Ethers	160
1. 2,3-Dimethyl-D-Glucose	160
2. 2,4-Dimethyl-D-Glucose	162
3. 2,5-Dimethyl-D-Glucose	164
4. 2,6-Dimethyl-D-Glucose	165
5. 3,4-Dimethyl-D-Glucose	167
6. 3,5-Dimethyl-D-Glucose	168
7. 3,6-Dimethyl-D-Glucose	168
8. 4,6-Dimethyl-D-Glucose	170
9. 5,6-Dimethyl-D-Glucose	171
IV. Trimethyl Ethers	172
1. 2,3,4-Trimethyl-D-Glucose	172
2. 2,3,5-Trimethyl-D-Glucose	174
3. 2,3,6-Trimethyl-D-Glucose	176
4. 2,4,6-Trimethyl-D-Glucose	179
5. 2,5,6-Trimethyl-D-Glucose	182
6. 3,4,6-Trimethyl-D-Glucose	182
7. 3,5,6-Trimethyl-D-Glucose	184
V. Tetramethyl Ethers	186
1. 2,3,4,6-Tetramethyl-D-Glucose	186
2. 2,3,5,6-Tetramethyl-D-Glucose	189

I. Introduction

In this review, methylation will connote the replacement of the primary or secondary alcohol groups of a sugar by methoxyl groups. The term will not be applied to the formation of methyl glycosides inasmuch as the reducing group of a cyclic sugar, although alcoholic in form, is in fact a cyclic semi-acetal group. The derived methyl glycoside does not behave as an ether and the original sugar may be regenerated from it by hydrolysis with aqueous acid.

The true methyl ethers of the monosaccharides have been of the utmost service in two respects. In the determination of constitution, the stability of the ether link under drastic conditions of temperature, acidity, alkalinity, oxidation and reduction has provided an ideal means of "labeling" free hydroxyl groups, and the ease with which the methyl ethers of the sugars may be purified, by high vacuum distillation or by crystallization, gives them great importance as reference compounds and as characterizing derivatives. On the other hand, the stability of the methyl ethers militates against their use as intermediates in the chemical synthesis of disaccharides and higher saccharides. For synthetic purposes recourse is had rather to the sugar esters and other derivatives which are hydrolyzable either by acid or alkali.

Constitutional studies have broadly followed the pattern of methylation of the unsubstituted alcoholic groups in a carbohydrate, hydrolytic or oxidative disruption of the molecule and identification of the resulting methylated fragments. In this way methyl ethers have been invaluable in the determination of the ring structures of the monosaccharides and in the elucidation of the constitutions of the more complex saccharides.

In this article, the authors have endeavored to summarize the methods of synthesis and the proofs of constitution of all the known methyl ethers of D-glucopyranose and D-glucofuranose. Acyclic glucose ethers are not considered in this review. Later articles will deal with monosaccharides other than glucose. It has not been possible to discuss in full all the reactions involved, but to offset this disadvantage the bibliography has been made as complete as possible and tables have been compiled of the physical properties of the methyl-D-glucoses and of their more important derivatives.

The most generally applicable method for the methylation of carbohydrates is that of Haworth, in which the methylating agents are dimethyl sulfate and concentrated sodium hydroxide solution. This method possesses the advantages that the solvent is water and that, under appropriate conditions, it is suitable for the direct methylation of reducing sugars. The earlier, well-known method of Purdie, which employs methyl iodide and silver oxide as reagents, cannot be used in an aqueous medium and requires that the reducing group of a sugar be first protected from oxidation by, for example, the formation of the methyl glycoside. It is the present custom only to use the Purdie reagents for the completion of methylation or for the methylation of sugars already partly substituted by other groups. Other methods of more recent date, which have advantages in certain circumstances, include treatment of the carbohydrate in liquid ammonia with sodium and then with methyl iodide (Muskat, K. Freudenberg) and the use of thallous hydroxide and dimethyl sulfate (Menzies, Hirst). Diazomethane cannot be used for the methylation of the alcoholic hydroxyl groups of a carbohydrate.

In choosing the most appropriate method for the methylation of a particular compound account has to be taken of several factors, namely, the presence of groups in the molecule which are susceptible to attack by one or more of the methylating reagents, the solubility of the compound in the different methylation media, the ease of isolation of the product, the relative efficiency of the reactions, and the relative costs of the various processes.

The two commonest groups liable to be modified by methylation are the reducing group and any ester groups which may be present in the molecule. Reducing sugars are disrupted under alkaline conditions, oxidized by silver oxide and converted to aldehyde–ammonias and 1-amino derivatives by liquid ammonia. Hence none of the above methods can be applied, without modification, to the methylation of a reducing sugar. It is possible in the case of the Haworth method to overcome this difficulty by employing in the initial stage an excess of dimethyl sulfate, which is acidic, until the sensitive reducing group has been methylated (*i.e.*, until the mixture fails to reduce Fehling's solution). An alternative procedure is to convert the sugar to the methyl glycoside by treatment with hydrogen chloride in dry methyl alcohol and after etherification by one of the above methods, to regenerate the reducing group by hydrolysis with aqueous mineral acid.

For the methylation of the free hydroxyl groups of a partially esterified carbohydrate, *e.g.*, of a partially acetylated compound, only the Purdie method is available. Although liquid ammonia does not react with esters at $-70°C.$, there is a danger of saponification if the temperature of the reaction mixture is allowed to rise before all the ammonia has been removed. The aqueous alkali employed in the methods of Haworth or Menzies will, of course, remove ester groups by hydrolysis.

For maximum efficiency, methylation should be a single-phase reaction and so those methods employing aqueous alkali are most suitable for water-soluble compounds and the Purdie method for non-polar compounds. A carbohydrate which is initially soluble in water (*e.g.*, starch, dextran) may be less soluble when partially methylated and for this reason an inert solvent (*e.g.*, acetone, dioxane, carbon tetrachloride) is often introduced during the later stages of a methylation with dimethyl sulfate. It is interesting to note that methyl α-D-glucopyranoside forms a trithallium derivative which is insoluble in water and consequently the introduction of more than three methoxyl groups by the Menzies method is extremely difficult. Compounds which are insoluble in both water and organic solvents (*e.g.*, cellulose) are best methylated by subjecting their acetates to simultaneous de-acetylation and methylation, using dimethyl sulfate and sodium hydroxide, in the presence of a solvent for the acetate. The liquid ammonia method seems to be less dependent on the solubility or insolubility of the carbohydrate; it is, for example, by far the best method of preparing trimethyl starch, although starch is insoluble in liquid ammonia.

As far as concerns the ease of isolation of the methylated product, the Purdie method possesses advantages, because the excess silver oxide and the silver iodide produced are easily removed by filtration and the excess methyl iodide by distillation. It is sometimes necessary to use a long purification procedure in order to free a methylated product obtained by one of the other methods from all traces of inorganic impurities (*e.g.*, sodium sulfate, sodium methyl sulfate, thallous sulfate or sodium iodide).

The relative costs of the various methods of methylation are not important in small-scale work, but must be taken into consideration when large quantities of reagents are employed. On this account the Purdie method has serious limitations, for neither methyl iodide nor silver oxide is a cheap commodity and, furthermore, six or more treatments may be necessary, each requiring a four-fold excess of silver oxide and a six-fold excess of methyl iodide.

For the preparation of partially methylated sugars direct methylation is not as a rule a suitable process, for the mixture of isomers thus produced can be fractionated only with difficulty. Instead, it is customary to introduce protecting substituents prior to methylation and to remove them afterwards. As will be seen later, this is the method used in the synthesis of mono-, di-, and tri-methylglucoses.

II. Monomethyl Ethers

1. *2-Methyl-D-Glucose*

Methods of Formation.—A monomethyl hexose, prepared by Pictet and Castan[1] from the so-called "α-glucosan" by the action of sodium methoxide, was claimed to be 2-methyl-D-glucose for the reason that it gave no phenylosazone under ordinary conditions. A similar claim was made as to the constitution of the non-crystalline methyl hexose prepared from methyl anhydro-α-D-glucoside mono-oleate by Irvine and Gilchrist,[2] but it is doubtful whether pure 2-methyl-D-glucose was isolated in either case because scission of anhydro-rings frequently leads to optical inver-

(1) A. Pictet and P. Castan, *Helv. Chim. Acta*, **3**, 645 (1920).
(2) J. C. Irvine and Helen S. Gilchrist, *J. Chem. Soc.*, **125**, 1 (1924).

sion. The low value for $[\alpha]_D$ recorded by Irvine and Gilchrist for the monomethyl hexose supports this view.

In 1928 and 1929, three methods, all involving the use of 3,4,6-triacetyl-D-glucose or its derivatives, were reported for the synthesis of 2-methyl-D-glucose. Glucosides of this sugar were produced by methylation and hydrolysis of ethyl 3,4,6-triacetyl-β-D-glucoside (from 1,2-anhydro-3,4,6-triacetyl-D-glucose),[3] 3,4,6-triacetyl-α-D-glucosyl chloride[4] and 3,4,6-triacetyl-D-glucose,[5] respectively. In each of these cases acid hydrolysis of the glucoside yielded a sirup, which formed a crystalline phenylhydrazone (m. p. 175–178°) but not a phenylosazone under the usual conditions. A warning that the location of an ether group introduced by methylation of a sugar acetate cannot necessarily be predicted from a knowledge of the original positions of the ester groupings was given in 1931 by Haworth, Hirst and Teece,[6] who obtained 2-methyl-β-D-glucose in crystalline form (m. p. 157–159°, identified by means of its phenylhydrazone) from methyl 2,3,4-triacetyl-α-D-glucoside, by treatment with methyl iodide and silver oxide and subsequent hydrolysis. During the methylation process the acetyl residue originally present on C2 had migrated to C6. The same crystalline 2-methyl-β-D-glucose was obtained by Oldham and Rutherford[7] from 3,5,6-tribenzoyl-D-glucofuranose via the sirupy α- and β-mixture of methyl 2-methyl-D-glucofuranoside.

Mercaptalation has been favored by several groups of workers as an initial step in the formation of 2-methyl-D-glucose, although Pacsu,[8] who first used this method, wrongly believed the crystalline product (m. p. 156–157°) to be the 4-methyl ether. The similarity between this compound and that of Brigl and Schinle,[9] who methylated 3,4,5,6-tetrabenzoyl-D-glucose ethyl mercaptal and then removed the benzoyl and mercaptal groups with sodium methoxide and mercuric chloride respectively, was noticed by Schinle,[10] who conclusively demonstrated that the two methyl glucoses were identical. Meanwhile Levene, Meyer and Raymond[11,12] had reached a similar conclusion, with which Pacsu[13]

(3) W. J. Hickinbottom, *J. Chem. Soc.*, 3140 (1928).
(4) T. Lieser, *Ann.*, **470**, 104 (1929).
(5) P. Brigl and R. Schinle, *Ber.*, **62B**, 1716 (1929).
(6) W. N. Haworth, E. L. Hirst and Ethel G. Teece, *J. Chem. Soc.*, 2858 (1931).
(7) J. W. H. Oldham and Jean K. Rutherford, *J. Am. Chem. Soc.*, **54**, 1086 (1932).
(8) E. Pacsu, *Ber.*, **58B**, 1455 (1925).
(9) P. Brigl and R. Schinle, *Ber.*, **63B**, 2884 (1930).
(10) R. Schinle, *Ber.*, **64B**, 2361 (1931).
(11) P. A. Levene, G. M. Meyer and A. L. Raymond, *Science*, **73**, 291 (1931).
(12) P. A. Levene, G. M. Meyer and A. L. Raymond, *J. Biol. Chem.*, **91**, 497 (1931).
(13) E. Pacsu, *Ber.*, **65B**, 51 (1932).

TABLE I

The Physical Properties of 2-Methyl-D-glucose and Some of Its Derivatives

Compound	Melting point, °C.	$[\alpha]_D$	Rotation solvent	References
2-Methyl-β-D-glucose	157–159	$+12.0 \to +66.0$	H_2O	6, 7, 9, 10
phenylhydrazone	175–178	$-12.3, -13.3$	C_5H_5N	3, 4, 5, 7
p-toluidide	150–151	—	—	14
1,3,4,6-tetrabenzoate	169–170	-6.2	$CHCl_3$	15
2-Methyl-D-gluconic acid	—	—	—	16
3,4,5,6-diisopropylidene-				
methyl ester	44	$+41$	H_2O	16
amide	139	$+39$	H_2O	16
γ-lactone	liquid	$+45 \to +37$	H_2O	16
Methyl 2-methyl-α-D-				
glucopyranoside	147–148	$+155$	H_2O	6
3,4,6-triacetate	120	$+145$	$CHCl_3$	6
Methyl 2-methyl-β-D-				
glucopyranoside	97–98	$-37.5, -41.9$	H_2O, CH_3OH	15
3,4,6-triacetate	74–75	$+5.9$	C_2H_5OH	5
3,4,6-tribenzoate	119–120	-40.4	$CHCl_3$	15
3,4,6-tritosylate	168–169	$+13.8$	$CHCl_3$	7
4,6-ethylidene-	122–123	-66.0	$CHCl_3$	17
4,6-benzylidene-	170–171	-69.2	$CHCl_3$	7, 18
Methyl 2-methyl-α,β-D-				
glucofuranoside	liquid	—	—	7, 19

himself in 1932 agreed. The true course of the synthesis effected by Pacsu[8] was as follows: D-glucose dibenzyl mercaptal → 5,6-isopropylidene-D-glucose dibenzyl mercaptal → 2-methyl-D-glucose dibenzyl mercaptal → 2-methyl-D-glucose.

Proof of Constitution.—Very little reliable evidence regarding the position of the ether group in 2-methyl-D-glucose is available from the methods of synthesis outlined above. In some cases[9,10,13] the structures

(14) Eleanor Mitts and R. M. Hixon, *J. Am. Chem. Soc.*, **66**, 483 (1944).
(15) J. W. H. Oldham, *J. Am. Chem. Soc.*, **56**, 1360 (1934).
(16) W. N. Haworth, E. L. Hirst and K. A. Chamberlain, *J. Chem. Soc.*, 795 (1937).
(17) D. J. Bell and R. L. M. Synge, *J. Chem. Soc.*, 833 (1938).
(18) J. W. H. Oldham and Mary A. Oldham, *J. Am. Chem. Soc.*, **61**, 1112 (1939).
(19) D. J. Bell, *J. Chem. Soc.*, 186 (1936).

of intermediate compounds isolated during the synthesis were in doubt until the structure of the final product was ascertained, while in others the labile character of an acetyl group precludes the use of a knowledge of the initial location of the ester groups as evidence of the position of the methyl group in the product.[3-6] The synthesis effected by Oldham and Rutherford[7] is more conclusive from a structural viewpoint than the others, since these workers protected the 3, 5 and 6 positions of the well-characterized 1,2-isopropylidene-D-glucofuranose by forming the comparatively stable tribenzoate, removed the isopropylidene residue by hydrolysis, methylated positions 1 and 2 and then debenzoylated the product by Zemplén's catalytic method. The sirupy methyl glucoside thus synthesized gave, on hydrolysis with aqueous acid, crystalline 2-methyl-β-D-glucose.

The methyl group was originally assigned to C2 because this partially methylated sugar readily formed a phenylhydrazone but not a phenylosazone.[3,4] More drastic conditions were necessary before the second phenylhydrazine residue could be introduced and in this process the methyl group was lost, the product being D-glucose phenylosazone.[5,6] Supplementary evidence was obtained by Levene, Meyer and Raymond[12] from a comparison of the relative rates of oxidation of this methylglucose and other 2-substituted glucose derivatives. These authors also showed that a 6-methyl ether was not present, inasmuch as oxidation gave rise to a monomethyl-saccharic acid, and that the 4 and 5 positions were not methylated, since changes in the optical rotation of a solution of the monomethyl-D-glucose in acidic methyl alcohol indicated the formation of pyranoside as well as furanoside. Since the physical properties of the monomethyl-D-glucose were quite distinct from those of 3-methyl-D-glucose, which was already known, Levene, Meyer and Raymond concluded that the allocation of the methyl group to C2 was fully justified.

2. *3-Methyl-D-Glucose*

Methods of Formation.—The introduction of a methyl group at position 3 in glucose is greatly facilitated by the orientation of the hydroxyl groups in this sugar, which precludes the formation of cyclic acetal derivatives involving C3. Such cyclic structures, especially the isopropylidene derivatives, have frequently been employed for the protection of the other hydroxyl groups during the synthesis of 3-methyl-D-glucose. The methylation of 1,2:5,6-diisopropylidene-D-glucofuranose provides the easiest route for the synthesis of 3-methyl-D-glucose. Whereas the methylating agents initially employed by Irvine and

Scott[20] and by others[21,22,23] were methyl iodide and silver oxide, Schmidt and Simon[24] preferred dimethyl sulfate in the presence of an aqueous solution of potassium hydroxide. On several occasions 1,2:5,6-diisopropylidene-D-glucofuranose has been treated with sodium or potassium in an anhydrous solvent, such as ether, benzene or liquid ammonia, giving the corresponding 3-sodio (or potassio) derivative, the metal being subsequently replaced by a methyl group by treatment with methyl iodide or dimethyl sulfate.[22,23,25–28]

In 1934 Vargha[29] methylated the so-called "1,2-isopropylidene-3-acetyl-6-trityl-D-glucofuranose" and obtained the 3-methyl derivative instead of the expected derivative of 5-methyl-D-glucose. The 3-methyl derivative was converted into 3-methyl-D-glucose by hydrolysis, first with alkali and then with acetic acid. Vargha pointed out that the observed acyl migration from C3 to C5 may have occurred during the formation of the so-called "1,2-isopropylidene-3-acetyl-6-trityl-D-glucofuranose" from the known 1,2-isopropylidene-3-acetyl-D-glucofuranose, or during the methylation process.

The protection of hydroxyl groups by the formation of cyclic acetals in the synthesis of 3-methyl-D-glucose has been effected with other reagents than acetone. For example, the 4,6-benzylidene and 4,6-ethylidene acetals of methyl D-glucopyranoside have each been employed for this purpose. Dewar and Fort[30] converted methyl 4,6-ethylidene-β-D-glucopyranoside 3-nitrate into the corresponding 2-benzoate, from which the nitrate grouping was eliminated by treatment with a mixture of iron and zinc powders in glacial acetic acid. Methylation of the product gave methyl 2-benzoyl-3-methyl-4,6-ethylidene-β-D-glucopyranoside. In a somewhat similar fashion, Bolliger and Prins[31] prepared 3-methyl-D-glucose via methyl 2-tosyl-4,6-benzylidene-α-D-glucopyranoside, which had been obtained from methyl 4,6-benzylidene-α-D-glucopyranoside by preferential tosylation.

(20) J. C. Irvine and J. P. Scott, *J. Chem. Soc.*, **103**, 564 (1913).
(21) J. C. Irvine and T. P. Hogg, *J. Chem. Soc.*, **105**, 1386 (1914).
(22) K. Freudenberg and R. M. Hixon, *Ber.*, **56B**, 2119 (1923).
(23) C. G. Anderson, W. Charlton and W. N. Haworth, *J. Chem. Soc.*, 1329 (1929).
(24) O. T. Schmidt and A. Simon, *J. prakt. Chem.*, **152**, 190 (1939).
(25) D. J. Loder and W. L. Lewis, *J. Am. Chem. Soc.*, **54**, 1040 (1932).
(26) I. E. Muskat, *J. Am. Chem. Soc.*, **56**, 693 (1934).
(27) I. E. Muskat, *J. Am. Chem. Soc.*, **56**, 2449 (1934).
(28) R. L. Sundberg, C. M. McCloskey, D. E. Rees and G. H. Coleman, *J. Am. Chem. Soc.*, **67**, 1080 (1945).
(29) L. v. Vargha, *Ber.*, **67B**, 1223 (1934).
(30) J. Dewar and G. Fort, *J. Chem. Soc.*, 496 (1944).
(31) H. R. Bolliger and D. A. Prins, *Helv. Chim. Acta*, **28**, 465 (1945).

Another method of synthesis of derivatives of 3-methyl-D-glucose lies in the scission of the anhydro-ring of 2,3-anhydro-D-allose derivatives with sodium methoxide, when the entry of the methoxyl group is accompanied by a Walden inversion on C3.[32]

The usual method for the crystallization of 3-methyl-D-glucose is that reported by Irvine and Scott,[20] who obtained the α-form as short rectangular plates from methyl alcohol solution and delicate prismatic needles of the β-form when acetone was added to the mother liquor. The β-form is unstable and changes slowly into the α-form when stored.[21]

Proof of Constitution.—When Irvine and Scott[20] first obtained a methyl-D-glucose by methylation and hydrolysis of diisopropylidene-D-glucose, it was clear that the methyl group in this compound was not located on C2 because of the ready formation of a phenylosazone without loss of the ether group. This belief has since been confirmed by the observation of Schmidt and Simon[24] that oxidation of the calcium salt of the methylgluconic acid, by treatment with hydrogen peroxide in the presence of ferric acetate, gave a monomethyl-D-arabinose. In order to ascertain the constitution of this particular methyl-D-glucose, it was necessary therefore to decide whether the methyl group was present on C3, C4, C5 or C6 and, since diisopropylidene-D-glucose was for two decades the only known source, the problem was inevitably linked with the determination of the positions of the acetone residues in this parent compound. Irvine and his school[20,21] at first believed that the methoxy group was in position 6 because the monomethyl glucosazone was identical with that obtained from the so-called "6-methyl-D-fructose," but it was noted that the structure assigned to the latter compound was itself open to doubt. By 1923 Karrer[33] and Freudenberg,[34,35] who respectively examined the action of potassium permanganate on and the formation of pyrazole derivatives from diisopropylidene-D-glucose, had reached the conclusion that the isopropylidene groups were attached to the 1,2 and 5,6 positions, so that the methyl glucose was known to be the 4- or 3-derivative, depending on whether the sugar ring in diisopropylidene-D-glucose was of the propylene oxide type, a structure favored by Irvine, or of the butylene oxide (furanose) type, as suggested by Freudenberg. Levene and Meyer[36] were able to decide between these two possibilities by oxidizing the methyl-D-glucose to the corresponding methyl-D-glucosaccharolactone (which incidentally confirmed that C6 did not carry

(32) S. Peat and L. F. Wiggins, *J. Chem. Soc.*, 1810 (1938).
(33) P. Karrer and O. Hurwitz, *Helv. Chim. Acta*, **4**, 728 (1921).
(34) K. Freudenberg and F. Brauns, *Ber.*, **55B**, 3233 (1922).
(35) K. Freudenberg and A. Doser, *Ber.*, **56B**, 1243 (1923).
(36) P. A. Levene and G. M. Meyer, *J. Biol. Chem.*, **60**, 173 (1924).

the methyl group), reduction of which gave 3-methyl-D-glucuronic acid and not 3-methyl-L-guluronic acid, as would have been the case had the original material been 4-methyl-D-glucose. In addition, the methylglucose was converted, by ascent of the series via the cyanohydrin, into a monomethyl-D-heptonolactone, which was dextrorotatory and which was therefore, according to Hudson's lactone rule, the 4-methyl-δ-heptonolactone and not the 5-methyl-γ-heptonolactone. The starting material must therefore have been 3-methyl-D-glucose. The formation from the methyl-D-glucose of both methyl furanosides and methyl pyranosides provides confirmation that C4 is not methylated.[37] The demonstration by Anderson, Charlton and Haworth[23] that monoisopropylidene-D-glucose was converted by methylation and hydrolysis into 3,5,6-trimethyl-D-glucofuranose, which in turn gave rise to 2,3,5,6,-tetramethyl-D-glucofuranose, provided conclusive proof that the ring system of the acetone derivatives of D-glucose was furanose (butylene oxide type) and that the monomethyl hexose derived therefrom was 3-methyl-D-glucose. (The accompanying Table II is on page 155.)

3. 4-Methyl-D-Glucose

Methods of Formation.—The introduction of a methyl grouping in position 4 of glucose has been accomplished along two distinct routes, involving (a) the 2,3:5,6-diisopropylidene derivative of the open chain D-glucose dibenzyl mercaptal and (b) the 2,3,6-triesters of methyl β-D-glucopyranoside. To Schinle[43] belongs the credit for the observation that the sirup produced from D-glucose dibenzyl mercaptal and acetone, in the presence of anhydrous copper sulfate, is a mixture of the 2,3:5,6-diisopropylidene and 5,6-monoisopropylidene derivatives. The sirup gives rise, by methylation and acid hydrolysis, to a mixture of 4-methyl-D-glucose dibenzyl mercaptal (from the diisopropylidene compound) and 2-methyl-D-glucose dibenzyl mercaptal (from the monoisopropylidene compound), from which each of the components can be isolated in crystalline form. An aqueous suspension of mercuric chloride liberates the corresponding methylglucoses from the methylated mercaptals. The 4-methyl-D-glucose thus isolated by Schinle did not crystallize, but it gave a phenylosazone (m. p. 159°), which was identical with the so-called 4,5,6-trimethyl-D-glucose phenylosazone of Pacsu.[8]

(37) P. A. Levene and R. T. Dillon, *J. Biol. Chem.*, **92**, 769 (1931).
(38) W. J. Heddle and E. G. V. Percival, *J. Chem. Soc.*, 249 (1939).
(39) K. Freudenberg, W. Dürr and H. v. Hochstetter, *Ber.*, **61B**, 1735 (1928).
(40) B. Helferich and O. Lang, *J. prakt. Chem.*, **132**, 321 (1932).
(41) K. Freudenberg and E. Plankenhorn, *Ber.*, **73B**, 621 (1940).
(42) J. K. Mertzweiller, D. M. Carney and F. F. Farley, *J. Am. Chem. Soc.*, **65**, 2367 (1943).
(43) R. Schinle, *Ber.*, **65B**, 315 (1932).

Table II
The Physical Properties of 3-Methyl-D-glucose and Some of Its Derivatives

Compound	Melting point, °C.	[α]D	Rotation solvent	References
3-Methyl-D-glucose				
α-form	161, 168	+104.5 → +55.5	H₂O	21, 23, 25, 29
β-form	133.5–135	+31.9 → +55.1	H₂O	20, 21
anilide	154–155	−108.5	CH₃OH	21
phenylosazone	178, 185	−105.5 → −41.6	C₂H₅OH	23, 29, 31, 38
1,2:5,6-diisopropylidene-	liquid	ca. −34	C₂H₅OH	23, 24, 26, 27
1,2-isopropylidene-5,6-dibenzoate	81–82	—	—	39
β-tetraacetate	95–96	−5.2	CHCl₃	40
β-tetrabenzoate	198–199	+3.6	CHCl₃	15
1,2,4,6-tetraazobenzoate	220–224	+163	CHCl₃	41, 42
3-Methyl-D-gluconic acid	liquid	−9.4	H₂O	24
sodium salt	—	+15.6	H₂O	24
phenylhydrazide	140–141	−1.7	H₂O	24
Methyl 3-methyl-α-D-glucopyranoside	—	—	—	—
4,6-benzylidene-	150–151	+119.5	C₂H₂Cl₄	31
2-tosyl-4,6-benzylidene-	156–157	+56.3	CHCl₃	31
Methyl 3-methyl-β-D-glucopyranoside	liquid	−26	H₂O	15, 30, 40
4,6-benzylidene-	166	—	—	32
2-acetyl-4,6-ethylidene-	101–103	−41.2	CHCl₃	30
2-tosyl-4,6-ethylidene-	123–124	−22.8	CHCl₃	30
2,4,6-tribenzoate	125–126	+14.7	CHCl₃	15
2,4,6-triacetate	90	−34.8	CHCl₃	40

The isolation of methyl 4-methyl-β-D-glucopyranoside proceeds quite smoothly from methyl β-D-glucopyranoside 2,3,6-triacetate, tribenzoate or trinitrate, by treatment with methyl iodide and silver oxide and, subsequently, with alkali in the case of the acetate[40] and benzoate,[44] or with a reducing agent in the case of the nitrate.[45] The real difficulty in such syntheses lies, of course, in the preparation of the parent triester. Helferich and Lang[40] prepared methyl β-D-glucopyranoside 2,3,6-triacetate from the isomeric 2,3,4-triacetate by treatment with potassium hydroxide in alcohol, a process which involved acyl migration, while Levene and Raymond[44] showed that partial benzoylation of the methyl β-D-gluco-

(44) P. A. Levene and A. L. Raymond, J. Biol. Chem., **97**, 763 (1932).
(45) D. J. Bell and R. L. M. Synge, J. Chem. Soc., 836 (1938).

pyranoside 2,3-dibenzoate of Ohle and Spencker[46] introduced a third benzoate grouping on C6. The methyl β-D-glucopyranoside 2,3,6-trinitrate used by Bell and Synge[45] was synthesized from methyl 4-acetyl-6-trityl-β-D-glucopyranoside 2,3-dinitrate by first replacing the trityl residue with a nitrate group (by the action of fuming nitric acid in chloroform) and then deacetylating the product by the catalytic method of Zemplén.

So far as the authors are aware, 4-methyl-D-glucose has not yet been isolated in crystalline form.

Proof of Constitution.—Schinle[43] suggested that the methylglucose, formed together with 2-methyl-D-glucose from D-glucose dibenzyl

TABLE III

The Physical Properties of 4-Methyl-D-glucose and Some of Its Derivatives

Compound	Melting point, °C.	$[\alpha]_D$	Rotation solvent	References
4-Methyl-D-glucose	liquid	+53, +61 (equil.)	H_2O	8, 49
phenylosazone	158–160	$-32.6 \to -15.5$	C_2H_5OH	43, 49
1,2,3,6-tetraacetate	liquid	—	—	49
4-Methyl-D-glucose dibenzyl				
mercaptal	96–98	−63.9	C_5H_5N	8, 43, 49
2,3,5,6-tetraacetate	69–70	—	—	43
4-Methyl-D-glucono-δ-lactone	liquid	+34 (equil.)	H_2O	49
Methyl 4-methyl-β-D-glucopyranoside	liquid	—	—	40
2,3,6-triacetate	106–108	ca. −34	$CHCl_3$	44, 45, 49
2,3,6-tribenzoate	liquid	ca. +58	—	44
2,3,6-trinitrate	liquid	—	—	45

mercaptal, was 4-methyl-D-glucose on the grounds that its phenylosazone, which retained the methoxyl group, was different from the known 3-, 5- and 6-methyl-D-glucose phenylosazones. An element of doubt arose, however, when it was shown by Levene and Raymond[47] that the so-called 5-methyl-D-glucose of Ohle and Vargha,[48] which had been used as a reference compound by Schinle, was really 6-methyl-D-glucose. The new methylglucose may, therefore, have been either the 4- or the 5-methyl derivative. The dubiety was largely removed by the synthesis of the

(46) H. Ohle and K. Spencker, *Ber.*, **61B**, 2387 (1928).
(47) P. A. Levene and A. L. Raymond, *J. Biol. Chem.*, **97**, 751 (1932).
(48) H. Ohle and L. v. Vargha, *Ber.*, **62B**, 2435 (1929).

same methylglucose from methyl 2,3,6-triacetyl-β-D-glucopyranoside. The presence of the pyranose ring in the latter substance precluded the formation of a 5-methyl ether, unless, as suggested by Munro and Percival,[49] a change in the sugar ring occurred during the methylation process. Such a change was highly unlikely and these authors proved that the monomethylglucose was indeed the 4-methyl derivative by converting it into 2,3,4,6-tetramethyl-D-glucono-δ-lactone (identified as its crystalline phenylhydrazide) and also into crystalline 2,3,4,6-tetramethyl-D-glucopyranose.

4. 5-Methyl-D-Glucose

Methods of Formation.—In the hexose series, the greater stability of the pyranose form (in which C5 is involved in the cyclization) in comparison with the furanose and aldehydo forms presented a special problem in the synthesis of 5-methyl-D-glucose and delayed the isolation of this compound until 1936, by which time all the methyl-D-glucoses that are isomeric with it were already known. Vargha[50] showed that, by a carefully controlled hydrolysis with aqueous acetic acid, the 5,6 acetone residue could be preferentially removed from 1,2:5,6-diisopropylidene-3-tosyl-D-glucofuranose and that partial benzoylation of the product gave the 6-benzoate, which, on treatment with methyl iodide and silver oxide, yielded 1,2-isopropylidene-3-tosyl-5-methyl-6-benzoyl-D-glucofuranose. Saponification and acetylation of this compound gave the first product in the series to be obtained crystalline, namely, 1,2-isopropylidene-3,6-diacetyl-5-methyl-D-glucofuranose, which served as a valuable intermediate for purification. The sirupy 5-methyl-D-glucose, generated by alkaline deacetylation followed by acidic scission of the isopropylidene group, gave a crystalline phenylosazone (m. p. approximately 128°). In the previous year[51] it had been reported that 5-methyl-D-glucose phenylosazone (m. p. 117°) is produced by methylation of D-glucose phenylosazone. The discrepancy between the two values recorded for the melting point is probably explained by Vargha's observation[50] that the melting point is not sharply defined.

Proof of Constitution.—The allocation of the methoxy group to C5 was made by Vargha[50] on the basis of the synthesis and also on the grounds that the phenylosazone was different from the known isomeric methylglucose phenylosazones and that the behavior of the reducing group was unusual. The rapidity with which the methylhexose reacted with Schiff's reagent, permanganate and Fehling's solution, thus simulat-

(49) J. Munro and E. G. V. Percival, *J. Chem. Soc.*, 873 (1935).
(50) L. v. Vargha, *Ber.*, **69B**, 2098 (1936).
(51) Elizabeth E. Percival and E. G. V. Percival, *J. Chem. Soc.*, **1398 (1935)**.

ing a true aldehyde rather than a sugar, together with the absence of mutarotation in aqueous solution, suggested that when there is no possibility of the formation of a pyranose ring in a reducing sugar the aldehydo form is favored rather than the furanose form. Since detosylation is

TABLE IV

The Physical Properties of 5-Methyl-D-glucose and Some of Its Derivatives

Compound	Melting point, °C.	$[\alpha]_D$	Rotation solvent	References
5-Methyl-D-glucose	liquid	−10.6	C_2H_5OH	50
phenylosazone	117, 128	−72 (30 min.) → −12 (equil.)	C_2H_5OH	50, 51
1,2-isopropylidene-	liquid	−13.1	$CHCl_3$	50
1,2-isopropylidene-3,6-diacetate	87	−15.2	$CHCl_3$	50
1,2-isopropylidene-3-tosyl-6-benzoate	liquid	−27.1	$CHCl_3$	50

frequently accompanied by a Walden inversion, it was possible that the 5-methylhexose isolated by Vargha was not a derivative of D-glucose, but this doubt was dispelled when the crystalline 1,2-isopropylidene-3,6-diacetyl-5-methyl-D-glucofuranose was converted, by saponification and methylation, into the known 1,2-isopropylidene-3,5,6-trimethyl-D-glucofuranose.[50]

5. *6-Methyl-D-Glucose*

Methods of Formation.—The route by which Helferich and Becker[52] first prepared 6-methyl-D-glucose has since been followed by other workers with the introduction of only minor modifications.[8,53,54] It involved tritylation of the primary alcohol group of methyl α-D-glucopyranoside, benzoylation of positions 2, 3 and 4, detritylation using dry hydrogen chloride in chloroform at 0° and methylation with silver oxide and methyl iodide to give crystalline methyl 2,3,4-tribenzoyl-6-methyl-α-D-glucopyranoside, from which sirupy 6-methyl-D-glucose was obtained by removal of the ester groups with sodium or ammonia in methanol and of the glucosidic methyl with aqueous acid. A sirupy methyl-D-glucose, which gave the same crystalline phenylosazone as that of Helferich and Becker,[52] was isolated by Ohle and Vargha[55] from "*iso*diacetone-D-glucose" by methylation and hydrolysis. A new synthesis,

(52) B. Helferich and Johanna Becker, *Ann.*, **440**, 1 (1924).
(53) B. Helferich and E. Günther, *Ber.*, **64B**, 1276 (1931).
(54) H. Ohle and K. Tessmar, *Ber.*, **71B**, 1843 (1938).
(55) H. Ohle and L. v. Vargha, *Ber.*, **62B**, 2425 (1929).

again involving the use of a cyclic diacetal, was reported in 1936 by Bell[56] and later employed by Freudenberg and Hüll.[57] In this case, 1,2-isopropylidene-3,5-benzylidene-6-acetyl-D-glucofuranose (prepared from 1,2-isopropylidene-6-acetyl-D-glucofuranose) was simultaneously deacetylated and methylated with methyl sulfate in aqueous alkali. The cyclic acetal residues were subsequently eliminated with hot aqueous mineral acid and the 6-methyl-D-glucose was isolated in crystalline form. A contrast to the above methods of formation, all of which entailed the methylation of compounds having C6 unsubstituted while the other hydroxyl groups were suitably protected, was provided by the synthesis accomplished by Ohle and Vargha,[48] wherein the opening with sodium methoxide of the anhydro-ring of 1,2-isopropylidene-5,6-anhydro-D-glucofuranose, prepared from 1,2-isopropylidene-6-tosyl-D-glucofuranose, gave 1,2-isopropylidene-6-methyl-D-glucofuranose and then, by hydrolysis, crystalline 6-methyl-D-glucose. Ohle and Vargha originally believed that ring scission had resulted in the attachment of the methoxyl grouping to C5 rather than C6, but this erroneous interpretation of the reaction was disproved three years later by Levene and Raymond.[47]

Proof of Constitution.—The work of Levene and Raymond,[47] Bell[56] and Ohle and Tessmar[54] clearly demonstrated that the same methyl-D-glucose, identified as its crystalline α-form, phenylosazone, β-tetraacetate and methyl β-D-glucoside triacetate, was isolated by each of the methods described and it is therefore permissible to draw conclusions regarding its constitution from each of these methods. Its synthesis from 1,2-isopropylidene compounds, prepared from the well-known 1,2-isopropylidene-D-glucofuranose, and its ability to form a phenylosazone without loss of the ether group have both distinguished it from 2-methyl-D-glucose.[48,55,56] Although its phenylosazone has a m. p. similar to that of 3-methyl-D-glucose phenylosazone, a comparison of its crystalline β-tetraacetate with the isomeric compound obtained from 3-methyl-D-glucose removed all doubt that it could be the 3-methyl ether.[53] Since the methylglucose has been synthesized from both furanose and pyranose derivatives of glucose, itself gives rise to furanosides and pyranosides,[47] and can be converted into both 2,3,4,6- and 2,3,5,6-tetramethyl-D-glucose,[47,48] it follows that neither C4 nor C5 is methylated. By this process of elimination the compound must be 6-methyl-D-glucose. This conclusion finds positive support in the proof by Ohle and Vargha[48,55] that the anhydro-ring of the 1,2-isopropylidene-anhydro-D-glucofuranose, an intermediate in their method of synthesis, was a 5,6 ring. Scission of this ring with sodium methoxide could give either 5-methyl-L-idose or

(56) D. J. Bell, *J. Chem. Soc.*, 859 (1936).
(57) K. Freudenberg and G. Hüll, *Ber.*, **74B**, 237 (1941).

6-methyl-D-glucose. The former is eliminated as a possibility by the fact that the methyl hexose is convertible into 2,3,4,6-tetramethyl-D-glucose by simple methylation. (The accompanying Table V is on pape 161.)

III. DIMETHYL ETHERS

1. *2,3-Dimethyl-D-Glucose*

Methods of Formation.—The methods used for the preparation of 2,3-dimethyl-D-glucose have all followed the same general pattern, involving the condensation of methyl α- or β- (and in one case[60] phenyl β-) D-glucopyranoside with an aldehyde, methylation of the resultant 4,6-cyclic acetal and removal of the glucosidic and acetal groups by hydrolysis with aqueous mineral acid. The aldehydes employed for this purpose are benzaldehyde,[60–64] acetaldehyde[65,66] and furfuraldehyde.[67] It was from methyl 4,6-benzylidene-α-D-glucopyranoside that Irvine and Scott[61] first isolated this dimethylglucose and, by fractional crystallization from ethyl acetate, separated the spherical aggregates of the α-form from the prismatic β-form. Both 2-methyl- and 3-methyl-D-glucose have been converted by similar routes into 2,3-dimethyl-D-glucose and its derivatives.[63,65,68]

The isolation of 2,3-dimethyl-D-glucose from the hydrolysis products of certain methylated polysaccharides has been an important factor in assigning structures to these polysaccharides. From trimethyl-starch it has been recovered in about 3% yield, together with 2,3,4,6-tetramethyl- and 2,3,6-trimethyl-D-glucopyranose, and arises from the points of linkage of the repeating chains of the amylopectin component.[57,69,70] From a dimethyl-starch the yield is considerably higher (75%).[71] Other sources are the methylated capsular polysaccharide of *Rhizobium radici-*

(58) B. Helferich, W. Klein and W. Schäfer, *Ber.*, **59B**, 79 (1926).
(59) B. Helferich and E. Himmen, *Ber.*, **62B**, 2136 (1929).
(60) C. M. McCloskey and G. H. Coleman, *J. Org. Chem.*, **10**, 184 (1945).
(61) J. C. Irvine and J. P. Scott, *J. Chem. Soc.*, **103**, 575 (1913).
(62) P. A. Levene and G. M. Meyer, *J. Biol. Chem.*, **65**, 535 (1925).
(63) K. Freudenberg, H. Toepffer and C. C. Andersen, *Ber.*, **61B**, 1750 (1928).
(64) T. H. Evans, I. Levi, W. L. Hawkins and H. Hibbert, *Can. J. Research*, **20B**, 175 (1942).
(65) B. Helferich and H. Appel, *Ber.*, **64B**, 1841 (1931).
(66) H. Appel, W. N. Haworth and (in part) E. G. Cox and F. J. Llewellyn, *J. Chem. Soc.*, 793 (1938).
(67) H. Bredereck, *Ber.*, **68B**, 777 (1935).
(68) J. W. H. Oldham and Jean K. Rutherford, *J. Am. Chem. Soc.*, **54**, 366 (1932).
(69) K. Freudenberg and H. Boppel, *Ber.*, **73B**, 609 (1940).
(70) C. C. Barker, E. L. Hirst and G. T. Young, *Nature*, **147**, 296 (1941).
(71) J. C. Irvine and J. Macdonald, *J. Chem. Soc.*, 1502 (1926).

TABLE V
The Physical Properties of 6-Methyl-D-glucose and Some of Its Derivatives

Compound	Melting point, °C.	$[\alpha]_D$	Rotation solvent	References
6-Methyl-α-D-glucose	143–145	+110 → +59, 55	H_2O	47, 48, 56, 57
phenylosazone	184–187	−75 → −44	C_2H_5OH	47, 53, 55, 56, 57
α-1,2,3,4-tetraacetate	120	+111.8	$CHCl_3$	53
β-1,2,3,4-tetraacetate	93, 96	+21	$CHCl_3$	47, 53
α-1,2,3,4-tetrabenzoate	154	+120.6	$CHCl_3$	54
β-1,2,3,4-tetrabenzoate	142–144	+66.9	$CHCl_3$	54
1,2,3,4-tetraazobenzoate	141–143	+ 180 (λ, 6252)	$CHCl_3$	57
1,2-isopropylidene-	71–72	−6.2	$CHCl_3$	47, 48
Methyl 6-methyl-α-D-glucopyranoside	liquid	+127.9	H_2O	58
2,3,4-tribenzoate	116–120	+116.4, +118.4	C_5H_5N	52, 53, 54
Methyl 6-methyl-β-D-glucopyranoside	133–135	−27.0	H_2O	59
2,3,4-triacetate	107–108	−12.4, −14.5	$CHCl_3$	47, 53, 59
2,3,4-tribenzoate	110–111	−22.5	C_5H_5N	54

colum[72] and the methylated dextrans synthesized from sucrose by *Leuconostoc mesenteroides*[73] and by *Leuconostoc dextranicum*.[74]

Proof of Constitution.—Most of the constitutional studies have been made on the dimethylglucose synthesized from methyl 4,6-benzylidene-α-D-glucopyranoside and, in general, specimens prepared by the alternative routes mentioned above have been identified by comparison with this standard through its excellent series of crystalline derivatives. The two crystalline dimethylglucoses obtained from the benzylidene compound by Irvine and Scott[61] were characterized as α- and β-forms of the same material by virtue of the fact that mutarotation in aqueous solution gave rise to the same equilibrium mixture in each case. At that time, the points of attachment of the benzylidene group were unknown, so that the course of synthesis afforded no evidence concerning the positions of the methyl groups other than that the carbon atom involved with C1 in the ring structure of methyl α-D-glucoside (now known to be C5) was not methylated. Irvine and Scott[61] noted that, under the usual conditions for osazone formation, the dimethylglucose formed only a phenyl-

(72) Elsa Schlüchterer and M. Stacey, *J. Chem. Soc.*, **776** (1945).
(73) Frances L. Fowler, Irene K. Buckland, F. Brauns and H. Hibbert, *Can. J. Research*, **15B**, 486 (1937).
(74) E. C. Fairhead, M. J. Hunter and H. Hibbert, *Can. J. Research*, **16B**, 151 (1938).

hydrazone, from which they deduced that C2 was methylated. This conclusion subsequently found support in the synthesis of the same dimethylglucose from 2-methyl-D-glucose.[68] A similar synthesis from 3-methyl-D-glucose established that the second methyl group was located on C3.[63,65] Confirmatory evidence that the hydroxyl groups on C4 and C6 were not substituted was provided by the oxidation of the dimethylglucose to a dimethyl-D-glucono-γ-lactone,[62] by the formation of a trityl ether[75,76] and by the liberation of formaldehyde from the terminal group when the dimethylglucose was treated with sodium periodate.[77,78] The structure as a whole was confirmed by the conversion of the dimethylglucose to both 2,3,4- and 2,3,6-trimethyl-D-glucopyranose[68,75,79] and by its oxidation with nitric acid to L-*threo*-dimethoxysuccinamide.[80] (The accompanying Table VI is on page 163.)

2. *2,4-Dimethyl-D-Glucose*

Methods of Formation.—The difficulties inherent in regulated synthesis of derivatives of 2,4-dimethyl-D-glucose, necessitating the protection of C3 and C6 of glucopyranose, were not surmounted until 1940, when Reeves, Adams and Goebel[83,84] prepared crystalline methyl 2,4-dimethyl-β-D-glucopyranoside from the known methyl 3-tosyl-2,4,6-triacetyl-β-D-glucopyranoside. This parent compound was treated with barium methoxide, or hydrogen chloride in dry methanol, under conditions such that it was deacetylated but not detosylated. The 6-trityl ether of the product was methylated, detritylated with hydrogen bromide in glacial acetic acid, and then treated with sodium amalgam in methanol to remove the tosyl group to give methyl 2,4-dimethyl-β-D-glucopyranoside. In a second method of synthesis,[84] 3-benzyl-D-glucose, prepared from 1,2:5,6-diisopropylidene-D-glucofuranose, was submitted to the same treatment as was methyl 3-tosyl-β-D-glucopyranoside. From the sirupy methyl 2,4-dimethyl-αβ-D-glucopyranoside the crystalline β-glucoside, identical with that already obtained, was isolated by seeding. The

(75) G. J. Robertson, *J. Chem. Soc.*, **737** (1933).
(76) J. W. H. Oldham and G. J. Robertson, *J. Chem. Soc.*, 685 (1935).
(77) R. E. Reeves, *J. Am. Chem. Soc.*, **63**, 1476 (1941).
(78) R. Jeanloz, *Helv. Chim. Acta*, **27**, 1509 (1944).
(79) J. C. Irvine and Jean K. Rutherford, *J. Am. Chem. Soc.*, **54**, 1491 (1932).
(80) S. Akiya, *J. Pharm. Soc. Japan*, **58**, 117 (1938); *Chem. Abstracts*, **32**, 4146 (1938).
(81) F. Smith, *J. Chem. Soc.*, 1035 (1940).
(82) D. S. Mathers and G. J. Robertson, *J. Chem. Soc.*, 696 (1933).
(83) R. E. Reeves, M. H. Adams and W. F. Goebel, *J. Am. Chem. Soc.*, **62**, 2881 (1940).
(84) M. H. Adams, R. E. Reeves and W. F. Goebel, *J. Biol. Chem.*, **140**, 653 (1941).

TABLE VI

The Physical Properties of 2,3-Dimethyl-D-glucose and Some of Its Derivatives

Compound	Melting point, °C.	$[\alpha]_D$	Rotation solvent	References
2,3-Dimethyl-D-glucose				
α-form	85–87	+81.9 → +48.3	Me_2CO	61
β-form	110, 121	+6.5 → +50.9	Me_2CO	60, 61
anilide	134	—	—	72
phenylhydrazone	amorph.	—	—	61
1,4,6-triazobenzoate	185, 209	+97.8 (λ, 6252)	$CHCl_3$	41, 69
2,3-Dimethyl-D-glucono-phenylhydrazide	167	—	—	64
2,3-Dimethyl-D-gluco-saccharamide	156	+28	H_2O	81
Methyl 2,3-dimethyl-α-D-glucopyranoside	83–85	+142.6, +150.2	H_2O	61, 67, 80
4,6-benzylidene-	122–123	+95.4, +97.0	Me_2CO	61, 63, 82
4,6-furylidene-	119–120	+98.4	$CHCl_3$	67
4,6-di-azobenzoate	143–144	+260.5 (λ, 6252)	$CHCl_3$	41
6-trityl-	169–170	+66.4	$CHCl_3$	73, 75, 76
4-acetyl-6-trityl-	153–154	+83.4	$CHCl_3$	75
4-benzoyl-6-trityl-	133	+56.0	Me_2CO	41
4-tosyl-6-trityl-	146–147	+66.3	$CHCl_3$	76
Methyl 2,3-dimethyl-β-D-glucopyranoside	62–64	−47.8	$CHCl_3$	15, 64, 82
4,6-benzylidene-	134, 141	−60.2	$CHCl_3$	32, 63, 64, 82
4,6-ethylidene-	109–111	−47.8	$CHCl_3$	65
4,6-dibenzoate	99–101	−8.2	$CHCl_3$	15, 82
4,6-dibenzene-sulfonate	119–120	−21.0	$CHCl_3$	68

crystalline α-isomer, characterized by conversion into the β-glucoside, proved to be identical with the methyl dimethyl-D-glucoside prepared a decade earlier by partial methylation of methyl 6-trityl-α-D-glucopyranoside.[84,85] A third method for the preparation of methyl 2,4-dimethyl-β-D-glucopyranoside, reported by Dewar and Fort,[86] involved the methylation with silver oxide and methyl iodide and subsequent denitration with a mixture of iron and zinc powders in glacial acetic acid, of methyl β-D-glucopyranoside 3,6-dinitrate. A shorter method for the preparation of the dinitrate considerably increased the value of this synthesis.[87]

(85) A. Robertson and R. B. Waters, *J. Chem. Soc.*, 1709 (1931).
(86) J. Dewar and G. Fort, *J. Chem. Soc.*, 492 (1944).
(87) J. Dewar, G. Fort and N. McArthur, *J. Chem. Soc.*, 499 (1944).

Catalytic reduction and methanolysis of the methylated capsular polysaccharide of Type III pneumococcus gives a mixture of methyl 2,3,6-trimethyl- and 2,4-dimethyl-αβ-D-glucopyranosides. The latter, which arises from glucuronic acid units in the polysaccharide, can be separated into crystalline α- and β-isomers identical with synthetic specimens.[84,88]

Proof of Constitution.—From the steps in its synthesis from methyl 3-tosyl-2,4,6-triacetyl-β-D-glucopyranoside, Reeves, Adams and Goebel[83] tentatively postulated that the crystalline product was methyl 2,4-dimethyl-β-D-glucopyranoside. They exercised commendable caution,

TABLE VII

The Physical Properties of 2,4-Dimethyl-D-glucose and Some of Its Derivatives

Compound	Melting point, °C.	$[\alpha]_D$	Rotation solvent	References
2,4-Dimethyl-D-glucose	liquid	—	—	84, 88
Methyl 2,4-dimethyl-α-D-glucopyranoside	79–81	+159, +186	Me$_2$CO	84, 88
Methyl 2,4-dimethyl-β-D-glucopyranoside	124	−16.3, −18.6	Me$_2$CO	83, 84, 86, 88
3-tosylate	liquid	−2.3	CHCl$_3$	84
3-tosyl-6-trityl-	amorph.	−1.0	CHCl$_3$	84
3-acetyl-6-nitrate	liquid	−5.7	CHCl$_3$	86
6-nitrate	liquid	−15.2	CHCl$_3$	86
3,6-dinitrate	liquid	−7.0	CHCl$_3$	86, 87

however, since none of the intermediate compounds was crystalline and the yield was only 2.5%. Their belief was substantiated when the same dimethyl methyl glucoside was synthesized from methyl 2-methyl-4,6-ethylidene-β-D-glucopyranoside[86] and when, after acidic hydrolysis, it was shown to yield 4-methyl-D-glucose phenylosazone.[84,86,88]

3. *2,5-Dimethyl-D-Glucose*

Methods of Formation.—The only 2,5-dimethyl derivatives in the glucose series of which the authors have any record are those of D-glucuronic acid and of 3,6-anhydro-D-glucose. D-Glucuronolactone with cold acidic methanol gives the γ-lactone of methyl β-D-glucofururonoside, which can be converted by treatment with methyl iodide and silver oxide into the sirupy γ-lactone of methyl 2,5-dimethyl-β-D-glucofururonos-

(88) R. E. Reeves and W. F. Goebel, *J. Biol. Chem.*, **139**, 511 (1941).

ide and thence to its crystalline amide.[89,90] There is no record of any attempt to reduce the methylated lactone to methyl 2,5-dimethyl-β-D-glucofuranoside. Derivatives of 2,5-dimethyl-3,6-anhydro-D-glucose have been prepared from 1,2-isopropylidene-5-methyl-3,6-anhydro-D-glucofuranose by hydrolytic separation of the acetone group and methylation.[91]

TABLE VIII
The Physical Properties of Derivatives of 2,5-Dimethyl-D-glucose

Compound	Melting point, °C.	[α]D	Rotation solvent	References
Methyl 2,5-dimethyl-β-D-gluco-fururonoside	—	—	—	—
γ-lactone	90–91	+2.0	H₂O	89
amide	95	—	—	90
2,5-Dimethyl-3,6-anhydro-D-glucose	liquid	+110 → +120	H₂O	91
anilide	96	+143 (equil.)	C₂H₅OH	91
2,5-Dimethyl-D-gluco-saccharamide	169–170	—	—	89

4. *2,6-Dimethyl-D-Glucose*

Methods of Formation.—The first fruitful attempt to effect the preferential methylation of the second and sixth positions of D-glucose was reported, in 1932, by Oldham and Rutherford.[7] By tosylation of methyl 2-methyl-β-D-glucopyranoside and treatment of the tri-*p*-toluenesulfonate with sodium iodide in acetone, they isolated methyl 2-methyl-3,4-ditosyl-6-desoxy-6-iodo-β-D-glucopyranoside and thence the 6-nitrate. Reductive scission of the nitrate radical and methylation with the Purdie reagents gave crystalline methyl 2,6-dimethyl-3,4-ditosyl-β-D-glucopyranoside. Unfortunately it was not possible to eliminate the tosyl groups and for six years this compound remained the only known derivative of 2,6-dimethyl-D-glucose. Following two unsuccessful attempts by Bell[19,92] to prepare the free dimethyl sugar, which resulted in so great a preponderance of accompanying isomers that it could be separated only as the ditosyl derivative of Oldham and Rutherford, Bell and Synge[17] succeeded in isolating it as a colorless glass from methyl β-D-glucopyranoside 3,4-dinitrate by methylation, reductive denitration and acidic

(89) R. E. Reeves, *J. Am. Chem. Soc.*, **62**, 1616 (1940).
(90) L. N. Owen, S. Peat and W. J. G. Jones, *J. Chem. Soc.*, 339 (1941).
(91) W. N. Haworth, L. N. Owen and F. Smith, *J. Chem. Soc.*, 88 (1941).
(92) D. J. Bell, *J. Chem. Soc.*, 175 (1935).

scission of the glucosidic group. A similar route in the α-series followed by Reeves,[93] who employed methyl α-D-glucopyranoside 3,4-dinitrate, prepared by a shorter method than that for the β-isomer, was carried only as far as methyl 2,6-dimethyl-α-D-glucopyranoside. In a second method, Reeves[93] used methyl 3,4-di-N-phenylcarbamyl-α-D-glucopyranoside, obtainable from the 2,6-dibenzoate of methyl α-D-glucopyranoside. Another source of 2,6-dimethyl derivatives is 1,2-isopropylidene-3,5-dibenzyl-6-methyl-D-glucofuranose, from which the acetone residue can be removed by hydrolysis, leaving the liberated C2 available for methylation.[57]

TABLE IX

The Physical Properties of 2,6-Dimethyl-D-glucose and Some of Its Derivatives

Compound	Melting point, °C.	$[\alpha]_D$	Rotation solvent	References
2,6-Dimethyl-D-glucose	liquid	+58.3, +63.3 (equil.)	H_2O	17, 57
1,3,4-triazobenzoate	207, 131	−172 (λ, 6252), −275	$CHCl_3$	57
2,6-Dimethyl-D-gluconic acid	—	—	—	—
phenylhydrazide	127–129	+48.6	C_2H_5OH	17
Methyl 2,6-dimethyl-α-D-glucopyranoside	liquid	+152, +156	H_2O	93
3,4-di-N-phenyl-carbamate	235–236	+87	$CHCl_3$	93
Methyl 2,6-dimethyl-β-D-glucopyranoside	50–52	−43.5	$CHCl_3$	17
3,4-ditosylate	156–159	−8.2	$CHCl_3$	7, 17, 19, 92
3,4-dinitrate	74–76	−13.7	$CHCl_3$	17

Proof of Constitution.—The constitution of the methyl 2,6-dimethyl-3,4-ditosyl-β-D-glucopyranoside originally isolated by Oldham and Rutherford[7] was clearly established by its mode of synthesis from methyl 2-methyl-β-D-glucopyranoside. Obviously one methyl group had remained in position 2 and there was no doubt from the well-known and generally applicable reactions involved that the other was on C6. Since the sirupy dimethylglucose of Bell and Synge[17] gave the same methyl ditosyl-β-D-glucoside, it must have been 2,6-dimethyl-D-glucose. Confirmation of this was afforded by Freudenberg and Hüll,[57] who showed that their dimethylglucose sirup, which had approximately the same equilibrium rotation in aqueous solution as that of Bell and Synge[17] and

(93) R. E. Reeves, *J. Am. Chem. Soc.*, **70**, 259 (1948).

which had been synthesized from a derivative of 6-methyl-D-glucose, was converted by treatment with phenylhydrazine into 6-methyl-D-glucose phenylosazone, with the concomitant removal of the methyl group on C2.

5. *3,4-Dimethyl-D-Glucose*

Methods of Formation.—The key to the synthesis of 3,4-dimethyl-D-glucose lay in the preparation of a parent pyranose compound suitably substituted in positions 2 and 6. Such a parent compound, namely methyl 2-benzoyl-6-trityl-β-D-glucopyranoside, was prepared, in 1944, by Dewar and Fort[30] from methyl 4,6-ethylidene-β-D-glucopyranoside 3-nitrate by benzoylation of C2, reductive removal of the nitrate group, acidic scission of the acetal ring and tritylation of the C6 thus liberated. With methyl iodide and silver oxide the parent compound gave rise to methyl 2-benzoyl-3,4-dimethyl-6-trityl-β-D-glucopyranoside, which was converted by the usual methods into crystalline 3,4-dimethyl-D-glucose. In an alternative route pursued by the same workers, involving the same essential stages as the above process, the methylation of the third and fourth positions was achieved in two stages and passed therefore through a series of 3-methyl-D-glucose derivatives. In collaboration with McArthur, Dewar and Fort[87] reported a third method of synthesis, whereby methyl β-D-glucopyranoside 2,3,4,6-tetranitrate, when treated successively with sodium iodide in acetone and with silver nitrate in acetonitrile, gave a mixture of the 2,6- and 3,6-dinitrates of methyl β-D-glucopyranoside. The former, purified through its crystalline diacetate, after treatment with the Purdie reagents and reductive denitration, yielded crystalline methyl 3,4-dimethyl-β-D-glucopyranoside, identical with that previously obtained by Dewar and Fort.[30]

Proof of Constitution.—Since one synthesis of this dimethylglucose was accomplished via a series of derivatives of 3-methyl-D-glucose,[30] whose identity was established by the conversion of one of these derivatives into the known crystalline triacetate of methyl 3-methyl-β-D-glucopyranoside, it follows that one of the two methyl groups must have been located on C3. From the course of the same synthesis it was clear that the second methoxy group was situated on one of the carbon atoms (4 and 6) vacated by the acetal group. Inasmuch as the dimethylglucose formed a trityl compound, it was probable that the C6 hydroxyl was free and that, therefore, the methyl group was on C4. The methyl dimethylglucoside isolated during this synthesis must have possessed a β-pyranoside structure, since it had been derived from methyl 4,6-ethylidene-β-D-glucopyranoside 3-nitrate by changes not involving the glucoside group and, as it differed from all the known methyl dimethyl-β-D-glucopyranosides, it must have been the only missing isomer, namely, methyl 3,4-dimethyl-β-D-glucopyranoside.

TABLE X

The Physical Properties of 3,4-Dimethyl-D-glucose and Some of Its Derivatives

Compound	Melting point, °C.	$[\alpha]_D$	Rotation solvent	References
3,4-Dimethyl-β-D-glucose	113	+64.9 → +94.8 (equil.)	H_2O	30
phenylosazone	126	—	—	94
Methyl 3,4-dimethyl-β-D-gluco-				
pyranoside	79–81	−11.9	$CHCl_3$	30, 87
2,6-dinitrate	62–63	+9.7	$CHCl_3$	87
2-benzoyl-6-trityl-	142	+36.4	$CHCl_3$	30
6-trityl-	liquid	+14.2	$CHCl_3$	30

6. *3,5-Dimethyl-D-Glucose*

The 3,5-dimethyl derivative is still included among the rapidly diminishing number of dimethylglucoses which have yet to be prepared.

7. *3,6-Dimethyl-D-Glucose*

Methods of Formation.—An examination of the sirupy mixture of methyl dimethyl-glucosides, produced when the addition compound of methyl β-D-glucopyranoside with boric acid was treated with methyl iodide and silver oxide, led to the isolation by Bell[92] of a specimen of 3,6-dimethyl-α-D-glucose. The efforts of Bell to discover alternative methods of synthesis which would avoid the simultaneous formation of isomeric dimethylglucoses were soon rewarded and, within a year, he was able to report that 3,6-dimethyl-α-D-glucose had been prepared by two other routes.[95] In the first, the fully-substituted 1,2-isopropylidene-6-acetyl-D-glucofuranose 3,5-dinitrate, formed when the known 1,2-isopropylidene-6-acetyl-D-glucofuranose was treated in chloroform solution with nitrogen pentoxide, was simultaneously deacetylated and denitrated in position 3 by means of dimethylamine in alcoholic benzene. Methylation of the liberated 3 and 6 positions yielded 1,2-isopropylidene-3,6-dimethyl-D-glucofuranose 5-nitrate and thence, by reductive denitration and acidic hydrolysis, 3,6-dimethyl-α-D-glucose. In the second place,

(94) Emma J. McDonald and R. F. Jackson, *J. Research Natl. Bur. Standards,* **24,** 181 (1940).
(95) D. J. Bell, *J. Chem. Soc.,* 1553 (1936).

the method used earlier for the preparation of 6-methyl-D-glucose by Levene and Raymond[47] was applied to 1,2-isopropylidene-3-methyl-D-glucofuranose. The 6-tosyl derivative, formed preferentially when this compound was treated with an equimolecular proportion of *p*-toluenesulfonyl chloride in pyridine, was transformed by sodium methoxide into the 5,6-anhydro compound and then into 1,2-isopropylidene-3,6-dimethyl-D-glucofuranose, which gave rise to the free dimethylglucose when heated with aqueous mineral acid. More recently, Percival and Duff[96] have been able to effect a similar synthesis, in which a 6-sulfate was used instead of a 6-tosyl derivative.

Proof of Constitution.—The later samples of this dimethylglucose have all been shown to have the same melting point and equilibrium rotation in aqueous solution as Bell's original specimen,[92] which itself

TABLE XI
The Physical Properties of 3,6-Dimethyl-D-glucose and Some of Its Derivatives

Compound	Melting point, °C.	$[\alpha]_D$	Rotation solvent	References
3,6-Dimethyl-α-D-glucose	113–116	+102.5 → +61.5 (equil.)	H_2O	92, 95, 96
1,2-isopropylidene-	liquid	−45.8	$CHCl_3$	95
Methyl 3,6-dimethyl-β-D-glucopyranoside	liquid	+55.4	C_2H_5OH	92
2,4-ditosylate	158–160	−22.8	$CHCl_3$	92
2,4-dibenzoate	155–156	−11.7	$CHCl_3$	92

was characterized by virtue of its conversion into methyl 2,4-ditosyl-3,6-dimethyl-β-D-glucopyranoside, a compound already synthesized by Oldham, but not reported in the literature. We have been unable to find any subsequent publication by Oldham on this subject. There is no doubt that the dimethyl sugar was in fact the 3,6-derivative, as the following facts show. Its synthesis from 1,2-isopropylidene-3-methyl-D-glucofuranose[95,96] demonstrates that one of the methyl groups must be located on C3; the other cannot be on C2 or C4 because these groups were not free in the parent compound. Since the original synthesis by Bell[92] from methyl β-D-glucopyranoside precludes the allocation of the second methyl group to C5, the only possible position is C6. Support for this conclusion is found in the similarity between the method of synthesis involving anhydro-ring formation[95] and the synthesis of 6-methyl-D-glucose by Levene and Raymond.[47]

(96) E. G. V. Percival and R. B. Duff, *Nature*, **158**, 29 (1946).

8. 4,6-Dimethyl-D-Glucose

Methods of Formation.—The first sample of 4,6-dimethyl-D-glucose, prepared by Haworth and Sedgwick[97] in 1926, came, together with tri- and tetramethylglucoses, from the incomplete methylation of glucose with methyl sulfate and sodium hydroxide. It was not characterized until about a decade later, when it was isolated by Bell and Synge[98] from another source.

The systematic synthesis of 4,6-dimethyl-D-glucose would not appear at first sight to present any great difficulty, since 2,3-substituted methyl D-glucopyranosides are readily available from 4,6-benzylidene (or ethylidene) methyl D-glucopyranoside. Nevertheless, Mathers and Robertson,[99] the first workers to attempt this type of synthesis, were unfortunate in their choice of 2,3-substituents, the dibenzoate and the di-*p*-toluenesulfonate, in that the former tended to undergo acyl migration, while saponification of the latter was accompanied by inversion of configuration. A satisfactory solution to the problem was furnished by using the 2,3-dinitrate[98] and the 2,3-dibenzyl derivative,[100,41] each of these groups being stable during the methylation stage and readily removable by reduction afterwards.

Proof of Constitution.—Reference to Table XII will show that all of the methods mentioned above as being suitable for the synthesis of 4,6-dimethylglucose have been correlated by means of crystalline derivatives. If the tosyl, benzyl and nitrate radicals can be regarded as being non-migratory (and there is no evidence to the contrary) then the course of the synthesis from 4,6-acetals (the constitution of which has been proved by their conversion into 2,3-dimethyl-D-glucose) leaves no doubt that the methyl groups occupy positions 4 and 6.

This conclusion has been substantiated by independent evidence afforded by examination of the dimethylglucose itself. As early as 1926, it was shown by Haworth and Sedgwick[97] to be a pyranose sugar by its conversion to 2,3,4,6-tetramethyl-D-glucopyranose. The demonstration by Bell and Synge[98] that, with 1% hydrogen chloride in methyl alcohol at room temperature, there was no evidence of furanoside formation indicated that C4 was methylated. Since the methyl dimethyl-ditosyl-β-D-glucoside can be recovered unchanged after being heated at 100°C. with sodium iodide in acetone, a process which replaces a 6-tosyloxy group by iodine, the second methyl group must be on C6.[98]

(97) W. N. Haworth and W. G. Sedgwick, *J. Chem. Soc.*, 2573 (1926).
(98) D. J. Bell and R. L. M. Synge, *J. Chem. Soc.*, 1711 (1937).
(99) D. S. Mathers and G. J. Robertson, *J. Chem. Soc.*, 1076 (1933).
(100) D. J. Bell and J. Lorber, *J. Chem. Soc.*, 453 (1940).

TABLE XII

The Physical Properties of 4,6-Dimethyl-D-glucose and Some of Its Derivatives

Compound	Melting point, °C.	[α]D	Rotation solvent	References
4,6-Dimethyl-α-D-glucose	156–158	+110 → +64	H_2O	41, 97, 98, 100
1,2,3-triazobenzoate	145	+551 (λ, 6252)	$CHCl_3$	41
Methyl 4,6-dimethyl-α-D-				
glucopyranoside	liquid	+157	$CHCl_3$	41, 100
2,3-dibenzoate	liquid	—	—	99
2,3-ditosylate	113–115	+55	$CHCl_3$	99, 100
2,3-di-azobenzoate	120	+405 (λ, 6252)	Me_2CO	41
2,3-di-*p*-nitrobenzoate	114	+203	Me_2CO	41
2,3-dibenzyl-	liquid	+32.9	$CHCl_3$	41, 100
Methyl 4,6-dimethyl-β-D-				
glucopyranoside	50–52	−28	$CHCl_3$	32, 98
2,3-ditosylate	146–149	−14.7	$CHCl_3$	98

9. 5,6-Dimethyl-D-Glucose

Methods of Formation.—When 5,6-dimethyl-D-glucose was first synthesized by Salmon and Powell in 1939[101] use was made of a new non-migratory substituent, namely the benzoxymethyl group ($PhCH_2 \cdot O \cdot CH_2$—), which was claimed to be more stable to alkali than a sugar nitrate and more readily introduced than the benzyl group. A benzoxymethyl derivative is formed when a sugar is treated in boiling ether with chloromethyl benzyl ether in the presence of either sodium or solid potassium hydroxide. By the mild hydrolysis with aqueous acetic acid of 1,2:5,6-diisopropylidene-3-benzoxymethyl-D-glucofuranose (prepared from diisopropylidene-D-glucose), the 5,6-isopropylidene group was preferentially removed, making these two positions available for methylation, which proved to be an extraordinarily difficult process requiring the use of both methyl sulfate—sodium hydroxide and the Purdie reagents. The 1,2-isopropylidene-3-benzoxymethyl-5,6-dimethyl-D-glucofuranose sirup was reduced with sodium in methyl alcohol to crystalline 1,2-isopropylidene-5,6-dimethyl-D-glucofuranose, from which 5,6-dimethyl-D-glucofuranose was obtained by acidic hydrolysis. The same dimethylglucose was isolated in the following year by Freudenberg and Plankenhorn[41] by a similar series of reactions in which benzyl was substituted instead of benzoxymethyl.

(101) M. R. Salmon and G. Powell, *J. Am. Chem. Soc.*, **61**, 3507 (1939).

Proof of Constitution.—If the reasonable assumption is made that the benzoxymethyl group is non-migratory, the product of the above synthesis could be only 5,6-dimethyl-D-glucose. Nevertheless, Salmon and Powell[101] sought confirmation of its structure from a periodate oxidation of the dimethyl sugar, which should yield dimethyl-D-glyceraldehyde. This was indeed produced and was identified, after oxidation with bromine water, as crystalline esters of dimethylglyceric acid. The dimethylglucose showed enhanced reducing power similar to that observed by Vargha[50] in the case of 5-methyl-D-glucose. It reduced Fehling's solution and neutral permanganate without the application of heat and gave a positive Schiff's test for aldehydes. In addition, it exhibited a very low constant rotation in aqueous solution. These properties were consistent with there being present an appreciable proportion of the aldehydo form.

TABLE XIII

The Physical Properties of 5,6-Dimethyl-D-glucose and Some of Its Derivatives

Compound	Melting point, °C.	$[\alpha]_D$	Rotation solvent	References
5,6-Dimethyl-D-glucose	liquid	+4.0	H_2O	101
	liquid	+3.7	Me_2CO	41
p-bromophenylosazone	156	—	—	101
1,2-isopropylidene-	56	−12.8	H_2O	101
3-benzoxymethyl-	liquid	—	—	101
3-benzyl-	liquid	−15.8	Me_2CO	41
3-carbanilyl-	88–89	−12.3	C_2H_5OH	101
1,2,3-tri-p-nitrobenzoate	90–120	+90.0	Me_2CO	41
1,2,3-triazobenzoate	192	+13.3 (λ, 6252)	Me_2CO	41

IV. Trimethyl Ethers

1. *2,3,4-Trimethyl-D-Glucose*

Methods of Formation.—Originally the synthesis of 2,3,4-trimethyl-D-glucopyranose entailed the treatment of methyl α-D-glucopyranoside with a deficiency of the methylating agent and fractionation of the resultant mixture of glucosides.[102-105] Such methods have now been entirely replaced by other routes, designed to eliminate the possibility of formation of isomeric trimethylglucoses. These involve the methylation of either a 6-substituted methyl D-glucopyranoside or levoglucosan.

(102) T. Purdie and R. C. Bridgett, *J. Chem. Soc.*, **83**, 1037 (1903).
(103) W. N. Haworth, *J. Chem. Soc.*, **107**, 8 (1915).
(104) J. C. Irvine and J. S. Dick, *J. Chem. Soc.*, **115**, 593 (1919).
(105) P. A. Levene and G. M. Meyer, *J. Biol. Chem.*, **48**, 233 (1921).

A suitable 6-substituent for this purpose is the trityl group, which can be readily removed under acid conditions when the methylation has been completed.[52,85,106–108]

Levoglucosan (1,6-anhydro-β-D-glucopyranose), which results from the vacuum distillation at elevated temperatures of D-glucose or its polymers, is a convenient parent compound for the synthesis of 2,3,4-trimethyl-D-glucose inasmuch as only two stages are required, namely methylation and scission of the anhydro-ring. The second stage has been effected by hydrolysis of the 2,3,4-trimethyl-levoglucosan with aqueous mineral acid[64,109] and by cold treatment in chloroform solution with fuming nitric acid and phosphorus pentoxide, the latter alternative giving the 1,6-dinitrate, which was subsequently denitrated with iron powder in glacial acetic acid.[110] In an early variation of this type of synthesis, reported by Irvine and Oldham,[111] the anhydro-ring of 2,3,4-triacetyl-levoglucosan was opened with phosphorus pentabromide prior to deacetylation and methylation, the bromine atoms thus introduced at C1 and C6 being eventually removed by prolonged heating at 150°C. with an excess of potassium acetate in methyl alcohol.

The naturally-occurring 1,6-linked glucose polymers furnish other sources of 2,3,4-trimethyl-D-glucose, which has been isolated from the hydrolysis products of the methyl ethers of gentiobiose,[112,113] melibiose,[114] raffinose[106,115] and dextran.[73,74,116]

Proof of Constitution.—2,3,4-Trimethyl-D-glucose has not been isolated in a crystalline form and comparisons have been confined to crystalline derivatives, usually the methyl β-D-glucoside. It is therefore appropriate at present to survey the evidence upon which the characterization of the β-glucoside, rather than the free sugar, is based.

The β-orientation of the glycosidic group is indicated by the ease with which the reducing trimethylsugar is generated by aqueous acid hydrolysis, the hydrolysis being accompanied by an increase in dextro-

(106) W. Charlton, W. N. Haworth and W. J. Hickinbottom, *J. Chem. Soc.*, 1527 (1927).

(107) W. N. Haworth, E. L. Hirst, E. J. Miller and A. Learner, *J. Chem. Soc.*, 2443 (1927).

(108) F. Smith, M. Stacey and P. I. Wilson, *J. Chem. Soc.*, 131 (1944).

(109) J. C. Irvine and J. W. H. Oldham, *J. Chem. Soc.*, **119**, 1744 (1921).

(110) J. W. H. Oldham, *J. Chem. Soc.*, **127**, 2840 (1925).

(111) J. C. Irvine and J. W. H. Oldham, *J. Chem. Soc.*, **127**, 2729 (1925).

(112) W. N. Haworth and B. Wylam, *J. Chem. Soc.*, **123**, 3120 (1923).

(113) G. Zemplén, *Ber.*, **57B**, 698 (1924).

(114) W. Charlton, W. N. Haworth and R. W. Herbert, *J. Chem. Soc.*, 2855 (1931).

(115) W. N. Haworth, E. L. Hirst and D. A. Ruell, *J. Chem. Soc.*, **123**, 3125 (1923).

(116) S. Peat, Elsa Schlüchterer and M. Stacey, *J. Chem. Soc.*, 581 (1939).

rotation.[109,114,116] With regard to the positions of the other three methyl radicals, the earliest fact established was that position 6 was not etherified. This was deduced from the observation that the reducing sirup formed by hydrolysis of the β-glucoside was oxidized by mild treatment with nitric acid to a trimethylsaccharic acid.[104,109,114,116] Direct oxidation of the methyl trimethyl-β-D-glucoside with alkaline permanganate yielded a monocarboxylic acid (methyl 2,3,4-trimethyl-β-D-glucuronoside) without simultaneous demethylation occurring, thus confirming that C6 was unsubstituted.[108] The synthesis of the methyl trimethyl-β-D-glucoside from methyl 6-desoxy-6-bromo-β-D-glucopyranoside, by a route designed to replace the bromine atom by a hydroxyl,[111] and its ability to form a trityl ether[106,107] and a mononitrate, which on treatment with sodium iodide in acetone gives rise to an iodohydrin,[110] all confirm that C6 does not carry a methoxyl grouping.

The additional evidence necessary to characterize the methyl trimethyl-β-D-glucoside as the 2,3,4-derivative was furnished by its conversion into 2,3,4,6-tetramethyl-D-glucopyranose.[109,116,117] If these deductions are correct, then drastic oxidation of the methyl trimethyl-β-D-glucoside with nitric acid would be expected to give rise to *i-xylo*-tri-methoxyglutaric acid and indeed this was demonstrated, in 1931, by Haworth and his school.[114] The synthesis of the same methyl trimethyl-β-D-glucoside from methyl 2,3-dimethyl-β-D-glucopyranoside, effected by Oldham and Rutherford,[68] has provided a useful correlation between the two series of methylglucoses. (The accompanying Table XIV is on page 175.)

2. *2,3,5-Trimethyl-D-Glucose*

Methods of Formation.—Although frequent references to 2,3,5-trimethyl-D-glucose were made in the literature published about 1920, the constitution assigned to this compound was based on the erroneous belief that the ring system of methyl α-D-glucoside was of the butylene oxide type. With the recognition that an amylene oxide (pyranose) ring was actually present, came the realization that the trimethylglucose in question was in fact the 2,3,4-derivative. The first specimen of authentic 2,3,5-trimethyl-D-glucose was prepared, in 1944, by Smith,[121] who methylated methyl 6-trityl-αβ-D-glucofuranoside, first with methyl sulfate in alkaline solution and then with methyl iodide in the presence of silver oxide, removed the protective trityl grouping by means of

(117) W. N. Haworth and Grace C. Leitch, *J. Chem. Soc.*, **121**, 1921 (1922).
(118) G. H. Coleman, D. E. Rees, R. L. Sundberg and C. M. McCloskey, *J. Am. Chem. Soc.*, **67**, 381 (1945).
(119) S. W. Challinor, W. N. Haworth and E. L. Hirst, *J. Chem. Soc.*, 258 (1931).
(120) F. Smith, *J. Chem. Soc.*, 1724 (1939).
(121) F. Smith, *J. Chem. Soc.*, 571 (1944).

TABLE XIV

The Physical Properties of 2,3,4-Trimethyl-D-glucose and Some of Its Derivatives

Compound	Melting point, °C.	$[\alpha]_D$	Rotation solvent	References
2,3,4-Trimethyl-D-glucopyranose	—	diverse	—	see text
anilide	145–146	—	—	116
1,6-dinitrate	86	+149.3	$CHCl_3$	75, 110
1,6-di-azobenzoate	165	−25 (λ, 6438)	$CHCl_3$	41, 57, 118
2,3,4-Trimethyl-D-gluconolactone	liquid	+80 → +32	H_2O	116
2,3,4-Trimethyl-D-glucuronic acid	—	—	—	—
β-methyl-uronide	137	−38	H_2O	108, 119
amide of α-methyl-	181–183	+138	H_2O	108, 120
amide of β-methyl-	193	−47	H_2O	120
2,3,4-Trimethyl-D-glucosaccharo-lactone	—	—	—	—
methyl ester	106–107	+146.5	C_6H_6	85, 114
Methyl 2,3,4-trimethyl-α-D-glucopyranoside	liquid	—	—	85, 102, 103, 104
6-trityl-	166–167	+88.9 (λ, 5461)	Me_2CO	85
Methyl 2,3,4-trimethyl-β-D-glucopyranoside	92–95	−22.9, −25.1	CH_3OH	109, 112, 115
6-desoxy-6-bromo-	24	−4.7	CH_3OH	111
6-nitrate	53–54	−5.2	$CHCl_3$	110
6-azobenzoate	122	−8 (λ, 6252)	Me_2CO	41

ethereal hydrogen chloride and finally eliminated the glucosidic grouping with hot aqueous mineral acid. The same sirupy trimethylglucose was synthesized two years later by the methylation and hydrolysis of 1,6-anhydro-β-D-glucofuranose, which is formed, together with levoglucosan (1,6-anhydro-β-D-glucopyranose), during the vacuum pyrolysis of starch.[122]

Proof of Constitution.—The method of synthesis from methyl 6-trityl-αβ-D-glucofuranoside furnished strong evidence that the trimethyl-D-glucose isolated by Smith[121] was substituted at positions 2, 3 and 5. This conclusion was confirmed by the observations (a) that the trimethyl-gluconolactone, produced when the sugar was oxidized with bromine

(122) R. J. Dimler, H. A. Davis and G. E. Hilbert, *J. Am. Chem. Soc.*, **68**, 1377 (1946).

water, was hydrated in aqueous solution at a rate characteristic of a γ-lactone (C4 not substituted), (b) that oxidation with nitric acid yielded a trimethylsaccharic acid (C1 and C6 not substituted), and (c) that the crystalline diamide of the trimethylsaccharic acid failed to give hydrazodicarbonamide when treated with Weerman's reagent and then with semicarbazide (C2 and C5 methylated). The trimethylglucose of Dimler, Davis and Hilbert[122] was characterized, not by its mode of formation from the glucosan, since the structure of this parent compound had not then been established, but by virtue of the fact that the lactone which was obtained from it yielded a crystalline phenylhydrazide identical with that derived from Smith's trimethylglucose.

TABLE XV

The Physical Properties of 2,3,5-Trimethyl-D-glucose and Some of Its Derivatives

Compound	Melting point, °C.	$[\alpha]_D$	Rotation solvent	References
2,3,5-Trimethyl-D-glucose 1,6-anhydro-	liquid 51–52	+17, −4.5 +18.9	H_2O Me_2CO	121, 122 122
2,3,5-Trimethyl-γ-D-gluconolactone phenylhydrazide	liquid 156–157	+62 (initially) +32, 38	H_2O CH_3OH	121 121, 122
2,3,5-Trimethyl-D-glucosaccharamide	213	+18	H_2O	121
2,3,5-Trimethyl-D-glucosaccharolactone methyl ester	77–78	−9.5 → +6	H_2O	90, 121

3. *2,3,6-Trimethyl-D-Glucose*

Methods of Formation.—The conversion of methyl 2,3-dimethyl-α (or β)-D-glucopyranoside into 2,3,6-trimethyl-D-glucose has been achieved by three routes, designed to protect C4 during the methylation of C6. In 1932, Oldham and Rutherford[68] reported that when an acetone solution of the 4,6-dibenzene-sulfonate of methyl 2,3-dimethyl-β-D-glucopyranoside was heated under pressure with sodium iodide the primary ester group was replaced by an iodine atom, which, in turn, could be exchanged for a nitrate radical and thence, by reductive hydrolysis, for a hydroxyl group. Methylation of the product yielded methyl 2,3,6-trimethyl-4-benzenesulfonyl-β-D-glucopyranoside, but the remaining ester group proved to be so stable that it was not possible to isolate 2,3,6-trimethyl-D-glucose itself. This difficulty was surmounted a few months later by Irvine and Rutherford,[79] whose starting material was the 4,6-dinitrate of methyl 2,3-dimethyl-αβ-D-glucopyranoside. An iodine atom was introduced at C6 in the usual way and was converted to the primary alcohol by saponification of the 6-acetate, formed by treatment of the iodo compound with silver acetate. The methyl 2,3-dimethyl-αβ-D-glucopyranoside 4-nitrate gave rise, when submitted to methylation, reductive denitration and acid hydrolysis, to crystalline 2,3,6-trimethyl-D-glucose. Although an attempt by Robertson[75] to utilize methyl 2,3-dimethyl-4-acetyl-α-D-glucopyranoside (prepared via the 6-trityl derivative) in the synthesis of 2,3,6-trimethyl-D-glucose did result in the isolation of the desired product, there was extensive contamination by its 2,3,4-isomer, acyl migration from C4 to C6 having occurred during the methylation stage.

A synthesis of a different type was effected in 1938 by Peat and Wiggins,[123] who found that treatment with methyl iodide and silver oxide of the mixture of anhydro methyl glycosides produced by the action of sodium methoxide on methyl 3-tosyl-αβ-D-glucoside gave *inter alia* methyl 2,6-dimethyl-3,4-anhydro-β-D-alloside, which was isolated in crystalline form. Scission of the anhydro-ring of this compound with boiling methyl alcoholic sodium methoxide introduced a third methyl group on C3 and, since this process was accompanied by a Walden inversion of the carbonium cation, the product was methyl 2,3,6-trimethyl-β-D-glucopyranoside.

This particular trimethylglucose is unique in that it was separated in crystalline form from the hydrolyzates of the methyl ethers of several naturally-occurring glucose polymers almost two decades before it was synthesized from glucose. These natural sources, which still furnish the most convenient routes for the preparation of 2,3,6-trimethyl-D-glucose, include maltose,[124-126] cellobiose,[127,128] lactose,[129-131] starch,[71,132] glycogen,[133,134] cellulose,[135-137] and lichenin.[138,139] The literature pub-

(123) S. Peat and L. F. Wiggins, *J. Chem. Soc.*, 1088 (1938).
(124) W. N. Haworth and Grace C. Leitch, *J. Chem. Soc.*, **115,** 809 (1919).
(125) J. C. Irvine and I. M. A. Black, *J. Chem. Soc.*, 862 (1926).
(126) C. J. A. Cooper, W. N. Haworth and S. Peat, *J. Chem. Soc.*, 876 (1926).
(127) W. N. Haworth and E. L. Hirst, *J. Chem. Soc.*, **119,** 193 (1921).
(128) P. Karrer and F. Widmer, *Helv. Chim. Acta*, **4,** 295 (1921).
(129) W. N. Haworth and Grace C. Leitch, *J. Chem. Soc.*, **113,** 188 (1918).
(130) J. C. Irvine and E. L. Hirst, *J. Chem. Soc.*, **121,** 1213 (1922).
(131) H. H. Schlubach and K. Moog, *Ber.*, **56B,** 1957 (1923).
(132) W. N. Haworth, E. L. Hirst and J. I. Webb, *J. Chem. Soc.*, 2681 (1928).
(133) A. K. Macbeth and J. Mackay, *J. Chem. Soc.*, **125,** 1513 (1924).
(134) W. N. Haworth, E. L. Hirst and J. I. Webb, *J. Chem. Soc.*, 2479 (1929).
(135) W. S. Denham and Hilda Woodhouse, *J. Chem. Soc.*, **111,** 244 (1917).
(136) J. C. Irvine and E. L. Hirst, *J. Chem. Soc.*, **123,** 518 (1923).
(137) K. Hess and W. Weltzien, *Ann.*, **442,** 46 (1925).
(138) P. Karrer and K. Nishida, *Helv. Chim. Acta*, **7,** 363 (1924).
(139) H. Granichstädten and E. G. V. Percival, *J. Chem. Soc.*, 54 (1943).

lished on the methyl ethers of these 1–4 linked glucopyranose polymers is voluminous and only a few of the vast number of papers on the subject are mentioned above.

Proof of Constitution.—The definitive synthesis of this crystalline trimethylglucose from 2,3-dimethyl-D-glucose served only to confirm the structure, which had already been assigned to the same compound derived by methylation and hydrolysis from natural D-glucose polymers.

As early as 1917, it was known that C2 and C3 in this trimethyl sugar each carried a methoxyl group, since (*a*) it failed to yield an osazone, and (*b*) the trimethyl-D-glucoheptonic acid derived from it by a cyanohydrin synthesis gave a lactone only with the concomitant loss of one of the methyl radicals.[135] The production of a dimethyl- and not a trimethyl-

Table XVI

The Physical Properties of 2,3,6-Trimethyl-D-glucose and Some of Its Derivatives

Compound	Melting point, °C.	$[\alpha]_D$	Rotation solvent	References
2,3,6-Trimethyl-α-D-glucose	121–123	+70 (equil.)	H_2O	130, 141, 143
β-1,4-diacetate	67–68	−8.7	$CHCl_3$	144
1-chloro-5-benzoate	122–123	−114.5	$CHCl_3$	145
diethyl mercaptal	71–72	−15	$CHCl_3$	146
diethyl mercaptal 4,5-dibenzoate	115–116	+61	$CHCl_3$	146
1,4-di-azobenzoate	172	+12.6 (λ, 6252)	Me_2CO	41, 42, 118
2,3,6-Trimethyl-D-gluconic acid	—	—	—	—
γ-lactone	29–30	+55 → +34	H_2O	143
δ-lactone	84	+99 → +42	H_2O	147
phenylhydrazide	145	—	—	143
Methyl 2,3,6-trimethyl-α-D-glucopyranoside	—	+149	CH_3OH	41
4-azobenzoate	90	+25 (λ, 6438)	$CHCl_3$	118
4-(3,5-dinitrobenzoate)	147	+56.3 (λ, 6252)	Me_2CO	41
Methyl 2,3,6-trimethyl-β-D-glucopyranoside	58–60	−48	$CHCl_3$	15, 68, 123, 125
4-benzenesulfonate	83–84	−35.6	$CHCl_3$	68
4-azobenzoate	95–96	−66 (λ, 6252)	Me_2CO	41, 118
Methyl 2,3,6-trimethyl-β-D-glucofuranoside	—	—	—	—
5-benzoate	55–56	−104.2	CH_3OH	148
5-tosylate	51–52	−62.1	CH_3OH	148

saccharic acid when the sugar was treated at 80°C. with nitric acid (density, 1.2) revealed that the third ether grouping was located on C6.[126,130] Although it was reported, in 1918, that the stability of the trimethylglucose towards permanganate characterized it as a member of the "normal" series,[127,129] it was not then realized that this implied the presence of a pyranose ring. More reliable evidence on this point has since been furnished by the conversion of the compound into the crystalline dextrorotatory 2,3,4,6-tetramethyl-D-glucopyranose.[127,133,140,141] The absence of an ether group on C4 of the trimethylglucose was verified when it was demonstrated that oxidation with bromine water yielded a γ-lactone, characterized by its rate of mutarotation in water and by its conversion to 2,3,5,6-tetramethyl-γ-D-gluconolactone,[141,142,143] and that treatment at room temperature with hydrogen chloride in methyl alcohol gave a methyl β-D-glucofuranoside, which could be transformed by methylation and hydrolysis into the sirupy levorotatory 2,3,5,6-tetramethyl-D-glucofuranose.[141]

A consideration of the individual carbon atoms of the sugar chain thus leads to the conclusion that this trimethylglucose is the 2,3,6-isomer and ample confirmation of this conclusion has been forthcoming. To mention only two cases: (a) oxidative degradation yields L-*threo*-dimethoxysuccinic acid,[143] and (b) the sugar does not condense with acetone.[130]

4. *2,4,6-Trimethyl-D-Glucose*

Methods of Formation.—Prominent among methods for the synthesis of 2,4,6-trimethyl-D-glucose are those based upon partial methylation of glucose by treatment with a deficiency of the methylating agents. Two factors seem to be responsible for the isolation of this compound in preference to its isomers, namely, the greater activity of positions 2,4 and 6 and the ease of crystallization of the α-form of the trimethyl sugar. It was by methods of this type that Haworth,[149] and Haworth and Sedgwick,[97] prepared the first samples of 2,4,6-trimethyl-D-glucose from sucrose and

(140) F. Micheel and K. Hess, *Ann.*, **449**, 146 (1926).
(141) J. C. Irvine and R. P. McGlynn, *J. Am. Chem. Soc.*, **54**, 356 (1932).
(142) W. Charlton, W. N. Haworth and S. Peat, *J. Chem. Soc.*, 89 (1926).
(143) H. C. Carrington, W. N. Haworth and E. L. Hirst, *J. Am. Chem. Soc.*, **55**, 1084 (1933).
(144) F. Micheel and K. Hess, *Ber.*, **60B**, 1898 (1927).
(145) K. Hess and F. Micheel, *Ann.*, **466**, 100 (1928).
(146) M. L. Wolfrom and L. W. Georges, *J. Am. Chem. Soc.*, **59**, 601 (1937).
(147) W. N. Haworth, S. Peat and J. Whetstone, *J. Chem. Soc.*, 1975 (1938).
(148) K. Hess and K. E. Heumann, *Ber.*, **72B**, 149 (1939).
(149) W. N. Haworth, *J. Chem. Soc.*, **117**, 199 (1920).

glucose, respectively, methyl sulfate and alkali being employed as methylating agents in each case. More recently Richtmyer[150] has used a similar technique to prepare 2,4,6-trimethyl-D-glucose from benzyl β- and phenyl β-D-glucopyranosides. When methyl α-D-glucopyranoside is treated, in aqueous solution, with thallous hydroxide there is precipitated a trithallium derivative, which with methyl iodide gives rise to four isomeric methyl trimethyl glucosides, the 2,4,6-isomer being in preponderance and being recoverable after acid hydrolysis as the crystalline reducing sugar.[151] A large excess of thallous hydroxide leads to complete methylation of the sugar.[151]

A prerequisite of the stepwise synthesis of 2,4,6-trimethyl-D-glucose is the isolation of a 3-substituted D-glucopyranose (the protective group being non-migratory), which will give rise on methylation to a 3-substituted methyl 2,4,6-trimethyl-αβ-D-glucopyranoside. Three protecting groups, namely benzyl,[152] nitrate[86] and tosyl,[18] have been used for this purpose. In the first two cases, the 3-substituent was removed, subsequent to methylation, by alkaline reduction and the glucosidic group by acid hydrolysis, thus yielding 2,4,6-trimethyl-D-glucose, but in the last case, the crystalline methyl 3-tosyl-2,4,6-trimethyl-α-D-glucopyranoside proved to be resistant to detosylation.

In 1938, Lake and Peat[153] reported that methyl 2,3-anhydro-4,6-dimethyl-β-D-mannopyranoside (prepared from methyl 2-tosyl-3,4,6-triacetyl-β-D-glucopyranoside by detosylation with sodium methoxide and subsequent methylation)[154] gave rise, when boiled with sodium methoxide in methyl alcohol, to almost equal proportions of methyl 2,4,6-trimethyl-β-D-glucopyranoside and methyl 3,4,6-trimethyl-β-D-altropyranoside. The former, after its separation by fractional distillation, yielded 2,4,6-trimethyl-D-glucose with hot mineral acid.

Polysaccharides from which 2,4,6-trimethyl-D-glucose can be prepared by methylation and hydrolysis include a polyglucose isolated from the cell wall of brewer's yeast,[155] laminarin[156] and the "hemicellulose" fraction of Iceland moss.[139]

Proof of Constitution.—Since 2,4,6-trimethyl-α-D-glucose is similar, both in its melting point and equilibrium rotation, to its more common 2,3,6-isomer, caution should be exercised in its identification. Fortunately, the two methyl β-D-glucosides have quite different properties

(150) N. K. Richtmyer, *J. Am. Chem. Soc.*, **61**, 1831 (1939).
(151) C. C. Barker, E. L. Hirst and J. K. N. Jones, *J. Chem. Soc.*, 1695 (1938).
(152) K. Freudenberg and E. Plankenhorn, *Ann.*, **536**, 257 (1938).
(153) W. H. G. Lake and S. Peat, *J. Chem. Soc.*, 1417 (1938).
(154) W. N. Haworth, E. L. Hirst and L. Panizzon, *J. Chem. Soc.*, 154 (1934).
(155) L. Zechmeister and G. Tóth, *Biochem. Z.*, **270**, 309 (1934).
(156) V. C. Barry, *Sci. Proc. Roy. Dublin Soc.*, **22**, 59 (1939).

TABLE XVII

The Physical Properties of 2,4,6-Trimethyl-D-glucose and Some of Its Derivatives

Compound	Melting point, °C.	$[\alpha]_D$	Rotation solvent	References
2,4,6-Trimethyl-α-D-glucose	123–126	$+111 \rightarrow +70$	CH_3OH	97, 150, 151, 152, 155
anilide	162–166	-113	CH_3OH	139
3-benzyl-	127–128	$+54.6$	$CHCl_3$	152
1,3-di-azobenzoate	115–120	$+190$ (λ, 6252)	Me_2CO	41, 118
2,4,6-Trimethyl-D-gluconamide	98–100	$+37.0$	$CHCl_3$	151, 153
Methyl 2,4,6-trimethyl-α-D-glucopyranoside	—	—	—	—
3-tosylate	123–124	$+53.6$	$CHCl_3$	18
Methyl 2,4,6-trimethyl-β-D-glucopyranoside	70–71	-27.4	$CHCl_3$	15, 86, 153
3-tosylate	104	$+2, -47$	$CHCl_3$	18, 152
Phenyl 2,4,6-trimethyl-β-D-glucopyranoside	108–109	-57.5	$CHCl_3$	150
Benzyl 2,4,6-trimethyl-β-D-glucopyranoside	94–95	-49.1	$CHCl_3$	150

and thus afford a convenient method of distinction. Consequently, the evidence given below on the proof of the constitution of 2,4,6-trimethyl-D-glucose has been taken only from those papers in which the compound has been shown to give a methyl β-D-glucoside having the melting point 70–71°C.

The early conclusion that this crystalline trimethyl-D-glucose was substituted in position 2, which was based on the observation that it did not form an osazone,[97] has been confirmed by the inability of the amide derived from it to show the Weerman test for an α-hydroxy amide[139,151,153] and by its synthesis from methyl 2-methyl-β-D-glucopyranoside.[18] The synthesis of the same trimethyl-D-glucose from 3-benzyl-D-glucose (see above), which had itself been prepared from the well-known 1,2:5,6-diisopropylidene-D-glucofuranose, afforded proof of the absence of a methoxyl group on C3.[152] The failure of the trimethylglucose to form either a furanoside when treated with cold dilute methyl alcoholic hydrogen chloride,[151,152] or a γ-lactone when oxidized with bromine water,[151] indicated that the second methoxyl group was located on C4.

Since the trimethylglucose has been shown to form a δ-lactone, characterized by its rapid mutarotation in aqueous solution,[151,153] and to give rise when methylated and hydrolyzed to 2,3,4,6-tetramethyl-D-glucopyranose,[139] it follows that C5 is unsubstituted. While it is clear that the third methoxyl group must be present on C6, little direct evidence is available to support this view. The syntheses of the trimethylglucose from methyl 2,3-anhydro-β-D-mannopyranoside (known to condense with benzaldehyde and therefore to carry free hydroxyl groups on C4 and C6)[153] and also from 3-benzyl-D-glucose[152] leave, however, no doubt that C6 is methylated.

5. *2,5,6-Trimethyl-D-Glucose*

The 2,5,6-trimethyl derivative of D-glucofuranose has not yet been prepared.

6. *3,4,6-Trimethyl-D-Glucose*

Methods of Formation.—In 1945, Sundberg, McCloskey, Rees and Coleman[28] were able to isolate 3,4,6-trimethyl-D-glucopyranose (which had hitherto been known only as a sirup) in the crystalline state, following a new method of synthesis based on methyl 3-methyl-4,6-benzylidene-β-D-glucopyranoside. The 2-benzyl derivative of this compound, prepared by heating with benzyl chloride in the presence of powdered potassium hydroxide, was subjected to mild acid hydrolysis such that the glucosidic group remained intact while the benzylidene residue was removed, thus liberating the alcohol groups on C4 and C6, these being subsequently etherified by treatment with methyl sulfate in the presence of alkali. Catalytic hydrogenation of the resulting crystalline methyl 2-benzyl-3,4,6-trimethyl-β-D-glucopyranoside gave the known crystalline methyl 3,4,6-trimethyl-β-D-glucopyranoside, from which the sirupy reducing sugar was generated by acid hydrolysis. Thick plates of the β-form were deposited at room temperature from a solution of the sirup in isopropyl ether and needles of the α-form from the mother liquor when this was cooled to 0°C.

Prior to this synthesis, 3,4,6-trimethyl-D-glucose (or its methyl β-D-glucoside) had been prepared by four different methods, each of which involved the shielding of C2 by the presence of either a double bond or an anhydro-ring. In 1922, Cramer and Cox[157] had shown that this particular trimethyl sugar was produced by the methylation and hydrolysis of the α-glucosan (1,2-anhydro-D-glucopyranose) of Pictet and Castan.[1] The same α-glucosan was involved indirectly in the synthesis effected by Haworth, Hirst and Panizzon[154] inasmuch as these

(157) M. Cramer and E. H. Cox, *Helv. Chim. Acta*, **5**, 884 (1922).

workers started from the methyl 3,4,6-triacetyl-β-ᴅ-glucopyranoside of Brigl,[158] a compound which is readily obtainable from α-glucosan. Haworth shielded C2 of the triacetylglucoside by tosylation, deacetylated and then methylated positions 3, 4, and 6. The most difficult stage of this method was the detosylation, which was finally effected with sodium methoxide after both alcoholic ammonia and aqueous potassium hydroxide had been used without success. Another anhydro-sugar, namely methyl 2,3-anhydro-4,6-dimethyl-β-ᴅ-alloside, has been shown by Peat and Wiggins[32] to give rise on treatment with sodium methoxide to a mixture of methyl 3,4,6-trimethyl-β-ᴅ-glucopyranoside and methyl 2,4,6-trimethyl-β-ᴅ-altropyranoside.

Hirst and Woolvin[159] preferentially methylated positions 3, 4 and 6 by the use of glucal, an unsaturated sugar derivative having a 1,2-olefinic bond, introduced by the reduction of triacetyl-α-ᴅ-glucopyranosyl bromide with zinc dust and acetic acid. By the action of an ethereal solution of perbenzoic acid on an aqueous solution of 3,4,6-trimethylglucal a sirupy product was obtained. This was shown to contain 3,4,6-trimethyl-ᴅ-glucopyranose, but, contrary to expectations, evidence of the presence of the isomeric 3,4,6-trimethyl-ᴅ-mannopyranose could not be found.

It has been reported by Granichstädten and Percival[139] that 3,4,6-trimethyl-ᴅ-glucose is one of the components of the mixture of methylated monosaccharides produced by methylation and hydrolysis of the "hemicellulose" fraction of Iceland moss.

Proof of Constitution.—Since the crystalline α- and β-forms of 3,4,6-trimethyl-ᴅ-glucose, isolated by Sundberg and coworkers,[28] were each obtained by hydrolysis of methyl 3,4,6-trimethyl-β-ᴅ-glucopyranoside,[32,154] the proof of the constitution of this glucoside obviously played an important part in the characterization of the trimethylglucose itself. Unfortunately, there is no such crystalline reference compound to correlate the earliest samples of sirupy 3,4,6-trimethyl-ᴅ-glucose[157,158] with the crystalline specimens isolated by Sundberg and coworkers.

It was shown by Haworth, Hirst and Panizzon[154] that hydrolysis and oxidation of the methyl trimethyl-β-ᴅ-glucoside gave a trimethylgluconolactone, which exhibited a rate of hydrolysis in aqueous solution similar to that of the δ-lactones of the glucose series, thus indicating the absence of a methoxyl group on C5. Furthermore, the amide derived from this lactone was degraded by the Weerman reagents (which incidentally proved that C2 was unsubstituted) to 2,3,5-trimethyl-ᴅ-arabofuranose, a sirup which was identified by its subsequent conversion to the known

(158) P. Brigl, *Z. physiol. Chem.*, **122**, 245 (1922).
(159) E. L. Hirst and C. S. Woolvin, *J. Chem. Soc.*, 1131 (1931).

crystalline 2,3,5-trimethyl-D-arabonolactone. The synthesis of the same methyl trimethyl-β-D-glucoside from methyl 2,3-anhydro-β-D-allopyranoside served to confirm that it was the 3,4,6-isomer.[32]

Even without this additional evidence drawn from earlier results, the synthesis of the crystalline forms of the trimethylglucose itself affords sufficient evidence to establish the constitution.[28] The parent compound was methyl 3-methyl-4,6-benzylidene-β-D-glucopyranoside and so it was evident that one of the methyl groups in the product was present on C3. The other two had clearly replaced the benzylidene residue (see above), which is known to favor the six-membered ring form bridging C4 and C6. Finally, the methyl β-D-glucopyranoside structure remained intact until the methylation stage had been completed, thus ensuring that C5 was not etherified.

TABLE XVIII

The Physical Properties of 3,4,6-Trimethyl-D-glucose and Some of Its Derivatives

Compound	Melting point, °C.	$[\alpha]_D$	Rotation solvent	References
3,4,6-Trimethyl-D-glucose				
α-form	76–77	+91.9 → +77.4	H_2O	28
β-form	97–98	+41.1 → +77.5	H_2O	28
phenylosazone	163–164	—	—	157
α-1,2-di-azobenzoate	162–164	342 (λ, 6438)	$CHCl_3$	118
3,4,6-Trimethyl-D-gluconolactone	liquid	+87 → +15	H_2O	154
phenylhydrazide	126	—	—	154
Methyl 3,4,6-trimethyl-α-D-glucopyranoside	—	—	—	—
2-azobenzoate	91	153 (λ, 6438)	$CHCl_3$	118
Methyl 3,4,6-trimethyl-β-D-glucopyranoside	51–52	−20, −16	$CHCl_3$	28, 32, 154
2-azobenzoate	113–114	67 (λ, 6438)	$CHCl_3$	118
2-tosylate	67	−16	$CHCl_3$	154
2-benzyl-	41–42	+9.9	$CHCl_3$	28

7. *3,5,6-Trimethyl-D-Glucose*

Methods of Formation.—Only one method has been used for the synthesis of 3,5,6-trimethyl-D-glucose, which has not yet been prepared in crystalline form. The method, published in 1913 by Irvine and Scott,[20] entails methylation of 1,2-isopropylidene-D-glucofuranose and hydrolysis with 0.5% hydrogen chloride in aqueous alcohol of the resultant 1,2-isopropylidene-3,5,6-trimethyl-D-glucofuranose. Modifications introduced

Table XIX
The Physical Properties of 3,5,6-Trimethyl-D-glucose and Some of Its Derivatives

Compound	Melting point, °C.	$[\alpha]_D$	Rotation solvent	References
3,5,6-Trimethyl-D-				
glucose	liquid	diverse	—	see text.
phenylosazone	70–72	—	—	23
1,2-isopropylidene-	liquid	−27.1, −29.5	C_2H_5OH	23, 48, 50, 105
1,2-trichloro-				
ethylidene-	113–114	−28	$CHCl_3$	161, 162
1,2-dichloro-				
ethylidene-	68	—	—	161
3,5,6-Trimethyl-D-				
gluconic acid	liquid	−6.3 → +5.4	$H_2O-C_2H_5OH$	62
sodium salt	—	+24.0	$H_2O-C_2H_5OH$	62
γ-lactone	44–45	+51.8 → +14.1	H_2O	147
amide	144	+34.0	H_2O	147
Methyl 3,5,6-trimethyl- α-D-glucofuranoside	liquid	+93	CH_3OH	165
Methyl 3,5,6-trimethyl- β-D-glucofuranoside	liquid	−87	CH_3OH	165

in the light of subsequent experiments include improvements in the isolation of 1,2-isopropylidene-D-glucofuranose by hydrolysis of the 1,2:5,6-diisopropylidene compound[160,161] and in the methylation stage, the latter being more efficiently accomplished with methyl sulfate and aqueous alkali than with the methyl iodide and silver oxide originally employed.[23,62,105,162]

Although 1,2-trichloroethylidene-3,5,6-trimethyl-D-glucofuranose has been prepared by methylation of 1,2-trichloroethylidene-D-glucofuranose,[161,162] there is no record of the chloral residue having been removed to generate the reducing trimethylglucose.

Proof of Constitution.—Irvine and his school[20,160,163] were not slow to recognize that the new trimethylglucose, prepared by them from isopropylidene-D-glucose, a compound not then characterized, differed from the normal series of glucose derivatives in the extreme ease of its oxida-

(160) J. C. Irvine and J. L. A. Macdonald, *J. Chem. Soc.*, **107**, 1701 (1915).
(161) H. W. Coles, L. D. Goodhue and R. M. Hixon, *J. Am. Chem. Soc.*, **51**, 519 (1929).
(162) W. Freudenberg and A. M. Vajda, *J. Am. Chem. Soc.*, **59**, 1955 (1937).
(163) J. C. Irvine and Jocelyn Patterson, *J. Chem. Soc.*, **121**, 2146 (1922).

tion with Fehling's solution and with neutral permanganate and in its very low equilibrium rotation. Their suggestion that it belonged to the anomalous class of "γ"-sugars has since been verified by its further methylation to methyl 2,3,5,6-tetramethyl-$\alpha\beta$-D-glucofuranoside[164] and thence, by hydrolysis, to 2,3,5,6-tetramethyl-D-glucofuranose,[23,48] which was identified, after oxidation to an aldonic acid, by means of the crystalline lactone and phenylhydrazide.[23] Since the trimethylglucose forms an osazone[23] and gives rise to a trimethyl-D-gluconamide, which is degraded by Weerman's reagent to a trimethyl-D-arabinose,[147] C2 must be unsubstituted. The only trimethyl-D-glucofuranose which fulfils these requirements is the 3,5,6-isomer.

V. Tetramethyl Ethers

1. *2,3,4,6-Tetramethyl-D-Glucose*

Methods of Formation.—Since reducing sugars are oxidized by silver oxide, with the formation of acid products,[166,167] silver oxide and methyl iodide can be used as methylating agents only when the reducing group is suitably protected. It was for this reason that Purdie and Irvine,[166] in their classical researches on the methyl ethers of D-glucose, chose to methylate Fischer's crystalline methyl α-D-glucoside (now known to be the pyranoside). On account of the insolubility of the glucoside in methyl iodide, it was necessary to use methyl alcohol as a solvent during the initial stages of the methylation; thereafter the product was soluble in methyl iodide. The sirupy methyl tetramethyl-α-D-glucoside was converted, by acid hydrolysis, into reducing, crystalline 2,3,4,6-tetramethyl-D-glucose. In spite of the fact that a large proportion of the methylating agents was wasted in the conversion of methyl alcohol into dimethyl ether, this method, which was extended to the β-series in 1905,[168] remained for more than a decade the only one for the preparation of tetramethylglucose. In 1915, Haworth[103] reported that methyl α-D-glucopyranoside can be methylated more conveniently and less expensively with methyl sulfate in the presence of 30% aqueous sodium hydroxide solution. Moreover, this method can be applied directly to reducing sugars, if alkaline conditions are avoided, by the use of excess methyl sulfate, until the sensitive reducing group has been methylated.

2,3,4,6-Tetramethyl-D-glucose is frequently isolated after hydrolysis of the methyl ethers of naturally-occurring glucose polymers (*e.g.* maltose,

(164) P. A. Levene and G. M. Meyer, *J. Biol. Chem.*, **70**, 343 (1926).
(165) P. A. Levene and G. M. Meyer, *J. Biol. Chem.*, **74**, 701 (1927).
(166) T. Purdie and J. C. Irvine, *J. Chem. Soc.*, **83**, 1021 (1903).
(167) T. Purdie and J. C. Irvine, *J. Chem. Soc.*, **85**, 1049 (1904).
(168) J. C. Irvine and A. Cameron, *J. Chem. Soc.*, **87**, 900 (1905).

cellobiose, gentiobiose, starch, cellulose and dextran). It arises from non-reducing terminal glucose units in the saccharides and its quantitative isolation forms the basis of the well-known "end group assay" method for the determination of chain-length.[169]

Proof of Constitution.—The determination of the positions of the methoxy groups in crystalline tetramethyl-D-glucose and the elucidation of the nature of the oxide ring in Fischer's methyl α-D-glucoside were essentially the same problem. The presence of a heterocyclic system in glucose, and in sugars in general, was originally postulated by Tollens to explain the asymmetry of the reducing carbon atom, as revealed by the formation of methyl α- and β-glucosides and by the phenomenon of mutarotation. Mainly on the basis of Baeyer's strain theory, Fischer formulated the sugars as five-membered ring structures (butylene oxide type), but there was no evidence that the conditions of strain in such a cyclic system would be analogous to those in cyclopentane. When the extensive researches of Hudson, which culminated in his famous "lactone rule," revealed with a high degree of probability that lactones derived from sugars by oxidation were γ-lactones (butylene oxide type), it was widely believed that Fischer's views had been confirmed. The possibility that oxidation of an unsubstituted sugar to a lactone, via an open-chain acid, might involve a change in the size of the ring was not regarded very seriously.

Such was the position until, in 1923, it was pointed out by Hirst and Purves[170] that recent work in another field[171,172] had shown that δ-hydroxy-aldehydes and -ketones react as ring structures (amylene oxide type). They considered that the properties of the sugars eliminated any possibility that they might contain ethylene or propylene oxide rings, but believed that it was still necessary to distinguish between butylene, amylene and hexylene oxide systems in the hexose series, and between butylene and amylene oxide structures in the pentose series. The oxidation of "normal" trimethylxylose with hot nitric acid (density, 1.20) to i-*xylo*-trimethoxyglutaric acid conclusively proved that, in this sugar at least, there was a six membered heterocyclic system of the amylene oxide type.[170] By similar methods, the same ring structure was shown to be present in the methyl ethers of the "normal" forms of arabinose[173] and rhamnose.[174]

(169) *Cf.* W. N. Haworth and E. L. Hirst, *Trans. Faraday Soc.*, **29**, 14 (1933).
(170) E. L. Hirst and C. B. Purves, *J. Chem. Soc.*, **123**, 1352 (1923).
(171) B. Helferich and T. Malkomes, *Ber.*, **55B**, 702 (1922).
(172) M. Bergmann and A. Miekeley, *Ber.*, **55B**, 1390 (1922).
(173) E. L. Hirst and G. J. Robertson, *J. Chem. Soc.*, **127**, 358 (1925).
(174) E. L. Hirst and A. K. Macbeth, *J. Chem. Soc.*, 22 (1926).

It was reported, in 1926, by Charlton, Haworth and Peat[142] that the lactones derived from a series of fully methylated sugars (and therefore possessing the same ring systems as the sugars) could be divided into two groups, according to their rates of rotational change in aqueous solution. The lactone from crystalline tetramethyl-D-glucose fell into the same group as those from the "normal" forms of trimethylxylose and trimethylarabinose. Since the latter sugars had already been shown to possess a pyranose (amylene oxide) structure, it seemed probable that crystalline tetramethylglucose was similarly constituted. Confirmation of this conclusion was obtained when the tetramethylglucose[175] (and later its derived lactone)[176] was shown to give rise, on oxidation with nitric acid (density, 1.42), to a mixture of L-*threo*-dimethoxysuccinic acid and i-*xylo*-trimethoxyglutaric acid, which were isolated as their crystalline amides. A butylene oxide (furanose) structure was thus no longer a possibility. Although this evidence did not eliminate a hexylene oxide ring, there had already been abundant proof (*e.g.*, uronic acid formation) that such a system was not present in methyl α-glucoside and hence in crystalline tetramethyl-glucose, which was therefore a pyranose sugar, namely 2,3,4,6-tetramethyl-D-glucopyranose.

TABLE XX

The Physical Properties of 2,3,4,6-Tetramethyl-D-glucose and Some of Its Derivatives

Compound	Melting point, °C.	$[\alpha]_D$	Rotation solvent	References
2,3,4,6-Tetramethyl-α-D-glucose	96	+92 → +84	H_2O	109
anilide	137–138	ca. +230	Me_2CO	177–180
p-toluidide	144	+157 → +53	CH_3OH	181
β-azobenzoate	125–126	−36 (λ, 6438)	$CHCl_3$	41, 118
2,3,4,6-Tetramethyl-D-gluconic acid	—	—	—	—
δ-lactone	liquid	+99 → +31	H_2O	142, 176, 182
phenylhydrazide	115	+42.1	C_2H_5OH	142, 176
amide	68–70	+60.4	Me_2CO	182, 183
Methyl 2,3,4,6-tetramethyl-α-D-glucopyranoside	liquid	+144	Me_2CO	167, 184
Methyl 2,3,4,6-tetramethyl-β-D-glucopyranoside	40–42	−17.4	C_2H_5OH	129, 167, 168

(175) E. L. Hirst, *J. Chem. Soc.*, 350 (1926).
(176) W. N. Haworth, E. L. Hirst and E. J. Miller, *J. Chem. Soc.*, 2436 (1927).
(177) J. C. Irvine and Agnes M. Moodie, *J. Chem. Soc.*, **93**, 95 (1908).

2. *2,3,5,6-Tetramethyl*-D-*Glucose*

Methods of Formation.—Since the direct methylation of D-glucose invariably leads to the formation of the crystalline pyranose form of the tetramethyl ether, it is essential that, in the preparation of 2,3,5,6-tetramethyl-D-glucofuranose, the ring system should be stabilized in the furanose form prior to methylation. The most convenient starting material is Fischer's methyl "γ"-D-glucoside[185] (now known to be a mixture of methyl α- and β-D-glucofuranoside), which is prepared by treatment of D-glucose with cold methyl alcoholic hydrogen chloride. When this glucoside mixture reacts with silver oxide and methyl iodide, there is produced a sirupy methyl tetramethyl-αβ-D-glucoside, which can be hydrolyzed very readily with aqueous acid to levorotatory 2,3,5,6-tetramethyl-D-glucofuranose.[186] Unlike its 2,3,4,6-isomer, this tetramethyl sugar has not yet been crystallized. It has frequently been prepared by methylation, followed by hydrolysis, of the methyl furanosides of partially methylated sugars, including 2,3,6- and 3,5,6-trimethyl-D-glucose.[187,188]

Proof of Constitution.—As has already been stated, it was recognized in 1926 that the fully methylated sugar lactones could be divided into two groups on the basis of their rates of hydration in aqueous solution.[142] One group, namely those most readily hydrated, was known to consist of δ-lactones, since it contained the lactones derived from the "normal" forms of trimethylxylose and trimethylarabinose, whose constitutions had already been established by methylation and oxidative degradation.[170,173] The more stable lactones of the other group were believed to be γ-lactones, and it was to this group that the lactone obtained by oxidation of the sirupy form of tetramethyl-D-glucose belonged. Since there was no possibility of a change in ring structure taking place during the oxidation of the methylated sugar to the lactone, it was deemed probable that sirupy tetramethyl-D-glucose was a butylene oxide (furanose) sugar. The synthesis, in the same year, of the sirupy tetramethyl-

(178) J. C. Irvine and R. Gilmour, *J. Chem. Soc.*, **93**, 1429 (1908).
(179) H. Pringsheim and A. Steingroever, *Ber.*, **59B**, 1001 (1926).
(180) M. L. Wolfrom and W. L. Lewis, *J. Am. Chem. Soc.*, **50**, 837 (1928).
(181) J. C. Irvine and A. Hynd, *J. Chem. Soc.*, **99**, 161 (1911).
(182) R. W. Humphreys, J. Pryde and E. T. Waters, *J. Chem. Soc.*, 1298 (1931).
(183) J. C. Irvine and J. Pryde, *J. Chem. Soc.*, **125**, 1045 (1924).
(184) J. C. Irvine and Agnes M. Moodie, *J. Chem. Soc.*, **89**, 1578 (1906).
(185) E. Fischer, *Ber.*, **47B**, 1980 (1914).
(186) J. C. Irvine, A.W. Fyfe and T. P. Hogg, *J. Chem. Soc.*, **107**, 524 (1915).
(187) F. Micheel and K. Hess, *Ann.*, **450**, 21 (1926).
(188) H. H. Schlubach and H. v. Bomhard, *Ber.*, **59B**, 845 (1926).

D-glucose and its lactone from 2,3,6-trimethyl-D-glucose provided conclusive proof that the ring system was not of the hexylene oxide type.[142,188] The final evidence necessary to characterize the tetramethylglucose in question as a furanose derivative was provided by Haworth, Hirst and Miller,[176] who demonstrated that oxidation of the tetramethylglucose with bromine water and of the resulting lactone with nitric acid yielded dimethoxysuccinic acid and oxalic acid, but not i-*xylo*-trimethoxyglutaric acid, the absence of which ruled out a pyranose structure.

TABLE XXI

The Physical Properties of 2,3,5,6-Tetramethyl-D-glucose and Some of Its Derivatives

Compound	Melting point, °C.	$[\alpha]_D$	Rotation solvent	References
2,3,5,6-Tetramethyl-D-glucose	liquid	−7.2, −11.1	H_2O	186–189
2,3,5,6-Tetramethyl-D-gluconic acid.	—	—	—	—
γ-lactone	26–27	+63 → +41	H_2O	23, 176, 182, 190
phenylhydrazide	136	—	—	23, 142, 143
amide	91	+39.2	H_2O	182
Methyl 2,3,5,6-tetramethyl-α-D-glucofuranoside	11	+106.5	CH_3OH	189
Methyl 2,3,5,6-tetramethyl-β-D-glucofuranoside	liquid	−72.7	CH_3OH	148

(189) W. N. Haworth, C. R. Porter and A. C. Waine, *J. Chem. Soc.*, 2254 (1932).

(190) H. D. K. Drew, E. H. Goodyear and W. N. Haworth, *J. Chem. Soc.*, 1237 (1927).

ANHYDRIDES OF THE PENTITOLS AND HEXITOLS

By L. F. Wiggins

The Imperial College of Tropical Agriculture, Trinidad, British West Indies

Contents

I. Introduction	191
II. Anhydrides of Hexitols	192
1. 1,4-Anhydro-D-mannitol (Mannitan)	192
2. 1,4- and 3,6-Anhydro-D-sorbitol	194
3. Styracitol and Polygalitol	198
4. Anhydrodulcitols	203
5. The 2,5-Anhydrohexitols	204
6. The 1,4:3,6-Dianhydrides of D-Mannitol, D-Sorbitol and L-Iditol	206
7. Neomannide and β-Mannide	217
8. Ethylene Oxide Derivatives of Hexahydric Alcohols	218
III. Anhydrides of Pentitols	220
1. 1,4-Anhydro- and 1,4:2,5-Dianhydroxylitol	220
2. 1,5-Anhydroxylitol, -D-Arabitol and -Ribitol	221
IV. Some Uses of Hexitol Anhydrides	222
V. Tables of Properties of Pentitol and Hexitol Anhydrides and Their Derivatives	225
Table III. Pentitol Anhydrides and Their Derivatives	225
Table IV. Hexitol Monoanhydrides and Their Derivatives	226
Table V. Hexitol Dianhydrides and Their Derivatives	227

I. Introduction

Although the polyhydric alcohols derived from sugars have been the subject of numerous investigations over many years, the inner ethers or anhydrides of these substances have been for the most part neglected, despite the fact that many of them were isolated and some description of their properties recorded before the advent of the twentieth century. In recent years, however, the anhydrides of polyhydric alcohols derived from sugars have received marked attention in several countries. Some indeed have achieved importance in chemical industry. Most interest has been centered on the anhydrides of pentitols and hexitols and this article will be restricted to a discussion of the chemistry of these substances.

Anhydro ring formation in the sugar series takes place when certain derivatives are subjected to alkaline hydrolysis. These derivatives contain groups, the hydrolysis of which leads to the transitory formation of a carbonium cation; for example, halogen, nitric or sulfonic acid esters. This same method is applicable to analogous derivatives of the sugar alcohols. Owing to the greater stability of the alcohols as compared with the sugars and sugar derivatives, a wider variety of methods of ring formation are applicable in this series. For example, many sugar alcohol anhydrides can be obtained by acid-catalyzed dehydration by heat. Isosorbide, isomannide, mannitan and sorbitan are obtained in this way. The principles of ring scission of anhydro compounds which have been elucidated during the past decade and which have been summarized by Peat,[1] also hold good for the anhydro glycitol series.

The formation of anhydro-glycitols and the investigation of their structure and that of the products of their ring scission provide fascinating problems in structural organic chemistry and this structural aspect of the chemistry of anhydrides of polyhydric alcohols forms the mainstay of this article.

II. Anhydrides of Hexitols

1. *1,4-Anhydro-D-mannitol (Mannitan)*

Crystalline mannitan was probably the first authentic monoanhydro-hexitol to be described. Bouchardat[2] prepared it by heating D-mannitol with hydrochloric acid under pressure and described it as having melting point 137° and $[\alpha]_D - 23.5°$. Both Bouchardat and Vignon[3] prepared also a non-crystalline mannitan, the latter author by heating D-mannitol with a quarter of its weight of water at a high temperature and under pressure. This non-crystalline anhydride was doubtless essentially the same as the mannitan of melting point 137°, but was contaminated with other anhydrides and with D-mannitol itself. The crystalline substance had the empirical formula $C_6H_{12}O_5$, and its structure was later shown to be that of 1,4-anhydro-D-mannitol (I) by van Romburgh and van der Burg.[4] This was accomplished through its conversion, by dehydration with formic acid, into the highly unsaturated and optically active 2-vinyldihydrofuran (II). This, on hydrogenation, gave racemic 2-ethyltetrahydrofuran (III) identical with that produced by the dehydration of Wohlgemuth's 1,4-hexanediol (IV). Although this series of transformations offered a good indication that mannitan was indeed 1,4-anhydro-D-mannitol, the proof was not rigid because (*a*) it did not demonstrate that the anhydride retained the D-mannitol configuration,

(1) S. Peat, *Advances in Carbohydrate Chem.*, **2**, 37 (1946).
(2) G. Bouchardat, *Ann. chim. et phys.*, [5] **6**, 100 (1875).
(3) L. Vignon, *Ann. chim. et phys.*, [5] **2**, 466 (1874).
(4) P. van Romburgh and J. H. N. van der Burg, *Proc. Acad. Sci. Amsterdam*, **25**, 335 (1922).

and (b) it did not preclude ring shift under the conditions of formation of the vinyldihydrofuran. At a much later date, however, Valentin[5] synthesized 3,6-anhydro-D-mannitol from methyl 3,6-anhydro-α-D-mannopyranoside and found it to be identical with mannitan. (Because of the symmetry of the mannitol molecule 3,6-anhydro- and 1,4-anhydro-D-mannitol are the same.) Further evidence for the allocation of the ring to C1 and C4 comes from the work of Hockett and coworkers[6] on the rate of oxidation of mannitan with lead tetraacetate. The anhydride rapidly consumes one and slowly consumes a second atomic proportion of oxygen. This is the behavior expected of a 1,4-anhydride of D-mannitol.

```
      CH₂─┐                                                    CH₂OH
      |   |                                                    |
    HOCH  |                                                    CH₂
      |   O        CH══CH          CH₂──CH₂                    |
    HOCH  |        |   \H          |    \H                     CH₂
      |   |        |    \          |     \                     |
     HC───┘        CH₂   C─CH══CH₂ CH₂    C─CH₂─CH₃            CHOH
      |             \   /           \    /                     |
     HCOH            \ /             \  /                      CH₂
      |               O               O                        |
     CH₂OH                                                     CH₃

       I              II                III                     IV
```

1,4-Anhydro-D-mannitol is also obtained by a novel route from 1,6-dibenzoyl-D-mannitol. When this compound is heated with p-toluenesulfonic acid in acetylene tetrachloride, it suffers partial dehydration and one of the products is a dibenzoyl derivative of mannitan. On debenzoylation of this substance it yields crystalline 1,4-anhydro-D-mannitol.[6,7] The formation of dibenzoyl-1,4-anhydro-D-mannitol has involved the migration of at least one benzoyl group and Hockett and coworkers[6] believe that the initial product is either 2,6- or 3,6-dibenzoyl-1,4-anhydro-D-mannitol.

The same anhydride of mannitol has been obtained recently by the deamination of 1-amino-1-desoxy-D-mannitol.[8] It is interesting to note that this method of formation of anhydro rings is of fairly general application. Thus 1,4-anhydro-D-sorbitol is obtained from 1-amino-1-desoxy-D-sorbitol. Similarly, methyl 2,3-anhydro-4,6-benzylidene-α-D-mannopyranoside and methyl 2,3-anhydro-4,6-benzylidene-α-D-allopyranoside[9]

(5) F. Valentin, *Collection Czechoslov. Chem. Communs.*, **8**, 35 (1936).
(6) R. C. Hockett, H. G. Fletcher, Jr., Elizabeth Sheffield, R. M. Goepp, Jr. and S. Soltzberg, *J. Am. Chem. Soc.*, **68**, 930 (1946).
(7) P. Brigl and H. Grüner, *Ber.*, **66**, 1945 (1933); **67**, 1582 (1934).
(8) V. G. Bashford and L. F. Wiggins, *Nature*, **165**, 566 (1950).
(9) L. F. Wiggins, *Nature*, **157**, 300 (1946).

are obtained by the deamination of methyl 3-amino- and 2-amino-4,6-benzylidene-α-D-altropyranoside, respectively.

2. *1,4- and 3,6-Anhydro-D-sorbitol*

Crystalline 1,4-anhydro-D-sorbitol is of much more recent origin than 1,4-anhydro-D-mannitol, inasmuch as its preparation was first recorded in 1946.[10] It was obtained by the restricted dehydration of D-sorbitol and has been given the trivial name arlitan. Two groups of workers simultaneously effected proof of its constitution. Hockett and coworkers[11] treated the anhydride with lead tetraacetate and since its rate of oxidation coincided with that of ethyl D-galactofuranoside and since one molecular proportion of formaldehyde was formed, the authors concluded that the ring must involve C1 and C4 of a hexitol chain. Assuming that no other carbon atoms are involved, sorbitan is either 1,4-anhydro-D-sorbitol or 1,4-anhydro-D-dulcitol (Walden inversion at C4 of the sorbitol molecule). They therefore synthesized 3,6-anhydro-D-dulcitol (enantiomorphous with 1,4-anhydro-L-dulcitol) but found it to be different from arlitan.

Soltzberg, Goepp and Freudenberg[10] proved the constitution of arlitan by a synthetic method. They obtained a liquid tetramethyl derivative of arlitan, characterized it by its boiling point, density and refractive index, and compared it with 2,3,5,6-tetramethyl-1,4-anhydro-D-sorbitol (V) synthesized by the following method. Tetramethyl-D-glucofuranose

```
      CH2────┐           CHOH    ┐           CH2OH
      |      |           |       |           |
      HCOCH3 |           HCOCH3  |           HCOCH3
      |      |O          |       |O          |
      CH3OCH |           CH3OCH  |           CH3OCH
      |      |           |       |           |
      HC─────┘           HC──────┘           HCOH
      |                  |                   |
      HCOCH3             HCOCH3              HCOCH3
      |                  |                   |
      CH2OCH3            CH2OCH3             CH2OCH3
         V                  VI                  VII
```

(VI) was hydrogenated to the corresponding sorbitol derivative (VII) in which the only groups unsubstituted were those situated at C1 and C4. Acid-catalyzed ring closure gave a product identical with the tetramethyl derivative of arlitan.

A further proof of the structure of arlitan has been given by Bashford

(10) S. Soltzberg, R. M. Goepp, Jr. and W. Freudenberg, *J. Am. Chem. Soc.*, **68**, 919 (1946).

(11) R. C. Hockett, Maryalice Conley, M. Yusem and R. I. Mason, *J. Am. Chem. Soc.*, **68**, 922 (1946).

and Wiggins.[12] Isosorbide has been shown to be 1,4:3,6-dianhydro-D-sorbitol (VIII), (see page 212). Therefore, if arlitan were treated with reagents having no tendency to change the original ring form and it were found that they convert it into 1,4:3,6-dianhydro-D-sorbitol, the transformation would provide additional proof that arlitan is 1,4-anhydro-D-sorbitol or 3,6-anhydro-D-sorbitol. This has been accomplished in the following way.

Arlitan was treated with p-toluenesulfonyl chloride, followed by acetic anhydride, in pyridine solution and the tosyltriacetylanhydro-D-sorbitol so obtained was then treated with sodium methoxide. The product of the reaction was 1,4:3,6-dianhydro-D-sorbitol. Accordingly,

```
     CH₂─┐         CH₂─┐         CH₂─┐         CH₂─┐
      |  │          |  │          |  │          |  │
     HCOH │        HCOH │        HCOH │        HCOH │
      |   │O        |   │O       |    │O        |   │O
   ┌─CH   │      ┌─CH   │        HOCH │        HOCH │
   │  |   │      │  |   │          |  │          |  │
   │  HC──┘      │  HC──┘          HC─┘         HC──┘
  O  |         O  |                 |         ┌─CH
   │ HCOH        │ HOCH            HCOTs      O |
   │  |          │  |               |         └─CH₂
   └─CH₂         └─CH₂             CH₂OH
     VIII          IX                X            XI
```

arlitan must be either 1,4- or 3,6-anhydro-D-sorbitol and since it is not identical with the known 3,6-anhydro-D-sorbitol of Fischer and Zach[13] it must have the former structure.

A secondary product of some interest was also obtained from the reaction of sodium methoxide with the tosyltriacetylanhydro-D-sorbitol. This was 1,4:3,6-dianhydro-L-iditol (IX). One possible explanation for the occurrence of this substance is that although tosylation of 1,4-anhydro-D-sorbitol proceeds preferentially at the hydroxyl group attached to C6, secondary reaction may also occur at C5 with the formation of 5-tosyl-1,4-anhydro-D-sorbitol (X). This on alkaline hydrolysis would yield, through Walden inversion at C5, 1,4:5,6-dianhydro-L-iditol (XI). If the ethylene oxide ring then rearranged to the more stable tetrahydrofuran structure, 1,4:3,6-dianhydro-L-iditol (IX) would result. In any case the isolation of this substance adds weight to the evidence which has already accrued that arlitan is 1,4-anhydro-D-sorbitol.

Evidence that this explanation for the formation of the L-iditol anhydride is reasonable comes from the fact that when crystalline 6-chloro-6-desoxy-1,4-anhydro-D-sorbitol (XII) is treated with sodium methoxide no 1,4:3,6-dianhydro-L-iditol is isolated.[12]

(12) V. G. Bashford and L. F. Wiggins, *J. Chem. Soc.*, 299 (1948).
(13) E. Fischer and K. Zach, *Ber.*, **45**, 2068 (1912).

Three benzylidene derivatives of 1,4-anhydro-D-sorbitol have been described. Soltzberg, Goepp and Freudenberg[10] described two, namely, (a) m. p. 136–140°, $[\alpha]_D$ + 33.72° and (b), m. p. 121–122° (no $[\alpha]_D$ quoted) and Bashford and Wiggins[8] described the third, (c) m. p. 151°, $[\alpha]_D$ + 15.4°. The structures of (a) and (b) have not yet been determined, but the latter authors have shown that (c) is 3,5-benzylidene-1,4-anhydro-D-sorbitol (XIII). It might be mentioned that the similarity in the physical properties of (a) and (c) and the fact that (a) has a 4° range in its melting point indicate that they are possibly one and the same substance. The monobenzylidene derivative (c) gave rise to a crystalline p-toluenesulfonate which readily suffered exchange with sodium iodide in acetone, a reaction which, taking place in the hexitol series, strongly indicates that the p-toluenesulfonyl group is attached to a primary carbon

XII XIII XIV XV

atom. This being the case, there are only three possible structures for the monobenzylidene derivative of 1,4-anhydrosorbitol, namely, 2,3-, 2,5- and 3,5-benzylidene-1,4-anhydro-D-sorbitol, (XIV), (XV) and (XIII), respectively. The first two structures are unlikely on steric grounds but had the compound (c) the structure XIV it would have reacted with lead tetraacetate. This was not the case so XIV is eliminated as a possible structure. It had been observed, however, that the 6-chloro-6-desoxy-3,5-benzylidene-1,4-anhydro-D-sorbitol (XVI) derived from 1,4:3,6-dianhydro-D-sorbitol (see page 197) on chlorination with thionyl chloride gave, not the expected 2,6-dichloro-2,6-didesoxy-3,5-benzylidene-1,4-anhydro-D-sorbitol (XVII), but bis(6-chloro-6-desoxy-3,5-benzylidene-1,4-anhydro-D-sorbitol) 2,2'-sulfite (XVIII). Therefore if the unknown benzylidene-1,4-anhydro-D-sorbitol under discussion, after subjection to the same chlorination reaction, gave this same sulfite, then its structure would thereby follow. The benzylidene-1,4-anhydro-D-sorbitol did in fact afford this same sulfite, so that its benzylidene grouping must be attached to C3 and C5 and the compound is represented by XIII.

6-Chloro-6-desoxy-3,5-benzylidene-1,4-anhydro-D-sorbitol (XVI) possesses an unusual property in that the benzylidene group is removed by distillation in alkaline or neutral solution. From the neutral solution it was possible to isolate 6-chloro-6-desoxy-1,4-anhydro-D-sorbitol as its triacetyl derivative. At the same time diacetyl-1,4:3,6-dianhydro-D-sorbitol was also formed. This removal of an acetal residue from a carbohydrate derivative by means of alkali is of very rare occurrence. Indeed the author is unaware of another example.

The 3,6-anhydro-D-glucose (XIX) of Fischer and Zach[13] on reduction with sodium amalgam or with hydrogen in the presence of Raney nickel[14] gives 3,6-anhydro-D-sorbitol (XX), a method of synthesis which establishes the ring structure of the substance since that of 3,6-anhydro-D-glucose has been proved by oxidation of the 1,2-isopropylidene-D-glucofuranose derivative (XXI), to 1,2-isopropylidene-D-xyluronic acid[15] (XXII). No proof of the configuration of the substance that has been named 3,6-anhydro-D-glucose has been given and the presence of the D-glucose configuration has been tacitly assumed hitherto.[1] 3,6-Anhydro-D-sorbitol has now been converted into the same 1,4:3,6-dianhydro-D-sorbitol as that obtained from 1,4-anhydro-D-sorbitol. Since the latter

(14) R. Montgomery and L. F. Wiggins, *J. Chem. Soc.*, 390 (1946).
(15) H. Ohle and H. Erlbach, *Ber.*, **62**, 2758 (1929).

has been proved to have the D-sorbitol configuration this series of transformations offers proof that the D-glucose configuration is retained in 3,6-anhydro-D-glucose.

Other derivatives of 3,6-anhydro-D-sorbitol will be discussed under section 6.

3. *Styracitol and Polygalitol*

Styracitol is one of the few naturally occurring anhydrohexitols. It was first isolated by Asahina[16] in 1907 by extracting the husks of the fruit of *Styrax obassia* with 60% alcohol. It was recognized[17] as a crystalline compound of m. p. 155° and $[\alpha]_D - 56.47°$, forming crystalline tetrabenzoate, tetracarbanilate and tetraacetate derivatives. Asahina and Takimoto[18] in 1931 reported that it also formed two dibenzylidene derivatives (*a*) m. p. 163–5°, $[\alpha]_D - 148.7°$ and (*b*) m. p. 192–193°, $[\alpha]_D - 80.47°$, as well as a diacetone derivative, m. p. 96–97°, $[\alpha]_D - 115.2°$.

Polygalitol was first isolated by Chodat[19] who obtained it from *Polygala amara*. Later Picard[20] obtained the same compound from *Polygala vulgaris* and Shinoda, Sato and Sato[21] from *Polygala tenuifolia*. This compound had the formula $C_6H_{12}O_5$ and was isomeric with styracitol. It had m. p. 142–143° and formed a crystalline tetraacetate, m. p. 73–75°. On oxidation with hydrogen peroxide followed by treatment with phenylhydrazine it gave phenyl-D-glucosazone, which was also obtained in a similar way from styracitol.[21a]

The subject of the constitution of styracitol and polygalitol has been the object of much controversy since their discovery. To styracitol has been ascribed the structure of 1,5-anhydro-D-mannitol and 1,5-anhydro-D-sorbitol alternatively from time to time during the past twenty years and since it was early recognized that polygalitol was epimeric with styracitol[21,21a] the corresponding epimeric configuration was ascribed to it. In 1930, Zervas[22] synthesized styracitol from the tetraacetyl-2-hydroxy-D-glucal (XXIII) of Maurer and Mahn[23] by hydrogenation followed by hydrolysis of the acetyl groups. Zervas assumed at this time that styracitol was 1,5-anhydro-D-sorbitol (XXIV). Asahina and Takimoto[18] claimed that this was the case because their oxidation of the

(16) Y. Asahina, *Arch. Pharm.*, **245**, 325 (1907).
(17) Y. Asahina, *Arch. Pharm.*, **247**, 157 (1909).
(18) Y. Asahina and H. Takimoto, *Ber.*, **64**, 1803 (1931).
(19) R. Chodat, *Arch. sci. phys. et nat.*, [3] **20**, 599 (1888).
(20) P. Picard, *Bull. soc. chim. biol.*, **9**, 692 (1927).
(21) J. Shinoda, S. Sato and D. Sato, *Ber.*, **65**, 1219 (1932).
(21a) See also M. Bergmann and L. Zervas, *Ber.*, **64**, 2032 (1931).
(22) L. Zervas, *Ber.*, **63**, 1689 (1930).
(23) K. Maurer and H. Mahn, *Ber.*, **60**, 1316 (1927).

tetramethyl derivative gave dextrorotary dimethoxysuccinic acid (see later formula XXVIIIa); this could only be obtained if styracitol had the D-sorbitol configuration.

Freudenberg and Rogers[24] then offered evidence that styracitol was 1,5-anhydro-D-mannitol (XXV), which was based on the difference in

```
     CH────┐           CH₂────┐          CH₂────┐
     ‖     │           │      │          │      │
     COAc  │           HCOH   │          HOCH   │
     │     │           │      │O         │      │
   AcOCH   │O          HOCH   │          HOCH   │O
     │     │           │      │          │      │
     HCOAc │           HCOH   │          HCOH   │
     │     │           │      │          │      │
     HC────┘           HC─────┘          HC─────┘
     │                 │                 │
     CH₂OAc            CH₂OH             CH₂OH
     XXIII             XXIV              XXV
```

configuration on C2 and C3 in XXIV and XXV. Criegee[25] had shown that cis-glycols are oxidized more rapidly with lead tetraacetate than trans-glycols and therefore this fact if applied to styracitol should differentiate between XXIV and XXV. Freudenberg and Rogers found that styracitol was oxidized twice as rapidly by lead tetraacetate as polygalitol, and therefore they represented styracitol by XXV and polygalitol by XXIV. However, Freudenberg and Sheehan[26] in a later paper published contrary evidence. Methylation of styracitol gave a tetramethyl derivative identical in constants with that recorded by Asahina and Takimoto. Tetramethyl-D-glucopyranose was hydrogenated to the corresponding sorbitol derivative (XXVI), and this was

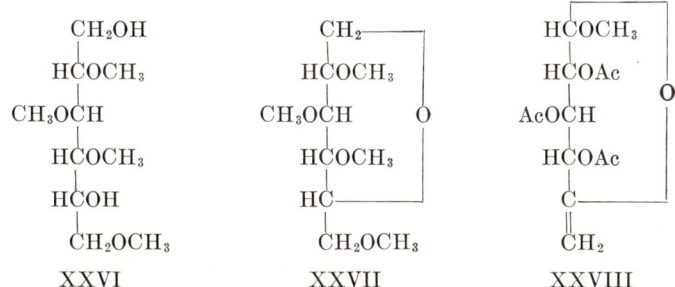

```
     CH₂OH              CH₂────┐          HCOCH₃─┐
     │                  │      │          │      │
     HCOCH₃             HCOCH₃ │          HCOAc  │
     │                  │      │          │      │O
   CH₃OCH               CH₃OCH │O       AcOCH    │
     │                  │      │          │      │
     HCOCH₃             HCOCH₃ │          HCOAc  │
     │                  │      │          │      │
     HCOH               HC─────┘          C──────┘
     │                  │                 ‖
     CH₂OCH₃            CH₂OCH₃           CH₂
     XXVI               XXVII             XXVIII
```

dehydrated by heating with sulfuric acid. The product on ultimate analysis was shown to be a tetramethyl-1,5-anhydrohexitol and since it had been derived by a process which appeared to offer no possibility of

(24) W. Freudenberg and E. F. Rogers, J. Am. Chem. Soc., **59**, 1602 (1937).
(25) R. Criegee, Ann., **495**, 211 (1932); **507**, 159 (1933).
(26) W. Freudenberg and J. T. Sheehan, J. Am. Chem. Soc., **62**, 558 (1940).

inversion of configuration, it was considered to be the D-sorbitol derivative XXVII. Table I shows that its constants agree closely with those of methylated styracitol, and hence they concluded that styracitol was 1,5-anhydro-D-sorbitol. However, when 2,3,4,6-tetramethyl-1,5-anhydro-D-mannitol was synthesized by a similar route, the product was not

TABLE I
Comparison of Methylated Derivatives (W. Freudenberg and Sheehan)[26]

	$[\alpha]_D$	n^{25}_D	Density
Tetramethylstyracitol	−36.5°	1.4520	1.0849
2,3,4,6-Tetramethyl-1,5-anhydro-D-sorbitol	−36.2	1.4518	1.0876
2,3,4,6-Tetramethyl-1,5-anhydro-D-mannitol	+30.6	1.4479	1.0435
Tetramethylpolygalitol	+67.6	1.4444	1.0571

identical with methylated polygalitol (see Table I). This indicated that configurational changes had occurred during the transformations and since earlier work[21,21a] had shown unequivocally that polygalitol is epimeric with styracitol, doubt is cast on the rigidity of this method of proof of the constitution of styracitol and polygalitol.

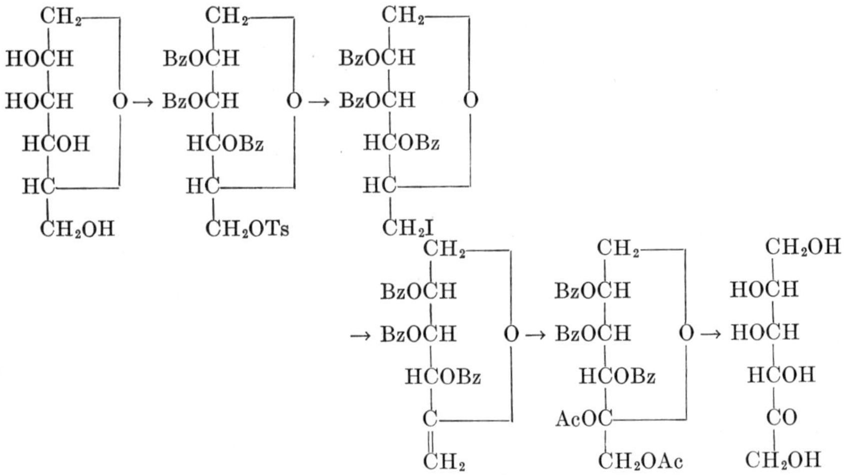

FIG. 1.—Transformation of styracitol to D-fructose.

An ingenious proof of constitution of styracitol has been devised by Zervas and Papadimitriou.[27] They prepared the 6-tosyl tribenzoyl derivative of styracitol and from it the 6-iodo-6-desoxy derivative, which on treatment with silver fluoride gave a compound possessing a double

(27) L. Zervas and Irene Papadimitriou, *Ber.*, **73**, 174 (1940).

bond between C5 and C6 analogous to the methyl 2,3,4-triacetyl-5,6-α-D-glucopyranoseenide (XXVIII) of Helferich and Himmen.[28] When this was oxidized with peracetic or perbenzoic acid and the product saponified and then treated with phenylhydrazine, phenyl-D-glucosazone was isolated. The identification of D-fructose (as phenyl-D-glucosazone) proves that styracitol is 1,5-anhydro-D-mannitol because had it been 1,5-anhydro-D-sorbitol, L-sorbose would have resulted after this series of transformations, which may therefore be represented by the sequence shown in Fig. 1.

Additional evidence was provided by Hockett, Dienes and Ramsden,[29] who compared the rates of oxidation with lead tetraacetate of styracitol and methyl α-D-mannopyranoside, and also of polygalitol and methyl

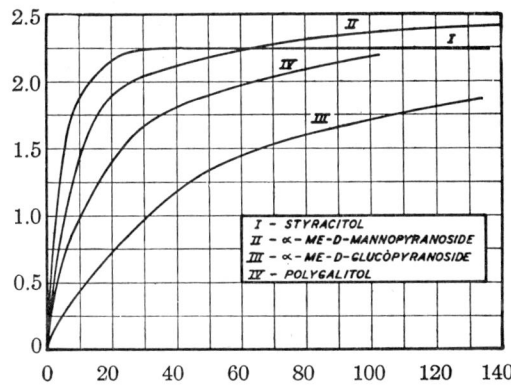

FIG. 2.—Oxidation of sugar derivatives by lead tetraacetate in dry acetic acid: 0.00025 mole substance in 100 cc.; ratio of Pb(OAc)$_4$ to substance, 15.2/1; water less than 0.05 %.

α-D-glucopyranoside. If styracitol is 1,5-anhydromannitol, it has the same *cis*-glycol group at C2, C3 as methyl α-D-mannopyranoside and its rate of oxidation should resemble that of the mannoside. Furthermore, if polygalitol is 1,5-anhydro-D-sorbitol it has a *trans*-glycol group at C2, C3 as has methyl α-D-glucopyranoside and its rate of oxidation should therefore resemble that of the latter substance. As shown in Fig. 2 the rate of oxidation of styracitol follows closely that of methyl α-D-mannopyranoside; that of polygalitol, however, lies between those of methyl α-D-mannopyranoside and methyl α-D-glucopyranoside, so that no safe conclusion may be drawn regarding it. Hockett and his coworkers also removed the objection to the acceptance of these structures which the

(28) B. Helferich and E. Himmen, *Ber.*, **62**, 2136 (1929).
(29) R. C. Hockett, Margaret Dienes and H. E. Ramsden, *J. Am. Chem. Soc.*, **65**, 1474 (1943).

work of Asahina[18] had presented. Asahina had oxidized tetramethylstyracitol with nitric acid and reported the formation of dextrorotary dimethoxy-succinic acid (XXVIIIa). This confers the D-sorbitol configuration on styracitol, as has been mentioned. Hockett and Conley[29a] repeated this work and isolated only oxalic acid and levorotary dimethoxy-succinic acid (XXVIIIb). This permits styracitol to have either the D-mannitol or D-sorbitol configuration.

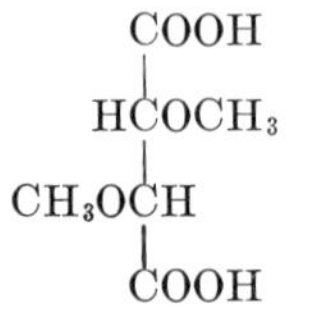

$$
\begin{array}{cc}
\text{COOH} & \text{COOH} \\
\text{HCOCH}_3 & \text{CH}_3\text{OCH} \\
\text{CH}_3\text{OCH} & \text{HCOCH}_3 \\
\text{COOH} & \text{COOH} \\
\text{XXVIIIa} & \text{XXVIIIb} \\
\text{L(+)-Dimethoxy-} & \text{D(−)-Dimethoxy-} \\
\text{succinic acid} & \text{succinic acid} \\
\text{(Dimethyl-L-threaric acid)} & \text{(Dimethyl-D-threaric acid)}
\end{array}
$$

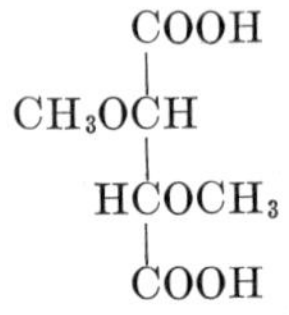

Further evidence that polygalitol and styracitol have an epimeric relationship has been provided by Richtmyer, Carr and Hudson,[30] who isolated both substances in the Zervas synthesis of styracitol. Moreover, they proved conclusively that polygalitol has the D-sorbitol configuration by synthesizing it from D-glucose, using reagents which affected only the reducing group. Acetobromo-D-glucose was converted into

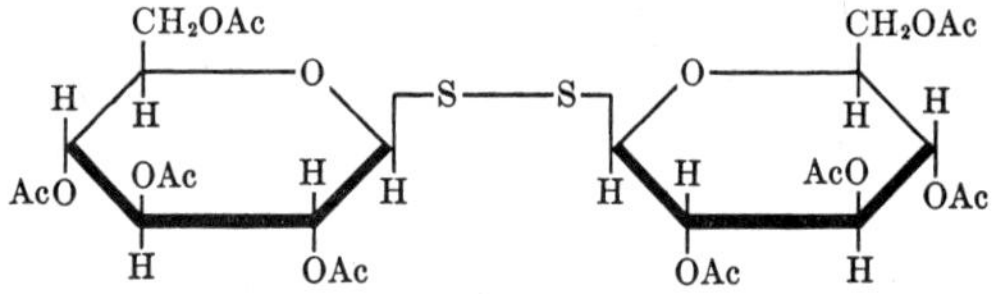

XXIX

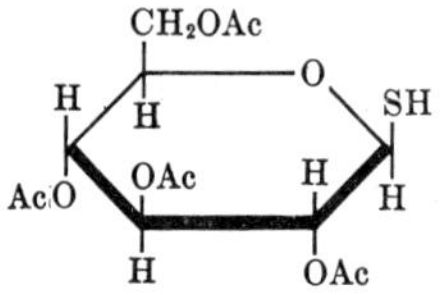

XXX 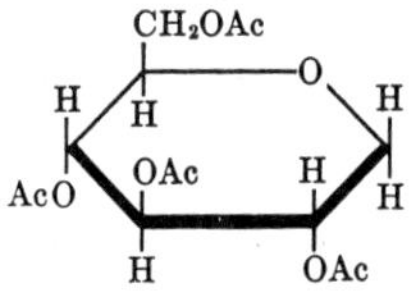 XXXI

octaacetyl-D-diglucopyranosyl disulfide (XXIX) and thence into tetraacetyl-D-glucothiose (XXX) which, on desulfurization with Raney nickel according to Bougault, Cattelain and Chabrier[31] gave a product identical with polygalitol tetraacetate (XXXI). The same compound was also

(29a) R. C. Hockett and Maryalice Conley, *J. Am. Chem. Soc.*, **66**, 464 (1944).

(30) N. K. Richtmyer, C. J. Carr and C. S. Hudson, *J. Am. Chem. Soc.*, **65**, 1477 (1943).

(31) J. Bougault, E. Cattelain and P. Chabrier, *Compt. rend.*, **208**, 657 (1939).

obtained on the desulfurization of XXIX and of ethyl tetraacetyl-D-glucopyranosyl xanthate.[31a] A similar unequivocal synthesis of styracitol was carried out subsequently by Fried and Walz[31b] through the reductive desulfurization of ethyl 1-thio-β-D-mannopyranoside tetraacetate, obtained in turn by acetylation of the reaction product from D-mannose, ethyl mercaptal and hydrochloric acid.

The aceritol of Perkin and Uyeda[32] was shown to be identical with polygalitol[24] (see also the later reference 103).

4. Anhydrodulcitols

Although several anhydrohexitols have been obtained from dulcitol or D-galactose, the constitution of few has been proved. Berthelot[33] obtained a liquid anhydride from dulcitol having the composition of a monoanhydrohexitol, but neither proof of its structure nor crystalline derivatives have been recorded. Maquenne[34] did obtain, however, crystalline bromo and chloro derivatives of an anhydrohexitol from

```
   CH₂─┐         CH₂─┐        C₁₀H₇·SCH ┐        CH₂OH
   |   |         |   |         |        |        |
   HCOH|         HOCH|          HCOH    |        HCOH
   |   |         |   |         |        |        |
   HOCH|  O      HOCH|  O       HOCH    |  O     ─CH ┐
   |   |         |   |         |        |        |   |
   HOCH|         HOCH|          HOCH    |        HOCH|
   |   |         |   |         |        |        |   | O
   HC──┘         HC──┘          HC──────┘        HCOH|
   |             |              |                 |   |
   CH₂OH         CH₂OH          CH₂OH            ─CH₂─┘
   XXXII         XXXIII         XXXIV             XXXV
```

dibromo and dichlorodulcitol but the constitution of these has not been determined so that it is not known whether they are derivatives of Berthelot's liquid anhydride or not. On the other hand Freudenberg and Rogers[24] obtained a 1,5-anhydrohexitol tetraacetate from tetraacetyl-2-hydroxy-D-galactal according to the method of Zervas.[22] This might be a derivative of 1,5-anhydro-D-dulcitol (XXXII) or of 1,5-anhydro-D-talitol (XXXIII). Recently, Fletcher and Hudson[35] carried out the reductive desulfurization of 2-naphthyl 1-thio-β-D-galactopyranoside tetraacetate (XXXIV) and obtained an anhydrohexitol tetraacetate

(31a) H. G. Fletcher, Jr., *J. Am. Chem. Soc.*, **69**, 706 (1947).
(31b) J. Fried and Doris E. Walz, *J. Am. Chem. Soc.*, **71**, 140 (1949).
(32) A. G. Perkin and Y. Uyeda, *J. Chem. Soc.*, **121**, 66 (1922).
(33) M. Berthelot, Chimie organique fondée sur la synthèse, vol. 2, Paris, p. 209 (1860).
(34) L. Maquenne, "Les Sucres," Carré & Naud, Paris, p. 125 (1900).
(35) H. G. Fletcher, Jr., and C. S. Hudson, *J. Am. Chem. Soc.*, **70**, 310 (1948).

which by its mode of preparation must be 1,5-anhydro-D-dulcitol tetraacetate. This had physical constants markedly different from those recorded by Freudenberg and Rogers[24] for the product of hydrogenation of tetraacetyl 2-hydroxy-D-galactal and this product is therefore probably 1,5-anhydro-D-talitol (XXXIII). That such is actually the case has been shown by Rosenfeld, Richtmyer and Hudson,[35a] who synthesized 1,5-anhydro-D-talitol by the reduction of 2,6-anhydro-D-altrose and found its physical constants to agree with those reported by Freudenberg and Rogers.

3,6-Anhydro-D-dulcitol (XXXV) has been prepared by a method which demonstrates its structure. Hockett and coworkers[11] prepared it from methyl 3,6-anhydro-α-D-galactopyranoside but both the anhydride itself and its tetraacetate were liquid. Carré[36] obtained another anhydro derivative of dulcitol by heating it with phosphoric acid. This also was a liquid but gave rise to a crystalline dibenzoate and a dicarbanilate and was a dianhydro derivative, but no constitutional studies have yet been carried out on this substance.

5. *The 2,5-Anhydrohexitols*

Brigl and Grüner[37] obtained several anhydrohexitol derivatives when they heated 1,6-dibenzoyl-D-mannitol with *p*-toluenesulfonic acid in acetylene tetrachloride. These include a dianhydrohexitol derivative (see section 6) and two monoanhydrohexitol derivatives, one of which was shown by Hockett and coworkers[38] to be 2,6 (or 3,6)-dibenzoyl-1,4-anhydro-D-mannitol. Brigl and Grüner had originally designated it as 1,6-dibenzoyl-2,4-anhydro-D-mannitol. The third product was also a dibenzoylanhydrohexitol. This was described as 1,6-dibenzoyl-2,5-anhydro-D-mannitol (XXXVI) by Brigl and Grüner.[37] Hockett, Zief and Goepp[39] have studied this compound and confirmed the presence of the 2,5-anhydro ring by reason of the behavior of the substance toward lead tetraacetate. The product of oxidation, however, was the inactive compound XXXVII, whereas if the 2,5-anhydride had possessed the D-mannitol configuration (XXXVI), a dissymmetric and therefore optically active product (XXXVIII) would have been obtained. This

(35a) D. A. Rosenfeld, N. K. Richtmyer and C. S. Hudson, *J. Am. Chem. Soc.*, **70**, 2201 (1948).

(36) P. Carré, *Compt. rend.*, **139**, 637 (1904).

(37) P. Brigl and H. Grüner, *Ber.*, **66**, 1945 (1933); **67**, 1582 (1934).

(38) R. C. Hockett, H. G. Fletcher, Jr., Elizabeth L. Sheffield, R. M. Goepp, Jr., and S. Soltzberg, *J. Am. Chem. Soc.*, **68**, 930 (1946).

(39) R. C. Hockett, M. Zief and R. M. Goepp, Jr., *J. Am. Chem. Soc.*, **68**, 935 (1946).

shows, provided that no racemization has occurred during the formation of the dialdehyde (XXXVII), that the 2,5-anhydrohexitol must be a D-sorbitol derivative and be represented by XXXIX. It is difficult to see how this 2,5-anhydro-D-sorbitol is obtained from D-mannitol. Its formation means that during the reaction optical inversion has occurred at C2. It may be that the *p*-toluenesulfonic acid present in the reaction mixture temporarily esterifies the hydroxyl at C2 and when it is

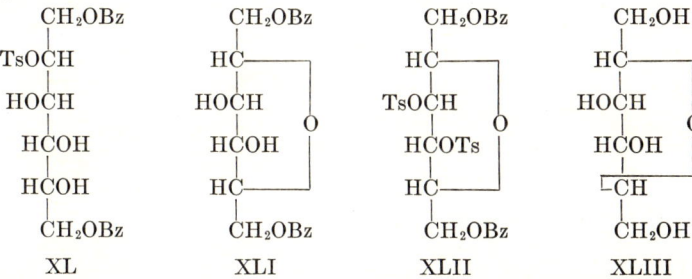

XXXVI XXXVII XXXVIII XXXIX

removed from the hexitol, concomitant Walden inversion and anhydro ring formation occur. It is interesting to note that tosylation of this 1,6-dibenzoyl-2,5-anhydro-D-sorbitol gave a ditosyl derivative identical with the anhydrohexitol derivative obtained by Müller and Vargha[40] by the tosylation of 1,6-dibenzoyl-D-mannitol. It may be that here the tosylation procedure has temporarily formed 1,6-dibenzoyl-2-tosyl-D-mannitol (XL) from which the tosyl group is split off with simultaneous

XL XLI XLII XLIII

Walden inversion and anhydro ring formation between C2 and C5 giving XLI, the formation of this being followed by complete tosylation to give XLII.

2,5-Anhydro-L-iditol (XLIII) has been obtained by Vargha[41] by a very interesting method from 1,6-ditosyl-2,4-benzylidene-D-sorbitol (XLIV). This compound, on treatment with one molecular proportion

(40) A. Müller and L. v. Vargha, *Ber.*, **66**, 1165 (1933).
(41) L. v. Vargha, *Ber.*, **68**, 1377 (1935).

of sodium methoxide, loses one tosyl group with the formation of 1-tosyl-2,4-benzylidene-5,6-anhydro-D-sorbitol (XLV). When this was hydrolyzed with 50% acetic acid it gave 1-tosyl-2,5-anhydro-L-iditol (XLVI) and its structure was demonstrated by Vargha and Puskás in 1943.[42] On treatment with one mole of tosyl chloride the compound afforded 1,6-ditosyl-2,5-anhydro-L-iditol (XLVII). The tosyloxy groups were shown to be attached to primary carbon atoms because they could readily be exchanged for iodine atoms by means of sodium iodide in acetone. The fact that the ring involved C2 and C5 was demonstrated by the fact that XLVII reacted with one mole of lead tetraacetate and gave rise to bis-(3-tosyl-L-glyceraldehyde) 2,2'-ether. In 1946, Vargha, Puskás and Nagy[43] obtained 2,5-anhydro-L-iditol itself by the sodium amalgam

CH_2OTs	CH_2OTs	CH_2OTs	CH_2OTs
HCO—	HCO—	HC—	HC—
HOCH CHPh	HOCH CHPh	HOCH O	HOC O
HCO—	HCO—	HCOH	HCOH
HCOH	HC	CH	CH
CH_2OTs	CH_2	CH_2OH	CH_2OTs.
XLIV	XLV	XLVI	XLVII

reduction of 1-tosyl-2,5-anhydro-L-iditol (XLVI). A liquid dianhydrohexitol of unknown constitution was also obtained. It is interesting to note, however, that the hydroxyl groups at C1 and C4 are thrown near enough together by the 2,5-anhydro ring in the iditol molecule for ring formation to take place between them, so that this dianhydrohexitol might well be 1,4:2,5-dianhydro-L-iditol.

6. *1,4:3,6-Dianhydrides of* D-*Mannitol,* D-*Sorbitol and* L-*Iditol*

Dianhydro-D-mannitol or isomannide was first prepared many years ago by Fauconnier[44] who obtained it by heating D-mannitol with hydrochloric acid. The dianhydrides of D-sorbitol and L-iditol were not prepared, however, until very recent times. Although isomannide is a crystalline substance forming well defined derivatives very little work has been carried out on it and its structure was not determined until 1945, but in that and the following years a flood of experiments having a

(42) L. v. Vargha and T. Puskás, *Ber.*, **76**, 859 (1943).

(43) L. v. Vargha, T. Puskás and E. Nagy, *Acta Bolyai*, **1**, 39 (1946); *J. Am. Chem. Soc.*, **70**, 261 (1948).

(44) A. Fauconnier, *Bull. soc. chim. France*, [2] **41**, 119 (1884).

bearing on the chemistry of this and allied substances appeared in the chemical literature.

Isomannide has the empirical formula $C_6H_{10}O_4$ and it forms crystalline dichloro, dibenzoyl, and ditosyl derivatives. Clearly it is a derivative of D-mannitol containing two anhydro rings and two free hydroxyl groups. The structural problem is to discover the position of these rings in the molecule. That the rings do involve C1 and C6 of the D-mannitol molecule is shown by the fact that isomannide suffers ring scission with hydrochloric acid, giving 1,6-dichloro-1,6-didesoxy-D-mannitol (XLVIII). Moreover, the reverse change can be effected, for on heating 1,6-dichloro-

1,6-didesoxy-D-mannitol under reduced pressure, hydrogen chloride is evolved and a high yield of isomannide is obtained. Isomannide cannot be 1,2:5,6-dianhydro-D-mannitol (XLIX) because the anhydro rings are stable to both hot sodium methoxide solution and methyl alcoholic ammonia, reagents which are known to cleave ethylene oxide rings with great ease. Isomannide also showed no reaction with lead tetra-acetate, therefore the two free hydroxyl groups which it contains cannot be vicinal. Taking these facts into account, and assuming that isomannide has the D-mannitol configuration, only three structures are possible, namely, 1,3:4,6- (L), 1,4:3,6- (LI) and 1,4:2,6-dianhydro-D-mannitol (LII). The first of these is sterically unlikely and furthermore the only authentic example of this type of structure in the carbohydrate series, namely, 3,5-anhydro-1,2-isopropylidene-D-xylofuranose (LIII) is unstable to hot sodium methoxide. Therefore it remains for distinction to be made between LI and LII. Isomannide forms a dichloro derivative and this on ring scission with hydrochloric acid gives rise to a tetrachloro-tetradesoxy-D-mannitol. Since isomannide itself gives 1,6-dichloro-1,6-didesoxy-D-mannitol when heated with hydrochloric acid, this tetrachloro derivative must be 1,x,y,6-tetrachloro-tetradesoxy-D-mannitol. On further chlorination with phosphorus pentachloride it affords the same hexachloro-hexadesoxy-D-mannitol as obtained from D-mannitol itself. Now if isomannide has structure LI it must give rise to 1,2,5,6-tetrachloro-tetradesoxy-D-mannitol (LIV), whereas if it has the structure LII it must give rise to 1,3,5,6-tetrachloro-tetradesoxy-D-mannitol (LV).

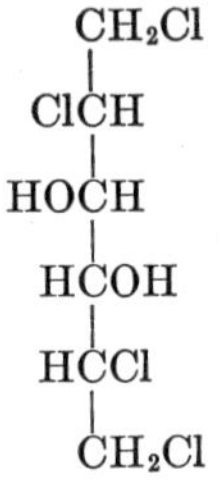

$$\begin{array}{cc}
CH_2Cl & CH_2Cl \\
ClCH & HOCH \\
HOCH & ClCH \\
HCOH & HCOH \\
HCCl & HCCl \\
CH_2Cl & CH_2Cl \\
LIV & LV
\end{array}$$

$$\begin{array}{ccc}
CH_2Cl & CH_2Cl & CH_2- \\
CH_3OCH & CH_3OCH & CH_3OCH \\
OCH & HOCH & -CH \\
HCO\!\!>\!\!CMe_2 & HCOH & HC- \\
HCOCH_3 & HCOCH_3 & HCOCH_3 \\
CH_2Cl & CH_2Cl & -CH_2 \\
LVI & LVII & LVIII
\end{array}$$

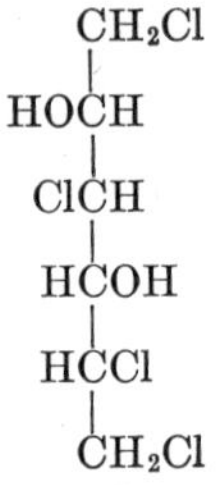

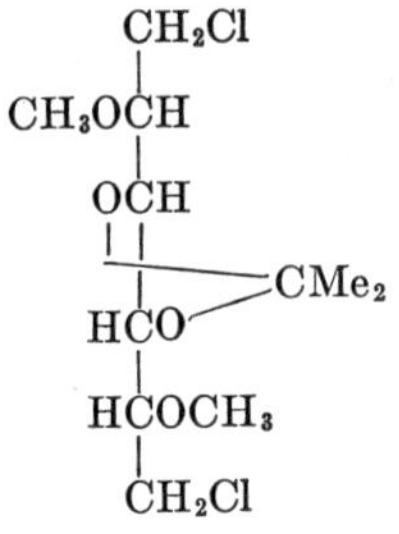

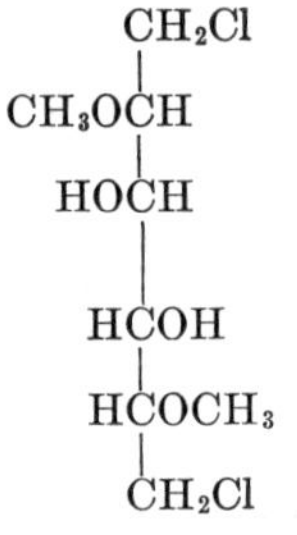

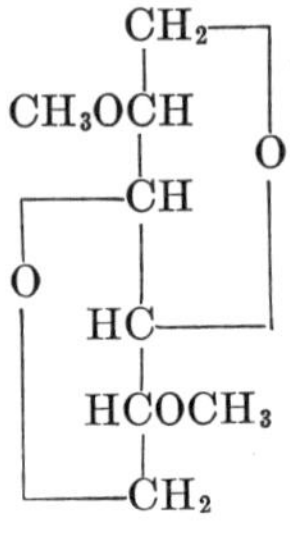

Wiggins[45] synthesized 1,2,5,6-tetrachloro-tetradesoxy-D-mannitol by an unequivocal route and found it to be identical with the tetrachloro-tetradesoxy-D-mannitol actually obtained from isomannide. Thus isomannide must be 1,4:3,6-dianhydro-D-mannitol (LI).

The dimethyl derivative of isomannide was synthesized by Wiggins as follows. 1,6-Dichloro-1,6-didesoxy-3,4-isopropylidene-D-mannitol was converted into 1,6-dichloro-1,6-didesoxy-2,5-dimethyl-3,4-isopropylidene-D-mannitol (LVI). This on mild acid hydrolysis gave 1,6-dichloro-1,6-didesoxy-2,5-dimethyl-D-mannitol (LVII) which with cold sodium methoxide furnished 2,5-dimethyl-1,4:3,6-dianhydro-D-mannitol (LVIII), identical with the product of methylation of isomannide itself. Further

(45) L. F. Wiggins, *J. Chem. Soc.*, 4 (1945).

evidence for the 1,4:3,6-allocation of the anhydro rings in isomannide was given by Hockett and Goepp and their coworkers.[46] They synthesized it by heating 1,4-anhydro-D-mannitol with sulfuric acid. Although the possibility of ring change exists in this procedure, it is strong evidence for the double tetrahydrofuran ring system in isomannide. They showed also that the hydroxyl groups in isomannide are secondary because of their slow rate of reaction with trityl chloride and likewise that of their ditosyl derivatives with sodium iodide. These workers also obtained isomannide dibenzoate by the method of Brigl and Grüner,[7] that is by heating 1,6-dibenzoyl-D-mannitol with p-toluenesulfonic acid in acetylene tetrachloride. This must involve migration of the benzoyl groups from the 1,6- to the 2,5-positions in the D-mannitol molecule. Thus, the dianhydro-D-mannitol dibenzoate first obtained by Brigl and Grüner is 1,4:3,6-dianhydro-D-mannitol 2,5-dibenzoate.

Montgomery and Wiggins[47] have studied the reaction of D-mannitol with hydrochloric acid in detail and the results throw some light on the mode of formation of isomannide from D-mannitol. When D-mannitol is heated under reflux with hydrochloric acid for several days, isomannide results in about 35–40% yield. But when D-mannitol is heated under pressure with fuming hydrochloric acid 1,6-dichloro-1,6-didesoxy-D-mannitol (40% yield) and very little isomannide are formed. On examination of the residues after separation of isomannide from the first reaction mentioned above, no fewer than two monoanhydrohexitol derivatives and three new dianhydrohexitol derivatives were encountered. The products isolated are summarized in Table II.

One of the dianhydrohexitols, B, was identified with 1,5:3,6-dianhydro-D-mannitol (neomannide, see below), but the other two, A and C, remain of unknown structure. An interesting observation concerning the dianhydride A has been made.[48] When 1,6-dichloro-1,6-didesoxy-D-mannitol was treated with sodium methoxide a new crystalline dianhydrohexitol was obtained. This had m. p. 118–119° but showed $[\alpha]_D + 33.6°$. It was therefore the enantiomorph of dianhydride A. Furthermore, when dichloro-didesoxy-D-mannitol is treated with sodium amalgam yet another dianhydride was isolated. This had m. p. 118–119° and $[\alpha]_D + 93.6°$ and is doubtless identical with the β-mannide of Siwoloboff.[49]

When D-mannitol was heated with fuming hydrochloric acid in a sealed tube and the main product, 1,6-dichloro-1,6-didesoxy-D-mannitol,

(46) R. C. Hockett, H. G. Fletcher, Jr., Elizabeth L. Sheffield, R. M. Goepp, Jr., and S. Soltzberg, *J. Am. Chem. Soc.*, **68**, 930 (1946).
(47) R. Montgomery and L. F. Wiggins, *J. Chem. Soc.*, 2204 (1948).
(48) L. F. Wiggins, *Nature 164*, 672 (1949).
(49) A. Siwoloboff, *Ann.*, **233**, 368 (1886).

Table II
Reaction Products of D-Mannitol and Boiling Hydrochloric Acid[47]

Monoanhydrides		Dianhydrides	
1,4-Anhydro-D-mannitol (Mannitan)	M. p. 145–147°	1,4:3,6-Dianhydro-D-mannitol (Isomannide)	M. p. 86–87°
Monochloro-desoxy-monoanhydrohexitol	M. p. 172–173°	—2,5-dimethanesulfonate	M. p. 104°
		Dianhydrohexitol (A)	M. p. 118–119° ($[\alpha]_D - 33.6°$)
		—dimethanesulfonate	M. p. 139–140°
		Dianhydrohexitol (B)	
		—dimethanesulfonate	M. p. 190–191°
		Dianhydrohexitol (C)	
		—dimethanesulfonate	M. p. 113–114°

removed, the residue was found to contain a little isomannide and in addition 2(5)-chloro-2(5)-desoxy-1,4:3,6-dianhydro-D-mannitol, two chlorinated anhydrohexitols, a monochloro derivative (probably 6-chloro-6-desoxy-1,4-anhydro-D-mannitol), a dichloro-didesoxy-anhydrohexitol and styracitol (1,5-anhydro-D-mannitol).

Summarizing, it appears that D-mannitol with boiling hydrochloric acid first suffers dehydration between C1 and C4 (because when the treatment is interrupted after only a few hours duration, 1,4-anhydro-D-mannitol is obtained), then between C3 and C6. The dehydration is probably achieved by the catalytic influence of the chloride ion. Dehydration is accompanied by the formation of smaller amounts of other ring compounds and by substitution of secondary hydroxyl groups by chlorine. With fuming hydrochloric acid, D-mannitol gives mainly 1,6-dichloro-1,6-didesoxy-D-mannitol and a little isomannide, although other ring products are also present and secondary substitution is more prolific. The following equilibrium may be regarded as representing the main reaction:

$$\text{HCl} + \text{Isomannide} \rightleftarrows \text{1,6-Dichloro-1,6-didesoxy-D-mannitol} + \text{H}_2\text{O}.$$

The presence of excess hydrogen chloride will favor the formation of 1,6-dichloro-1,6-didesoxy-D-mannitol whereas the elimination of hydrogen chloride by boiling a solution of D-mannitol with hydrochloric acid will favor the formation of isomannide. The limited yield of either substance is due to the formation of non-interconvertible products.

Overend, Montgomery and Wiggins[50] noted an interesting reaction between isomannide and phosphorus tribromide. None of the expected 2,5-dibromo-2,5-didesoxy-1,4:3,6-dianhydro-D-mannitol (LIX) was isolated; instead, ring scission took place and 1,6-dibromo-1,6-didesoxy-D-mannitol (LX) was formed. Since no such ring scission occurs with hydrobromic acid under similar conditions it was probable that the phosphorus tribromide itself had effected the ring breakdown by direct attack on the oxide rings.

```
   CH₂─┐              CH₂Br              CH₂OH              CH₂OH
   │   │              │                  │                  │
  BrCH  │            HOCH               HCOH               HCOCH₃
   │    O            │                  │                  │
  ┌─CH  │            HOCH            ┌─CH               ┌─CH
  │ │   │            │               │ │                │ │
  │ HC─┘             HCOH            │ HCOH             │ HCOH
 O │               O │              O │               O │
  │ HCBr             HCOH            │ HCOH             │ HCOCH₃
  │ │                │               │ │                │ │
  └─CH₂              CH₂Br           └─CH₂              └─CH₂

   LIX               LX               LXI                LXII


       CH₂─┐                       CH₂─┐
       │   │                       │   │
      HCOCH₃ │                    HCOH  │
       │    O                      │    O
     ┌─CH   │                    ┌─CH   │
     │ │    │                    │ │    │
     │ HC──┘                     │ HC──┘
    O │                         O │
     │ HCOCH₃                    │ HCOH
     │ │                         │ │
     └─CH₂                       └─CH₂

      LXIII                       LXIV
```

Sorbitol forms a crystalline dianhydride, namely isosorbide; this in contrast to isomannide was not described fully until recent times. Montgomery and Wiggins[51] fully described it in 1946 but other authors had referred to it previously. In 1927 Muller and Hoffman[52] described a liquid dianhydro-D-sorbitol which was probably an impure form of isosorbide. In 1940, Bell, Carr and Krantz[53] described experiments with a crystalline dianhydro-D-sorbitol. Montgomery and Wiggins[51]

(50) W. G. Overend, R. Montgomery and L. F. Wiggins, *J. Chem. Soc.*, 2201 (1948).
(51) R. Montgomery and L. F. Wiggins, *J. Chem. Soc.*, 390 (1946).
(52) J. Muller and U. Hoffman, U. S. Pat., 1,757,468 (1930).
(53) F. K. Bell, C. J. Carr and J. C. Krantz, Jr., *J. Phys. Chem.*, **44**, 862 (1940).

prepared isosorbide by boiling a mixture of hydrochloric acid and D-sorbitol in a similar manner to that used in the preparation of isomannide. In the D-sorbitol case, however, a much higher yield of dianhydride was isolated (70-75%). Isosorbide and isomannide are very similar in physical properties and general behavior. Both exhibit similar solubility, both are stable to alkaline hydrolytic reagents such as sodium methoxide and ammonia at fairly high temperatures and both are resistant to oxidation by lead tetraacetate, so that it was expected that they would possess similar structures. This was proved by Montgomery and Wiggins by the application of synthetical methods. In the first place isosorbide itself was synthesized from 3,6-anhydro-D-sorbitol (LXI) through the formation of the 1-tosyl derivative and its subsequent alkaline hydrolysis. This showed that at least one tetrahydrofuran ring was present. Hockett and coworkers[54] synthesized isosorbide by heating both 1,4-anhydro-D-sorbitol and 3,6-anhydro-D-sorbitol with sulfuric acid, thus showing the presence of both the 1,4- and the 3,6-anhydro rings in isosorbide. In this method, however, the possibility of acid-catalysed ring change is present though this possibility is minimized by the fact that similar treatment of 1,5-anhydro-D-sorbitol does not give isosorbide. A second synthetic approach devised by Montgomery and Wiggins provides more certain evidence. Haworth, Owen and Smith[55] prepared 2,5-dimethyl-3,6-anhydro-D-glucofuranose and this, on catalytic hydrogenation, gave 2,5-dimethyl-3,6-anhydro-D-sorbitol (LXII), a compound in which only the hydroxyl groups at C1 and C4 are free. The 1-tosyl derivative was prepared and the tosyl group removed by alkaline hydrolysis. The product was 2,5-dimethyl-1,4:3,6-dianhydro-D-sorbitol (LXIII) identical in all respects with the product of methylation of isosorbide. Therefore the latter must be represented by LXIV.

The rings in 1,4:3,6-dianhydro-D-sorbitol have different stabilities. Montgomery and Wiggins[56] treated the anhydride with hydrochloric acid under pressure and isolated some 1,6-dichloro-1,6-didesoxy-D-sorbitol and much 6-chloro-6-desoxy-1,4-anhydro-D-sorbitol, both isolated as their benzylidene derivatives. No 1-chloro-1-desoxy-3,6-anhydro-D-sorbitol was found. This experiment demonstrates that the 1,4-anhydro ring is more stable than the 3,6-ring in this substance. This is in agreement with the fact that partial dehydration of D-sorbitol in the 1,4-position can be preferentially effected and 1,4-anhydro-D-sorbitol isolated.

(54) R. C. Hockett, H. G. Fletcher, Jr., Elizabeth L. Sheffield and R. M. Goepp, Jr., *J. Am. Chem. Soc.*, **68**, 927 (1946).
(55) W. N. Haworth, L. N. Owen and F. Smith, *J. Chem. Soc.*, 88 (1941).
(56) R. Montgomery and L. F. Wiggins, *J. Chem. Soc.*, 237 (1948).

The simultaneous preparation of both isosorbide and isomannide from sucrose has been achieved.[57] This process entailed the hydrogenation of sucrose to a mixture of D-mannitol and D-sorbitol and the subjection of this mixture to dehydration in the presence of acid catalysts followed by fractional distillation.[58]

1,4:3,6-Dianhydro-L-iditol (isoidide)[59,60] has been prepared by methods similar to those for obtaining isomannide and isosorbide. It is formed, however, in much higher yield than either of these dianhydrides. Its structure as 1,4:3,6-dianhydro-L-iditol has been demonstrated by its formation from both 1,4:3,6-dianhydro-D-mannitol and 1,4:3,6-dianhydro-D-sorbitol. This transformation was accomplished by first dehydrogenating isomannide or isosorbide with Raney nickel to give a product which must contain the diketo structure LXV and then by hydrogenating the product. Dianhydro-L-iditol must therefore have structure (LXVI).

$$\begin{array}{cc}
\text{CH}_2\text{—} & \text{CH}_2\text{—} \\
| & | \\
\text{CO} & \text{HCOH} \\
| \quad\quad\text{O} & | \quad\quad\text{O} \\
\text{—CH} & \text{—CH} \\
| & | \\
\text{HC—} & \text{HC—} \\
\text{O} \quad | & \text{O} \quad | \\
\text{CO} & \text{HOCH} \\
| & | \\
\text{—CH}_2 & \text{—CH}_2 \\
\text{LXV} & \text{LXVI}
\end{array}$$

Wiggins[60] has obtained 1,4:3,6-dianhydro-D-iditol by a novel route which also proves its constitution. Considering D-mannitol (LXVII), it is readily perceived that if the configuration of the hydrogen and hydroxyl groups at C3 and C4 could each be reversed then D-iditol (LXVIII) would result. Furthermore, it is now recognized that when a tosyl group is removed under alkaline conditions from a sugar molecule containing a suitably placed hydroxyl group then Walden inversion and concomitant anhydro ring formation take place at the carbon atom originally carrying the tosyl group. If tosyl groups could be introduced into the D-mannitol molecule at C3 and C4 alkaline hydrolysis of such a ditosyl ester would give rise to Walden inversion at C3 and C4 and the formation of two tetrahydrofuran rings between C1 and C4, and C3 and C6, forming 1,4:3,6-dianhydro-D-iditol. This transformation was

(57) R. Montgomery and L. F. Wiggins, *J. Chem. Soc.*, 433 (1947).
(58) L. F. Wiggins, *Advances in Carbohydrate Chem.*, **4,** 299 (1949).
(59) H. G. Fletcher, Jr., and R. M. Goepp, Jr., *J. Am. Chem. Soc.*, **67,** 1042 (1945); **68,** 939 (1946).
(60) L. F. Wiggins, *J. Chem. Soc.*, 1403 (1947).

successfully accomplished through the 3,4-ditosyl-1,2:5,6-diisopropylidene-D-mannitol of Brigl and Grüner[61] (LXXIX).

No authentic 1,4:3,6-dianhydro derivatives of any other hexitols have yet been described (see however Table V).

CH$_2$OH | HOCH | HOCH | HCOH | HCOH | CH$_2$OH	CH$_2$OH | HOCH | HCOH | HOCH | HCOH | CH$_2$OH	Me$_2$C(O—CH$_2$, O—CH) | TsOCH | HCOTs | HCO—CMe$_2$—OCH$_2$
LXVII	LXVIII	LXIX

LXX (isomannide, bicyclic representation)

LXXI, LXXII, LXXIII (fused bis-tetrahydrofuran representations)

The ring system in these dianhydro hexitols is of interest and worthy of some discussion. The formula of isomannide (LXX) based on the Fischer projection formula for sugars does not convey the real character of the molecule and the author has chosen to write these substances as two fused tetrahydrofuran rings. Scale models show this to be a more exact representation. Thus isomannide is written as LXXI, isosorbide as LXXII, and L-isoidide as LXXIII.

Scale model studies show that with mannitol, sorbitol and iditol the

(61) P. Brigl and H. Grüner, *Ber.*, **67**, 1969 (1934).

fused tetrahydrofuran ring systems are easily formed, but when the series dulcitol, talitol and allitol is examined it is found that the closure of a second hydrofuran ring is much more difficult to effect. The photograph in Fig. 3 demonstrates these relations.

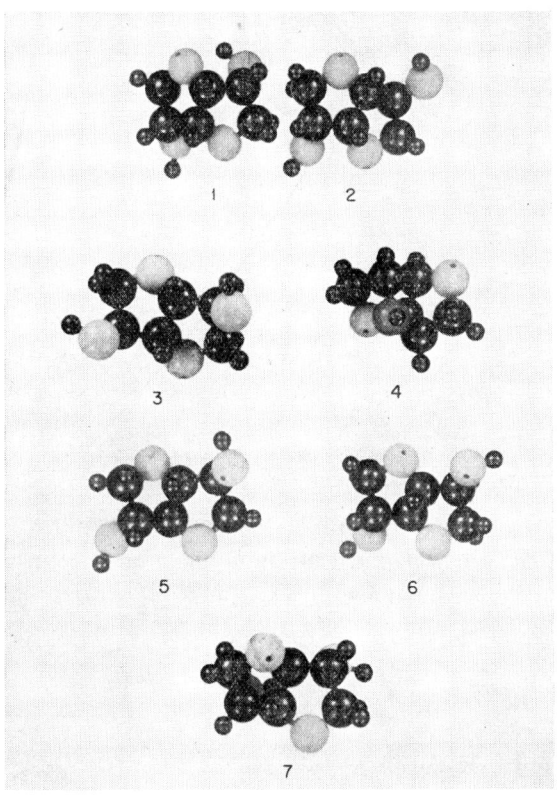

1. 1,4:3,6-Dianhydro-D-mannitol.
2. 1,4:3,6-Dianhydro-D-sorbitol (L-gulitol).
3. 1,4:3,6-Dianhydro-L-iditol.
4. 2,5 Imino-2,5-didesoxy-1,4:3,6-dianhydro-D-mannitol.
5. 1,4:3,6-Dianhydro-D-dulcitol.
6. 1,4:3,6-Dianhydro-D-allitol.
7. 1,4:3,6-Dianhydro-D-altritol (D-talitol).

FIG. 3.

The stereochemistry of the hexitols affects the manner of ring fusion, as is seen in Figs. 3 and 4. In mannitol, sorbitol (gulitol) and iditol, the 1,4:3,6-dianhydro rings are *cis*-oriented, whereas in the dulcitol, allitol and altritol (talitol) series the 1,4:3,6-rings are *trans*-oriented. This makes the ring system in this latter series almost planar although in the case of dianhydrodulcitol and allitol there is considerable strain in the molecules.

Some interesting facts have come to light from a study of the amino derivatives of isomannide, isosorbide and isoidide.[62] The diamines of isomannide (LXXIV) and isosorbide (LXXV) were made by Montgomery and Wiggins[63] by treating the corresponding ditosyl derivatives with

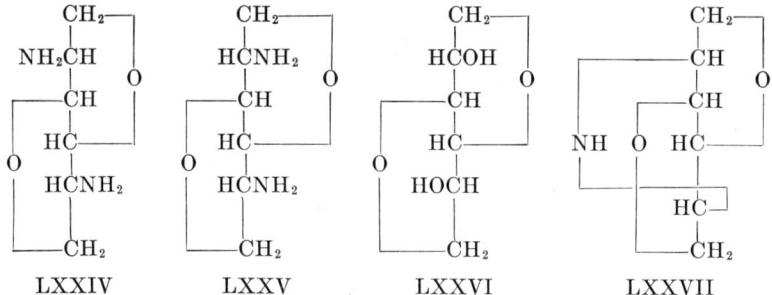

Fig. 4. (The rings in heavy type are in the plane of the paper)

methyl alcoholic ammonia. When however 2,5-ditosyl-1,4:3,6-dianhydro-L-iditol was treated with ammonia[62] no diamino derivative was isolated; instead a compound containing an imino group was obtained.

The structure of this imine presents an interesting problem. All the properties of the substance point to its being a derivative of a simple six-carbon compound containing a secondary amino group. A study of scale models shows, however, that it is quite impossible to form an imine bridge across C2 and C5 in either the dianhydrosorbitol or iditol series, but it is possible to form such an imine bridge in the case of 1,4:3,6-dianhydromannitol (see fig. 3). Since inversion of configuration on

(62) V. G. Bashford and L. F. Wiggins, *J. Chem. Soc.*, 371 (1950).
(63) R. Montgomery and L. F. Wiggins, *J. Chem. Soc.*, 393 (1946).

replacement of the tosyloxy groups by an imino group is possible, and in the absence of further evidence, the structure 2,5-imino-2,5-didesoxy-1,4:3,6-dianhydro-D-mannitol (LXXVII) has been tentatively assigned to this compound.

Since no other stereoisomer but that of iditol, mannitol or sorbitol can exist in the *cis*-fused ring series of 1,4:3,6-dianhydrohexitols and since amination experiments on the ditosyl derivatives of each isomer have given three different amino derivatives, it is reasonable to assume that the two diamines which have been isolated are of the same configuration as their parent dianhydrohexitols. These two diamines LXXIV and LXXV on treatment with nitrous acid suffered deamination as expected but instead of obtaining in one case isomannide and in the other isosorbide, only one dianhydrohexitol was isolated and that was dianhydro-L-iditol. The deamination must of course lead to the transitory carbonium cation LXXVIII and this on hydroxylation can take on any of the configurations, L-iditol, D-mannitol or D-sorbitol. That it preferentially takes on the configuration of L-iditol indicates that this is in some ways a more stable structure than the others. This behavior is paralleled

LXXVIII   LXXIX   LXXX   LXXXI

by the epimerization of 2,4:3,5-dimethylene-D-gluco- (LXXIX) and D-mannosaccharic acids[64] (LXXX) because, when the dimethyl esters of these acids are heated with barium hydroxide, both give rise to 2,4:3,5-dimethylene-L-idosaccharic acid (LXXXI).

7. *Neomannide and β-Mannide*

Two other dianhydrohexitols are known. One was prepared over sixty years ago by Siwoloboff[65] and was named β-mannide and the other, neomannide, was prepared by Hockett and Sheffield[66] in 1946. The

(64) W. N. Haworth, W. G. M. Jones, M. Stacey and L. F. Wiggins, *J. Chem. Soc.*, 61 (1944).
(65) A. Siwoloboff, *Ann.*, **233**, 368 (1886).
(66) R. C. Hockett and Elizabeth L. Sheffield, *J. Am. Chem. Soc.*, **68**, 937 (1946).

latter compound was synthesized from styracitol (1,5-anhydro-D-mannitol), by means of the Fischer-Zach reaction. This involved the formation of 6-tosyl-tribenzoyl-styracitol (LXXXII) and its alkaline hydrolysis, which led to a second ring closure in the mannitol molecule between C3 and C6, forming 1,5:3,6-dianhydro-D-mannitol (LXXXIII) (neomannide). This, written in the Haworth sugar formula may be represented by LXXXIV. Its structure, as shown by LXXXIII, is proved by the fact that neomannide does not react with lead tetraacetate. By its method of preparation neomannide could only be 1,5:4,6-, 1,5:2,6- or 1,5:3,6-dianhydro-D-mannitol and only the latter structure would show this inactivity towards the Criegee reagent.

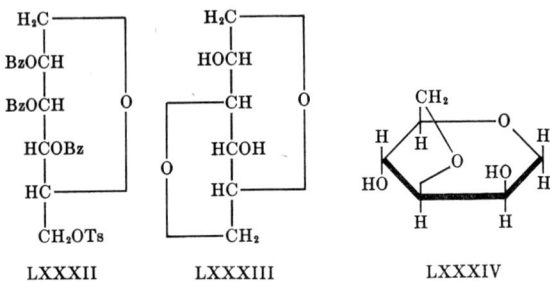

LXXXII   LXXXIII   LXXXIV

β-Mannide was obtained by treating 1,6-dichloro-1,6-didesoxy-D-mannitol with sodium amalgam. It was described as a crystalline compound of m. p. 118° and $[\alpha]_D + 94°$. The dianhydride obtained by Montgomery and Wiggins from the products of the action of hydrochloric acid on mannitol had precisely the same melting point as this, but the specific rotation was $[\alpha]_D - 33.6°$, so there can be no question of identity of the β-mannide with this material. A very interesting fact has come to light inasmuch as the action of sodium methoxide on 1,6-dichloro-1,6-didesoxy-D-mannitol gives a dianhydride of m. p. 118° and $[\alpha]_D + 33.6°$, which is obviously the enantiomorph of the material obtained amongst the products of the action of hydrochloric acid on D-mannitol. Recently a repetition of Siwoloboff's preparation of β-mannide has been achieved by the author and confirmation obtained of the constants ascribed to it in the older work. Thus, the interesting fact remains that two distinct dianhydrides have been obtained by the action of alkaline reagents on 1,6-dichloro-1,6-didesoxy-D-mannitol, but as yet no information about their structure is available.

8. *Ethylene Oxide Derivatives of Hexahydric Alcohols*

Ethylene oxide rings have as yet been introduced only at the ends of hexitol molecules by hydrolysis of suitable substituents attached to primary carbon atoms, so no question of Walden inversion has arisen.

In point of fact, only one instance of the occurrence of Walden inversion in the hexitol series has been recorded, namely that involved in the formation of 1,4:3,6-dianhydro-L-iditol by inversion at C3 and C4 of D-mannitol. Since it is probable that the sugar ring exerts a directive influence on the scission of ethylene oxide rings in sugar anhydrides such as methyl 2,3-anhydro-α-D-alloside,[67,68] a study of ethylene oxide rings asymmetrically situated in hexitols will be of importance because it will enable an examination to be made of the stereochemical effect of the absence of a sugar ring on the way in which an ethylene oxide ring undergoes scission.

The 5,6-anhydro derivatives of D-sorbitol and D-mannitol[41,69] behave in like manner to the 5,6-anhydro-1,2-isopropylidene-D-glucofuranose of Freudenberg and coworkers.[70] Thus, 5,6-anhydro-1,3:2,4-diethylidene-D-sorbitol (LXXXV) readily suffers ring scission with ammonia to give 6-amino 6-desoxy-1,3:2,4-diethylidene-D-sorbitol (LXXXVI)[69] or with sodium methoxide to form 6-methyl-1,3:2,4-diethylidene-D-sorbitol.[42] 1,2:3,4-Diisopropylidene-5,6-anhydro-D-mannitol gives rise

| LXXXV | LXXXVI | LXXXVII |

to 6-amino-6-desoxy-1,2:3,4-diisopropylidene-D-mannitol[69] (LXXXVII) which is hydrolyzed to the same amino-D-mannitol obtained by Roux[71] through the reduction of D-mannose oxime.

The most interesting ethylene oxide derivative of the hexitols encountered hitherto is 1,2:5,6-dianhydro-D-mannitol. This was obtained in the form of its 3,4-isopropylidene (LXXXVIII) and 3,4-ethylidene (LXXXIX) derivatives.[72] The isopropylidene compound was synthesized from 1,6-dichloro-1,6-didesoxy-D-mannitol through its conversion first into 1,6-dichloro-1,6-didesoxy-3,4-isopropylidene-D-mannitol. This,

(67) S. Peat and L. F. Wiggins, *J. Chem. Soc.*, 1810 (1938).
(68) F. H. Newth, W. G. Overend and L. F. Wiggins, *J. Chem. Soc.*, 10 (1946).
(69) L. F. Wiggins, *J. Chem. Soc.*, 388 (1946).
(70) K. Freudenberg, H. Toepfer and C. C. Anderson, *Ber.*, **61**, 1750 (1928).
(71) E. Roux, *Compt. rend.*, **138**, 503 (1904).
(72) L. F. Wiggins, *J. Chem. Soc.*, 384 (1946).

on dehalogenation by means of cold sodium methoxide, gave the dianhydride LXXXVIII. An alternative synthesis through 1,6-ditosyl-3,4-isopropylidene-D-mannitol was also achieved. The structure of this dianhydride follows from its method of synthesis and the fact that the rings are extremely labile. The alternative structure 1,5:2,6-dianhydro-3,4-isopropylidene-D-mannitol is very unlikely on steric considerations and moreover would contain the comparatively stable tetrahydropyran rings. The synthesis of LXXXIX was accomplished through the

```
        CH₂                    CH₂
       /                      /
      O                      O
       \                      \
        CH                     CH
        |                      |
        OCH                    OCH
        | |                    | |
        | >CMe₂                | >CHMe
        HCO                    HCO
        |                      |
        HC                     HC
         \                      \
          \                      \
           O                      O
          /                      /
        CH₂                    CH₂
       LXXXVIII               LXXXIX
```

corresponding 1,6-dichloro-1,6-didesoxy-3,4-ethylidene-D-mannitol. The ethylene oxide rings in these compounds were extremely labile and could be opened with $N/100$ hydrochloric acid. In fact, ring scission of LXXXVIII was effected in preference to the removal of the isopropylidene group. Scission with alkaline reagents was accomplished with like ease with formation of D-mannitol derivatives substituted at the primary hydroxyl groups. Neither secondary substitution nor Walden inversion was encountered. Ring scission with dibasic acids, for example phthalic acid, or with diamines was accompanied by a vigorously exothermic reaction and polymeric products resulted.

III. Anhydrides of Pentitols

1. *1,4-Anhydro- and 1,4:2,5-Dianhydroxylitol*

Until 1945 no anhydrides of xylitol or any other pentitol had been described. In that year a patent was granted to Grandel[73] to make anhydrides of pentitols by heating them with a catalyst, followed by vacuum distillation. Xylitol on being heated with methionic acid followed by vacuum distillation gave monoanhydroxylitol and a dianhydro-

(73) F. Grandel (to Alien Property Custodian), U. S. Pat. 2,375,915 (1945).

xylitol. Grandel gave no indication of the structure of these compounds. Carson and Maclay[74] obtained a crystalline monoanhydroxylitol (m. p. 37–38°) by heating xylitol with either sulphuric or benzenesulfonic acid. They proved that it was 1,4-monoanhydro-DL-xylitol (XC for the D enantiomorph) by studying its oxidation with sodium metaperiodate. The anhydride consumes one mole of oxidant but liberates no formaldehyde or formic acid and affords on further oxidation with hypobromite in the presence of strontium carbonate, strontium DL-hydroxymethyl-diglycolate (XCI for the D enantiomorph). These facts could only be explained if the anhydroxylitol possessed the tetrahydrofuran ring structure.

$$\begin{array}{cccc}
\text{CH}_2{-} & \text{CH}_2{-} & \text{CH}_2{-} & \text{CH}_2{-} \\
\text{HCOH} & \text{CO} & \text{HCO} & \text{HCOH} \\
\text{HOCH} & \text{HOCH} & \text{HOCH} & \text{HOCH} \\
\text{HC}{-} & \text{CO} & \text{HC}{-} & \text{HCOH} \\
\text{CH}_2\text{OH} & \text{HC}{-} & \text{CH}_2 & \text{CH}_2{-} \\
 & \text{CH}_2\text{OH} & & \\
\text{XC} & \text{XCI} & \text{XCII} & \text{XCIII}
\end{array}$$

Although no proof has been offered, the dianhydride of xylitol that was obtained by Grandel[73] most probably has the meso 1,4:2,5-ring structure that is represented by XCII.

2. *1,5-Anhydroxylitol, -D-Arabitol and -Ribitol*

Following the Zervas[22] synthesis of styracitol, Fletcher and Hudson[75] prepared triacetyl-2-hydroxy-D-xylal and from it by hydrogenation with palladium catalyst, a 1,5-anhydropentitol. This could conceivably have either the xylitol or lyxitol configuration. Only the latter however would be optically active. Since the anhydropentitol that was obtained was optically inactive it was the meso 1,5-anhydroxylitol (XCIII). Further proof of this configuration was obtained by the fact that the same compound was obtained by reductive desulfurization of phenyl 1-thio-β-D-xylopyranoside triacetate, a process involving no change on any carbon atom except C1. Furthermore, during periodate oxidation of the anhydride, two moles of the oxidant were consumed and one mole of formic acid was liberated. Only the tetrahydropyran ring structure could account for these facts.

(74) J. F. Carson and W. D. Maclay, *J. Am. Chem. Soc.*, **67**, 1808 (1945).
(75) H. G. Fletcher, Jr., and C. S. Hudson, *J. Am. Chem. Soc.*, **69**, 921 (1947).

The extension of the new technique of reductive desulfurization to other aryl 1-thiopentosides has resulted in the preparation of the remaining two possible 1,5-anhydropentitols, namely, 1,5-anhydro-D-arabitol and the meso 1,5-anhydroribitol. Fletcher and Hudson[76] obtained

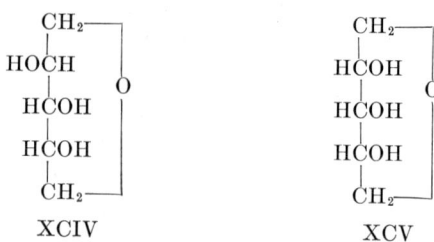

XCIV   XCV

1,5-anhydro-D-arabitol (XCIV) by the reductive desulfurization of either 2'-naphthyl 1-thio-α-D-arabinopyranoside triacetate or phenyl 1-thio-α-D-arabinopyranoside triacetate. Later, Jeanloz, Fletcher and Hudson[77] obtained the third isomeric 1,5-anhydropentitol namely, 1,5-anhydroribitol (XCV). It was obtained by condensation of tribenzoyl-β-D-ribopyranosyl bromide with potassium 2-thionaphtholate, forming 2'-naphthyl 2,3,4-tribenzoyl-1-thio-β-D-ribopyranoside which gave the anhydroribitol on being heated with Raney nickel in alcohol and subsequent debenzoylation.

IV. Some Uses of Hexitol Anhydrides

The anhydrides of hexahydric alcohols have already achieved marked usefulness in chemical industry and many of their derivatives show definite promise of value in industry and in medicine.

Perhaps the most important uses for these materials lie in their transformation into wetting and emulsifying agents and into drying oil ingredients. Numbers of long chain saturated and unsaturated fatty acid derivatives of hexitol anhydrides are marketed for these purposes. These are not pure hexitan or hexide derivatives but contain these substances as the fundamental constituents. The uses to which these substances are being put varies from laundry preparations to promoters for bacterial culture. Bloor[78] was probably the first to make long chain fatty acid esters of hexitans and hexides. In 1912 he described mannitan mono- and dilaurate and dianhydro-D-mannitol dilaurate and similar derivatives with stearic acid. Esters of D-sorbitol with unsaturated fatty acids were described in patent literature in 1937[79] but the products

(76) H. G. Fletcher, Jr., and C. S. Hudson, *J. Am. Chem. Soc.*, **69**, 1672 (1947).
(77) R. Jeanloz, H. G. Fletcher, Jr., and C. S. Hudson, *J. Am. Chem. Soc.*, **70**, 4052 (1948).
(78) W. R. Bloor, *J. Biol. Chem.*, **11**, 421 (1912).
(79) I. G. Farbenindustrie A. G., British Pat. 350,992 (1937).

were probably derivatives of D-sorbitol anhydrides. In recent years the hydroxypolyoxyalkylene derivatives of these partially esterified hexitol anhydrides have been made by the treatment of the ester with ethylene oxide in the presence of sodium methoxide.[80]

Some of the uses of these and similar compounds may now be enumerated. It has been found that hydroxypolyoxyalkylene derivatives of sorbitan mono-oleate (commercial name, Tween 80) favors the growth of tubercle bacilli and permits the multiplication of very small inocula.[81] Moreover, Tween 80 also nullifies the antitubercular effect of such substances as 5-amino-2-butoxypyridine[82] and of ethyl chaulmoograte.[83]

The partial esters of sorbitan or mannitan with fatty acids act as useful rust inhibitors when added to certain lubricating oils.[84] Thus, sorbitan monoleate in lubricating oils and in lithium greases shows this property.[85]

The hydroxypolyoxyalkylene ether of mannitan monolaurate has been used as a dispersing agent for laundry waxes for starching.[86] The corresponding sorbitan derivative has been described as a good emulsifier for dimethyl phthalate[87] and a similar derivative of sorbitan monostearate as a useful emulsifying and whipping agent for certain food products.[88] Useful emulsifying agents are also prepared either directly from anhydrohexitols or from the hexitols themselves by heating with coconut fatty acids.[89] Mannitan monolaurate has exceptional wetting properties and can be used as an insecticide against aphids or better in conjunction with rotenone.[90] Carotene is rendered water-soluble by the hydroxypolyoxyalkylene derivative of sorbitan monooleate.[91]

The products of esterification of D-sorbitol by linseed fatty acids are excellent drying oils and these products are doubtless derivatives of anhydrosorbitol.[92]

(80) W. C. Griffen (to Atlas Powder Co., Inc.), U. S. Pat. 2,380,166 (1945).
(81) B. D. Davis and R. J. Dubos, *J. Exptl. Med.*, **83**, 409 (1946); **86**, 215 (1947).
(82) H. S. Forrest, P. D'Arcy Hart and J. Walker, *Nature*, **160**, 94 (1947).
(83) H. Bloch, H. Erlenmeyer and E. Suter, *Experientia*, **3**, 199 (1947).
(84) T. E. Sharp (to Standard Oil Co. of Indiana), U. S. Pat. 2,398,193 (1946).
(85) G. M. Hain, D. T. Jones, R. L. Merker and W. A. Zisman, *Ind. Eng. Chem.*, **39**, 500 (1947); D. C. Atkins, H. R. Baker, C. M. Murphy and W. A. Zisman, *ibid.*, **39**, 491 (1947).
(86) W. C. Griffen (to Atlas Powder Co., Inc.), U. S. Pat. 2,374,931 (1945).
(87) F. A. Morton, *Proc. New Jersey Mosquito Exterm. Assoc.*, **33**, 69 (1946).
(88) N. F. Johnston (to R. F. Vanderbilt Co., Inc.), U. S. Pat. 2,422,486 (1947).
(89) K. R. Brown (to Atlas Powder Co. Inc.), U. S. Pat. 2,322,820 (1943).
(90) R. W. Roth and L. Pyenson, *J. Econ. Entomol.*, **34**, 474 (1941).
(91) R. M. Tomarelli, J. Charney and F. W. Bernhart, *Proc. Soc. Exptl. Biol. Med.*, **63**, 108 (1946).
(92) J. D. Brandner, R. H. Hunter, M. D. Brewster and R. E. Bonner, *Ind. Eng. Chem.*, **37**, 809 (1945).

Resinous products, described as odorless and tasteless resins, may be made by heating hexahydric alcohols with abietic acid or certain resins and a dibasic acid such as adipic acid. This product is also doubtless an anhydrohexitol derivative. The simple ethers such as methyl, ethyl or butyl ethers of hexitans or hexides may be useful as plasticizers.[93] Haworth and Wiggins described the dimethyl and diethyl ethers of 1,4:3,6-dianhydro-D-mannitol and D-sorbitol for the same purpose.[94] The alkenyl ethers are of interest because they themselves slowly polymerize and are also very high boiling liquids. 2,5-Diallyl-1,4:3,6-dianhydro-D-sorbitol (XCVI) polymerizes only in contact with gaseous oxygen and then only slowly. This substance, however, polymerizes about twice as fast as the corresponding methallyl ether.[95] The dianhydrides of D-mannitol and D-sorbitol formed diacrylate and dimethacrylate derivatives on being treated with acrylyl or methacrylyl chloride. Thus, 2,5-dimethacrylyl-1,4:3,6-dianhydro-D-mannitol (XCVII) and 2,5-diacrylyl-1,4:3,6-dianhydro-D-sorbitol (XCVIII) were obtained. These polymerized very rapidly on being heated even without a catalyst and also homogeneously copolymerized with methyl methacrylate forming very hard glass-like resins.[96]

(93) K. R. Brown (to Atlas Powder Co., Inc.), U. S. Pat. 2,420,519 (1947).
(94) W. N. Haworth and L. F. Wiggins, British Pat. 599,048 (1948).
(95) Hilda Gregory and L. F. Wiggins, *J. Chem. Soc.*, 1405 (1847).
(96) Hilda Gregory, W. N. Haworth and L. F. Wiggins, British Pat. 586,141 (1947); *J. Chem. Soc.*, 488 (1946).

The acetates of the 1,4:3,6-dianhydrohexitols may also be useful as plasticizers; both 2,5-diacetyl-1,4:3,6-dianhydro-D-mannitol and the analogous D-sorbitol derivative have been described.[97]

The hexitol anhydrides also have medicinal uses.[97a] Carr and Krantz[98] found that 1,4:3,6-dianhydro-D-mannitol was quite non-toxic to man and was a valuable diuretic. The dinitrates of both isomannide and isosorbide[99] have possible uses as agents for lowering blood pressure, though Burn and Stephenson[100] found that the D-sorbitol derivative was twice as active as the D-mannitol derivative in this respect. 2,5-Diamino-2,5-didesoxy-1,4:3,6-dianhydro-D-mannitol and -D-sorbitol have been prepared and their sulfanilamido derivatives obtained.[101] These however showed no outstanding bacteriostatic activity.

V. Tables of Properties of Pentitol and Hexitol Anhydrides and Their Derivatives

The following tables summarize the properties of well-authenticated derivatives of anhydropentitols and anhydrohexitols.

Table III
Pentitol Anhydrides and Their Derivatives

Substance	Melting point, °C.	Boiling point, °C.	[α]D	Rotation solvent	References
1,4-Anhydro-D,L-xylitol	37–38	145–165/0.01 mm.	0	—	74
—, 2,3,5-tri-(phenylcarbamyl)-	193–194.5	—	0	—	74
—, 2,3,5-tribenzoyl-	79–80	—	0	—	74
—, 2,3-diacetyl-5-trityl-	134–135	—	0	—	74
1,5-Anhydroxylitol	116–117	—	0	—	75
—, 2,3,4-triacetyl-	122–123	—	0	—	75
—, 2,3,4-tribenzoyl-	146–147	—	0	—	75
1,5-Anhydroribitol	127–128	—	0	—	77
—, 2,3,4-triacetyl-	133–134	—	0	—	77
—, 2,3,4-tribenzoyl-	156–157	—	0	—	77
1,5-Anhydro-D-arabitol	96–97	—	−98.6	H₂O	76
—, 2,3,4-triacetyl-	58	—	−74.2	CHCl₃	76
—, 2,3,4-tribenzoyl-	120–121	—	−220	CHCl₃	76

(97) W. N. Haworth and L. F. Wiggins, British Pat. 619,500 (1949).
(97a) Reviewed by C. J. Carr and J. C. Krantz, Jr., *Advances in Carbohydrate Chem.*, **1,** 175 (1945).
(98) C. J. Carr and J. C. Krantz, Jr., *Proc. Soc. Exptl. Biol. Med.*, **39,** 577 (1938); J. C. Krantz, Jr., U. S. Pat. 2,143,324 (1939).
(99) J. C. Krantz, Jr., C. J. Carr, S. E. Forman and F. W. Ellis, *J. Pharmacol.*, **67,** 187 (1939).
(100) J. H. Burn and R. P. Stephenson, private communication.
(101) R. Montgomery and L. F. Wiggins, *J. Chem. Soc.*, 393 (**1946**).

Table IV
Hexitol Monoanhydrides and Their Derivatives

Substance	Melting point, °C.	Boiling point, °C.	[α]D	Rotation solvent	References
1,4-Anhydro-D-sorbitol (Arlitan)	115–116	—	− 21.9	H_2O	10
—, 2,3,5,6-tetramethyl-	—	245–250 170–174/14 mm.	− 42.99	EtOH	10
—, 2,3,5,6-tetraacetyl-	52–54	—	+ 47.5	$CHCl_3$	102
—, 2,3,5,6-tetramethanesulfonyl-	122	—	− 3.5	Me_2CO	12
—, 6-p-tosyl-	—	—	+ 10.05	$CHCl_3$	12
—, 2,3,5-tribenzoyl-	161.5–163	—	+ 35.1	$CHCl_3$	11
—, 6-chloro-6-desoxy-	108–109	—	− 14.0	Me_2CO	56
—, 2,3,5-triacetyl-6-chloro-6-desoxy-	81–82	—	− 21.9	$CHCl_3$	56
—, 3,5-benzylidene-	154–155	—	+ 15.4	Me_2CO	12
—, 3,5-benzylidene-6-p-tosyl-	125.5	—	+ 9.1	$CHCl_3$	12
—, 3,5-benzylidene-6-chloro-6-desoxy-	154–155	—	− 17.4	$CHCl_3$	56
—, 3,5-benzylidene-6-iodo-6-desoxy-	144	—	—	—	12
—, monobenzylidene (a)-	136–140	—	+ 33.72	EtOH	10
—, monobenzylidene (b)-	121–122	—	—	—	10
Bis (6-chloro-6-desoxy-3,5-benzylidene-1,4-anhydro-D-sorbitol) 2,2′-sulfite	106–107	—	+ 46.8	$CHCl_3$	12
1,5-Anhydro-D-sorbitol (Polygalitol)	141–142	—	+ 42.5	H_2O	103
—, 2,3,4,6-tetraacetyl- (dimorphous)	65–67; 73–74	—	+ 38.9	$CHCl_3$	103
3,6-Anhydro-D-sorbitol	113	—	− 7.47	H_2O	13, 14
—, 2,5-dimethyl	70–71	—	− 15.6	$CHCl_3$	51
2,5-Anhydro-D-sorbitol	—	—	—	—	—
—, 1,6-dibenzoyl-[a]	137–138	—	+ 3.2	EtOH	37, 39
—, 1,6-dibenzoyl-3,4-di-p-tosyl-	142	—	+ 56.1	$CHCl_3$	40
5,6-Anhydro-D-sorbitol	—	—	—	—	—
—, 1-p-tosyl-2,4-benzylidene-	137	—	+ 4.0	C_5H_5N	41
—, 1,3:2,4-diethylidene-	136	—	− 7.3	$CHCl_3$	69
1,4(3,6)-Anhydro-D-mannitol (Mannitan)	145–148	—	− 23.8	H_2O	5
—, 2,3,5,6-tetraacetyl-	—	—	+ 23.0	AcOH	2
—, 5,6-benzylidene-	143–144	—	—	—	6
—, 2,3-dibenzoyl-5,6-benzylidene-[b]	161–162	—	—	—	6
	162	—	+ 33.6	$CHCl_3$	7
—, 2,3:5,6-dibenzylidene-	122–123 125–126	—	—	—	6
—, 2,6 (or 3,6)-dibenzoyl-[c]	142–145	—	− 48.4	$CHCl_3$	5
5,6-Anhydro-D-mannitol	—	—	—	—	16
—, 1,2:3,4-diisopropylidene-	—	100–110/0.2 mm. (bath)	+ 12.4	$CHCl_3$	69
1,5-Anhydro-D-mannitol (Styracitol)	155	—	− 56.5	H_2O	17
—, 2,3,4,6-tetraacetyl- (dimorphous)	58; 66–67	—	− 20.9	EtOH	104
—, 2,3,4,6-tetramethyl-	—	149–151/24 mm.	− 36.6	None	18
—, 2,3,4,6-tetrabenzoyl-	142	—	−150.4	$CHCl_3$	17
—, 2,3,4,6-tetranitro-	106	—	− 31.8	Me_2CO	17
—, 2,3,4,6-tetramethanesulfonyl	171–172	—	—	—	47
—, diisopropylidene-	96–97	—	−115.2	EtOH	18
—, dibenzylidene (a)-	163–165	—	−148.7	$CHCl_3$	18
—, dibenzylidene (b)-	192–193	—	− 80.5	$CHCl_3$	18
3,6-Anhydro-D-galactitol	—	—	+ 16.1	H_2O	11
—, 1,2,4,5-tetraacetyl-	—	—	+ 16.9	EtOH	11
1,5-Anhydro-D-galactitol	114–115	—	+ 76.6	H_2O	35
—, 2,3,4,6-tetraacetyl-	75–76	—	+ 49.1	$CHCl_3$	35
1,5-Anhydro-D-talitol	—	—	− 11.4	H_2O	35a
—, 2,3,4,6-tetraacetyl-	106–107	—	− 16.2	$CHCl_3$	35a
2,5-Anhydro-L-iditol	111–113	—	+ 12.6	H_2O	43
—, 1,3,4,6-tetraacetyl-	—	124–140/0.003 mm.	+ 13.2	$CHCl_3$	43
—, 1-p-tosyl-	146	—	+ 3.8	C_5H_5N	41
—, 1,6-di-p-tosyl-	146	—	+ 6.67	C_5H_5N	42
—, 1,6-diiodo-1,6-didesoxy-	110–111	—	+ 97.1	$CHCl_3$	42

[a] Brigl and Grüner[6] designated this compound 1,6-dibenzoyl-2,5-anhydro-D-mannitol. [b] Designated 1,6-dibenzoyl-2,4-anhydro-3,5-benzylidene-D-mannitol by Brigl and Grüner. [c] Designated 1,6-dibenzoyl-2,4-anhydro-D-mannitol by Brigl and Grüner.

(102) H. G. Fletcher, Jr., and Catherine M. Sponable, *J. Am. Chem. Soc.*, **70**, 3943 (1948).

(103) N. K. Richtmyer and C. S. Hudson, *J. Am. Chem. Soc.*, **65**, 64 (1943).

(104) Y. Asahina, *Ber.*, **45**, 2367 (1912).

Table V
Hexitol Dianhydrides and Their Derivatives

Substance	Melting point, °C.	Boiling point, °C.	$[\alpha]_D$	Rotation solvent	References
1,4:3,6-Dianhydro-D-sorbitol (Isosorbide)	61–63 61.9–64	160–165/10 mm. 160–175/2 mm.	+ 43.9 + 44.8	H_2O H_2O	51 54
—, 2,5-dimethyl-	—	93–95/0.1 mm. (bath)	+ 92.9	$CHCl_3$	51
—, 2,5-dibenzoyl-	102–103 101.5–102	— —	+ 24.5 + 23.1	$CHCl_3$ $CHCl_3$	51 54
—, 2,5-ditrityl-	92–94	—	+ 44.3	$CHCl_3$	54
—, 2,5-di-p-tosyl-	98.3–99.6 101–102	— —	+ 57.4 + 57.8	$CHCl_3$ $CHCl_3$	54 51
—, x-iodo-x-desoxy-x'-p-tosyl-	90.8–91.3	—	+ 52.73	$CHCl_3$	54
—, 2,5-dimethanesulfonyl-	122–123	—	+ 77.7	$CHCl_3$	12
—, 2,5-diacetyl-	60–61	—	+133.6	$CHCl_3$	56
—, 2,5-diallyl-	—	157–161/20 mm.	+ 93.4	$CHCl_3$	95
—, 2,5-dimethallyl-	—	157/15 mm.	+ 37.1	$CHCl_3$	95
—, 2,5-dichloro-2,5-didesoxy-	—	105/15 mm.	+ 55.8	$CHCl_3$	50
—, 5-chloro-5-desoxy(?)-	63–64	200–220/12 mm.	+ 48.0	$CHCl_3$	50
—, 2,5-dinitro-	52 71	— —	— —	— —	99 105
—, 2,5-diacrylyl-	57	—	+130.9	$CHCl_3$	96
—, 2,5-dimethacrylyl-	—	—	+ 78.7	$CHCl_3$	96
—, 2,5-diethyl-	—	132–133/12 mm.	+ 66.4	$CHCl_3$	94
—, 2,5-diamino-2,5-didesoxy-	—	105–110/0.1 mm. (bath)	+ 43.6	H_2O	101
—, 2,5-diamino-2,5-didesoxy-, oxalate	253–254	—	—	—	101
—, 2,5-diamino-2,5-didesoxy-, picrate	200 (dec.)	—	—	—	101
—, 2,5-diamino-2,5-didesoxy-, hydrochloride	320	—	—	—	101
—, 2,5-diamino-2,5-didesoxy-, sulfate	330	—	—	—	101
—, 2,5-diamino-2,5-didesoxy-, dimethylene-D-glucosaccharate	220–221 (dec.)	—	—	—	101
—, 2,5-diamino-2,5-didesoxy, disalicylidene	166–167	—	—	—	101
—, 2,5-di-(N⁴-acetylsulfanilamido)-2,5-didesoxy-	263–264	—	+ 31.4	Me_2CO-H_2O	101
—, 2,5-disulfanilamido-2,5-didesoxy-	239–240	—	+ 49.2	Me_2CO	101
—, 2,5-di-(p-nitrobenzenesulfonamido)-2,5-didesoxy-	216–217	—	+ 56.0	Me_2CO	101
1,4:3,6-Dianhydro-D-mannitol (Isomannide)	87	274	+ 91.4	H_2O	44
—, 2(5)-acetyl-	—	185–187/25 mm.	—	—	44
—, 2,5-diacetyl-	—	197/28 mm.	—	—	44
—, 2,5-di-(phenylcarbamyl)-	243	—	—	—	106
—, 2,5-diformyl-	115	—	—	—	44
—, 2,5-dichloro-2,5-didesoxy-	49 67	143/43 mm. —	— + 93.5	— $CHCl_3$	44 45

Hexitol Dianhydrides and Their Derivatives (Continued)

Substance	Melting point, °C.	Boiling point, °C.	$[\alpha]_D$	Rotation solvent	References
—, 2-chloro-2-desoxy-5-methanesulfonyl-	115–117	—	+ 66.4	$CHCl_3$	47
—, 2(5)-monomethyl-	44–45	—	—	—	44
—, 2,5-diethyl-	—	135–136/10 mm.	+126.8	$CHCl_3$	92
—, 2-chloro-2-desoxy-5-phenylcarbamyl(?)-	163	—	—	—	107
—, 2,5-dicrotyl-	—	171/10 mm.	—	—	94
—, 2,5-dibenzoyl-	133	—	+225.7	$CHCl_3$	77
—, 2,5-ditrityl-	92–94	—	+ 44.3	$CHCl_3$	6
—, 2,5-ditosyl-	{ 88.5–90 93–94	—	+ 92.2	$CHCl_3$	6 101
—, 2,5-dimethanesulfonyl-	104	—	+138.7	$CHCl_3$	47
—, 2,5-diiodo-2,5-didesoxy-	61–62	—	—	—	6
—, 2,5-diamino-2,5-didesoxy-	59–62	150/0.01 mm. (bath)	+ 33.6	$CHCl_3$	101
—, 2,5-diamino-2,5-didesoxy-, adipate	189	—	—	—	101
—, 2,5-diamino-2,5-didesoxy-, picrate	227–228 (dec.)	—	—	—	101
—, 2,5-diamino-2,5-didesoxy-, hydrochloride	280–300 (dec.)	—	—	—	101
—, 2,5-diamino-2,5-didesoxy-, sulfate	310	—	—	—	101
—, 2,5-diamino-2,5-didesoxy, dimethylene-galactosaccharate	246–247	—	—	—	101
—, 2,5-diamino-2,5-didesoxy, disalicylidene	188–189	—	—	—	101
—, 2,5-di-(N^4-acetylsulfanilamido)-2,5-didesoxy-	278–279	—	—	—	101
—, 2,5-disulfanilamido-2,5-didesoxy-	227–228	—	—	—	101
—, 2,5-di-(*p*-nitrobenzenesulfonamido)-2,5-didesoxy-	213–214	—	+ 7.5	Me_2CO	101
1,5:3,6-Dianhydro-D-mannitol (neomannide)	109.1–110.9	—	+ 6.4	H_2O	66
—, 2,5-demethanesulfonyl-	190	—	+ 11.5	Me_2CO	47

(105) L. F. Wiggins, Unpublished data.
(106) P. Carré, *Ann. chim. et phys.*, [8] **5,** 429 (1905).
(107) P. Carré and P. Mauclère, *Compt. rend.*, **192,** 1567 (1931).

ACTION OF CERTAIN ALPHA AMYLASES

By Mary L. Caldwell and Mildred Adams*

Columbia University, Takamine Laboratory,
New York, N. Y. Clifton, N. J.

CONTENTS

I. Foreword.. 229
II. Introduction... 229
III. Beta Amylases.. 231
IV. Alpha Amylases... 234
    1. Pancreatic Amylase...................................... 235
        a. Chemical Nature and Properties...................... 235
        b. Conditions Which Favor Activity.................... 236
        c. Activity Measurements and Ratios................... 238
        d. Extent of Hydrolysis of Starch...................... 239
        e. Stages of Very Slow Rates of Change................ 241
        f. Products... 245
        g. Action with Methylated Dextrins.................... 250
    2. Amylase of *Aspergillus oryzae*........................... 250
        a. Chemical Nature and Properties...................... 250
        b. Extent of Hydrolysis of Starch...................... 251
        c. Stages of Very Slow Rates of Change................ 253
        d. Products... 253
    3. Alpha Amylase of Malted Barley.......................... 255
        a. Occurrence.. 255
        b. Chemical Nature and Properties...................... 256
        c. Extent of Hydrolysis of Starch...................... 257
        d. Products... 259
    4. Other Alpha Amylases.................................... 265
        a. Salivary Amylase.................................... 265
        b. Bacterial Amylases.................................. 265
V. Discussion and Summary..................................... 266

I. Foreword

The manuscript for this chapter was submitted for publication in August, 1947, and reviews the literature to that date. Reference has also been made to a few researches of subsequent date in this active field.

II. Introduction

Amylases are hydrolytic enzymes which catalyze the hydrolysis of starches and glycogens. In addition to their intrinsic importance in

* Now with Bureau of Human Nutrition and Home Economics, U. S. Department of Agriculture, Beltsville, Maryland.

nature, in medicine and in industry they are of great value as tools for advancing our information about the substrates upon which they act. Amylases represent one of nature's means of breaking down these complex substances and by their use the chemist avoids the complicating influences of more drastic procedures.

At present, there are recognized two main types of amylases; the dextrinogenic, liquefying or alpha amylases and the saccharogenic or beta amylases. The former cause rapid fragmentation of starches to reducing dextrins which give no color with iodine (dextrinogenic action) and rapid decrease in the viscosity of starch pastes (liquefying action). The products formed by several dextrinogenic amylases have been found to exhibit falling or alpha mutarotation, hence the name alpha amylases is also applied to them. The saccharogenic amylases are characterized by their ability to split maltose rapidly from starches and glycogens without markedly disrupting the rest of the molecule. The dextrins which remain after the action of saccharogenic amylases have relatively high molecular weights, are practically non-reducing and retain the property of giving color with iodine. In those cases which have been examined, the reaction mixtures of saccharogenic amylases have been found to exhibit rising or beta mutarotation, hence the name beta amylases.

The terms dextrinogenic and saccharogenic are widely used at present as described above although neither is ideal. The dextrinogenic amylases exhibit so-called saccharogenic as well as dextrinogenic activities because they cause the formation of reducing substances and sugars; the saccharogenic amylases cause the formation of residual dextrins as well as of maltose. The terms alpha and beta amylases are more convenient.

No attempt is made here to survey the extensive literature on amylases. A number of recent reviews have discussed the general nature and properties of these enzymes[1-5] and this report will be devoted mainly to a summary of our present information about the action of some of the alpha amylases.

Interpretation of the results of investigations to determine the mechanism of the action of amylases has frequently been difficult because of the lack of sufficient knowledge concerning the substrates on which these enzymes act. It is now generally accepted that starches are made

---

(1) C. S. Hanes, *New Phytologist*, **36**, 101 (1937).

(2) Eric Kneen, in "Chemistry and Industry of Starch," edited by R. W. Kerr, Academic Press, Inc., New York, pp. 289–313 (1944).

(3) Mary L. Caldwell and Mildred Adams, in "Enzymes and Their Role in Wheat Technology," edited by J. A. Anderson, Interscience, New York, pp. 23–88 (1946).

(4) R. H. Hopkins, *Advances in Enzymol.*, **6**, 389 (1946).

(5) W. F. Geddes, *Advances in Enzymol.*, **6**, 415 (1946).

up of two types of components. These include glucosidic chains in which the glucose residues are united by 1,4 α-D-glucopyranosidic linkages, the so-called straight chain components or amyloses,[6] and those in which the linkages are for the most part 1,4 α-D-glucopyranosidic linkages but in which some of the glucose units are also joined by 1,6 α-D-glucopyranosidic linkages, the branched chain components or amylopectins.[6] The linear components are believed to assume helical forms under certain conditions.[7]

Glycogens are very highly branched polyglucoses, made up largely of 1,4 α-D-glucopyranosidic chains but with very frequent branchings caused by 1,6 α-D-glucopyranosidic linkages.[8]

Recent advances in our information of starches and glycogens make possible clearer interpretations of the results obtained with amylases. This better understanding should lead in turn to a more effective use of amylases as tools for further progress. For detailed information about starches and glycogens the reader is referred to a number of excellent reviews.[6,8–10]

III. Beta Amylases

No attempt is made in this chapter to review in detail our present information concerning the beta amylases. However, because it is impossible to consider adequately the properties of the alpha amylases without reference to beta amylases, it seems desirable to describe briefly a few of the characteristic properties of the beta amylases.

The beta amylases with which we are familiar at present are all of plant origin. They are characteristic of ungerminated grains such as wheat, barley, rice and oats and also are present in appreciable quantities in soy beans and in sweet potatoes.[3] Present evidence indicates that they cause the liberation of maltose from the non-reducing ends of the glucosidic chains of amyloses, amylopectins and of glycogens, and that they do not hydrolyze 1,6 α-D-glucopyranosidic linkages and probably not the 1,4 α-D-glucopyranosidic linkages adjacent to them.[1,3,6] They also do not appear to hydrolyze short straight chain saccharides such as maltotriose.[11,12]

(6) K. H. Meyer, *Advances in Colloid Sci.*, **6**, 1 (1942).
(7) R. E. Rundle and D. French, *J. Am. Chem. Soc.*, **65**, 558 (1943).
(8) K. H. Meyer, *Advances in Enzymol.*, **3**, 109 (1943).
(9) Dexter French, in "Chemistry and Industry of Starch," edited by R. W. Kerr, Academic Press, Inc., New York, pp. 113–176 (1944).
(10) T. J. Schoch, *Advances in Carbohydrate Chem.*, **1**, 247 (1945).
(11) K. Myrbäck and Elsa Leissner, *Arkiv Kemi, Mineral. Geol.*, **17A**, No. 18 (1943).
(12) K. Myrbäck, *J. prakt. Chem.*, **162**, 29 (1943).

Beta amylases hydrolyze amyloses completely to fermentable sugar provided conditions are favorable for the amylase and that retrogradation of the amylose during the hydrolysis is prevented.[6,13,14] They hydrolyze glycogens and the amylopectin components of starches only partially.[15] It is not known with certainty how many glucose units are left unhydrolyzed in amylopectins and glycogens because of the approach to a 1,6 α-D-glucopyranosidic linkage. Reports of approximately fifty percent hydrolysis of typical amylopectins from potato and corn starches[16] indicate that about half of the D-glucose of these molecules is in their outer free-end branches. The products of the action of beta amylases on amylopectins and glycogens are maltose and dextrins with high molecular weights and of negligible reducing action.[1,3,6,17]

The residual beta dextrins are readily hydrolyzed by alpha amylases, by acid or by steam and by certain glucosidases,[1,6,18] with the breaking of inner glucosidic chains between the branching points and the liberation of free branches made up of 1,4 α-D-glucopyranosidic chains. These partially hydrolyzed residual dextrins are again susceptible to hydrolysis by beta amylase until their free branches have again been removed.[1,6,18] Thus, beta and alpha amylases supplement each other and a trace of alpha amylase in a beta amylase preparation will cause much more extensive hydrolysis of starch and of glycogen than is brought about by beta amylase alone.[1,3,6]

The data[19] summarized in Figure 1 show that the extent of the hydrolysis of soluble potato starch by barley beta amylase reaches a limit which is independent of the concentration of the amylase. The data are typical of the action of beta amylases on unfractionated starches, when the hydrolyses are carried out at or near pH 4.5.[1,3,6,19,20] Under these conditions, the hydrolysis of unfractionated starches usually ceases when 60 to 64% of the maltose theoretically obtainable from the substrate has been formed. The exact value of the limit obviously will depend upon the concentration of amylopectin in the starch and upon its structure.

Blom and coworkers[21] have reported that after pretreatment at pH 3.4 and under suitable conditions of action, beta amylase does not

(13) K. H. Meyer, P. Bernfeld and J. Press, *Helv. Chim. Acta*, **23**, 1465 (1940).
(14) W. Z. Hassid and R. M. McCready, *J. Am. Chem. Soc.*, **65**, 1157 (1943).
(15) K. H. Meyer and Maria Fuld, *Helv. Chim. Acta*, **24**, 375 (1941).
(16) K. H. Meyer and P. Bernfeld, *Helv. Chim. Acta*, **23**, 875 (1940).
(17) C. O. Beckmann and Q. Landis, *J. Am. Chem. Soc.*, **61**, 1495, 1504 (1939).
(18) K. H. Meyer and P. Bernfeld, *Helv. Chim. Acta*, **25**, 399 (1942).
(19) G. A. van Klinkenberg, *Z. physiol. Chem.*, **212**, 173 (1932).
(20) Mary L. Caldwell and Susie E. Doebbeling, *J. Biol. Chem.*, **110**, 739 (1935).
(21) J. Blom, Agnete Bak and B. Braae, *Z. physiol. Chem.*, **241**, 273 (1936); **256**, 197 (1938).

hydrolyze starch beyond 53% theoretical maltose. Hopkins[4,22] has confirmed these results. At present, there appears to be no adequate explanation for the influence of hydrogen ion activity on the limit of hydrolysis of starch by beta amylase.

The characteristic of beta amylases of reaching a limit in the hydrolysis of starches gives the investigator a simple criterion for ascertaining whether his amylase preparation has been freed from alpha amylase, which usually accompanies beta amylase in nature. If the hydrolysis of Lintner's soluble potato starch at pH 4.5 ceases when approximately 64%

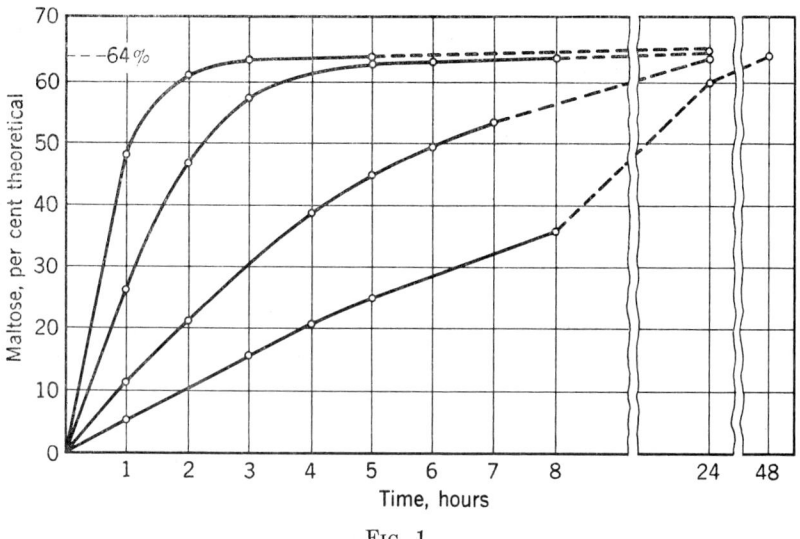

FIG. 1.

of the theoretical maltose has been formed, if this extent of hydrolysis is not increased by the addition of more of the amylase solution and if the reaction mixture retains the property of giving a blue-violet color with iodine due to the presence of the high molecular weight residual dextrins, the investigator can be confident that the beta amylase solution is not appreciably contaminated with traces of alpha amylase.[1,3,6,19,20]

In nature, amylases act hand in hand with each other and with other carbohydrases in bringing about the rapid breakdown of starches and glycogens to sugars which can be utilized readily by the living cell. Much important and practical information has been accumulated by the use of extracts and crude precipitates in which more or less natural mixtures of enzymes have been studied in the laboratory. However, the

(22) R. H. Hopkins, R. H. Murray and A. R. Lockwood, *Biochem. J.*, **40**, 507 (1946).

results of such investigations are hard to evaluate in terms of the action of a single enzyme and difficult to compare because the systems used by different investigators may have contained different kinds and proportions of collaborating enzymes and different kinds and proportions of electrolytes or of other impurities which may have influenced the reactions.

If amylases are to be used as tools for the detailed study of the breakdown and structure of their substrates it is obviously important to separate them from other enzymes and from other naturally associated constituents which may influence the results. It is then equally important to study the properties of the purified amylase and to supply it with the chemical environment necessary to protect it from inactivation and to enable it to act efficiently. With beta amylases this ideal has often been approached. Beta amylases from several sources have been prepared by selective inactivation of other enzymes that accompany them in nature[23] and highly active products have been obtained by extensive purification.[20,24-26] Balls and his associates have recently reported the crystallization of beta amylase from sweet potato.[27]

IV. Alpha Amylases

The general status of the alpha amylases with respect to the above ideal is not so satisfactory. The difficulties involved in the purification of these amylases are increased because there are no simple criteria by which the investigator can make sure that the alpha amylase under consideration is not contaminated by beta amylase or by some other glucosidase.[1,3,6,18,28] However, evidence is accumulating to indicate that these alpha amylases have important individual characteristics which will differentiate them from one another. Such characteristics should make these enzymes increasingly useful in the study of their substrates and thus amply repay continued efforts at their purification.

While alpha amylases from many sources are known, the present discussion will deal mainly with the alpha amylase of malted barley, pancreatic amylase and the amylase of *Aspergillus oryzae*. These

(23) E. Ohlsson, *Compt. rend. soc. biol.*, **87**, 1183 (1922); *Compt. rend. trav. lab. Carlsberg*, **16**, No. 7, 1 (1926); *Z. physiol. Chem.*, **189**, 17 (1930).

(24) K. Sjöberg and E. Eriksson, *Z. physiol. Chem.*, **139**, 118 (1924).

(25) H. Lüers and E. Sellner, *Wochschr. Brau.*, **42**, 97 (1925).

(26) G. A. van Klinkenberg, *Z. physiol. Chem.*, **209**, 253 (1932).

(27) A. K. Balls, R. R. Thompson and M. K. Walden, *J. Biol. Chem.*, **163**, 571 (1946); **173**, 9 (1948).

(28) E. Kneen, R. M. Sandstedt and C. M. Hollenbeck, *Cereal Chem.*, **20**, 399 (1943).

amylases are among those which have been most highly purified and most intensively studied and represent widely different origins, from animal, plant and fungus sources. A brief summary of the outstanding characteristics of each of these amylases will be followed by a discussion of their similarities and differences and by suggested interpretations of the results so far obtained.

1. *Pancreatic Amylase*

a. Chemical Nature and Properties.—Pancreatic amylase is protein. It has been purified from hog glands by a number of different procedures[29,30,31] and obtained in crystalline form.[32,33,34] Meyer and his coworkers[33,34] report that after extensive purification, a 4 to 5% aqueous solution of the amylase crystallizes readily in prisms or in fine needles.[33,34] Electrophoresis at various pH values of the crystallized product shows a single component. The authors conclude that the crystals must, therefore, be considered as pure amylase. The light absorption of the aqueous solution shows a maximum at 280 mμ and a small peak at 291 mμ.[34]

Pancreatic amylase is very labile and sensitive to its chemical environment. Its lability is accelerated by purification and by such factors as dilution of its aqueous solutions, dialysis of its aqueous solutions against water, unfavorable hydrogen ion activities and unfavorable temperatures.[29–31,35,36] The loss of amylase activity in solutions of pancreatic amylase increases with increasing temperature and is very rapid between 50° and 60°. The inactivation of pancreatic amylase in aqueous solution may be retarded by the addition of certain anions, of which the chloride ion is outstanding;[37–39] by the addition of certain cations, of which

(29) R. Willstätter, E. Waldschmidt-Leitz and A. R. F. Hesse, *Z. physiol. Chem.*, **126**, 143 (1923).
(30) H. C. Sherman, Mary L. Caldwell and Mildred Adams, *J. Am. Chem. Soc.*, **48**, 2947 (1926).
(31) H. C. Sherman, Mary L. Caldwell and Mildred Adams, *J. Biol. Chem.*, **88**, 295 (1930).
(32) Mary L. Caldwell, Lela E. Booher and H. C. Sherman, *Science*, **74**, 37 (1931).
(33) K. H. Meyer, Ed. H. Fischer and P. Bernfeld, *Helv. Chim. Acta*, **30**, 64 (1947).
(34) K. H. Meyer, Ed. H. Fischer and P. Bernfeld, *Arch. Biochem.*, **14**, 149 (1947).
(35) H. C. Sherman, E. C. Kendall and E. D. Clark, *J. Am. Chem. Soc.*, **32**, 1073 (1910).
(36) H. C. Sherman, *Proc. Nat. Acad. Sci. U. S.*, **9**, 81 (1923).
(37) H. C. Sherman, Mary L. Caldwell and Mildred Adams, *J. Am. Chem. Soc.*, **50**, 2529 (1928).
(38) H. C. Sherman, Mary L. Caldwell and Mildred Adams, *J. Am. Chem. Soc.*, **50**, 2535 (1928).
(39) H. C. Sherman, Mary L. Caldwell and Mildred Adams, *J. Am. Chem. Soc.*, **50**, 2538 (1928).

the calcium ion is outstanding;[40,41] by the addition of certain amino acids;[36,42,43] or by the presence of substrate.[30,41]

Considerable evidence indicates that the loss of activity of pancreatic amylase in dilute aqueous solution and the acceleration of this loss upon dialysis are due largely to hydrolysis of the protein molecule.[30,36] This loss of amylase activity is accompanied by decrease in protein nitrogen[30,33,36] and by the appearance of amino acids in the dialyzates.[36] It is assumed that the favorable influence of certain amino acids in decreasing the loss of amylase activity is due to their retarding influence upon this hydrolysis.[36] So far, the loss of pancreatic amylase activity, observed during dialysis of its solutions, has not been reversed by the union of the dialyzed solution with its dialyzate[30,33,36,41] as is possible under suitable conditions with certain enzymes which are composed of protein complexes with more or less readily dialyzable non-protein components. While the characteristic activity of pancreatic amylase must be considered primarily a property of the protein molecule as a whole, there is no doubt that it also depends upon the presence and arrangement of the primary amino groups of the molecule. When these groups were blocked or removed by such reagents as ketene, formaldehyde, or nitrous acid, corresponding decreases in the amylase activities were observed.[44] No evidence was obtained to indicate that sulfhydryl groups were present in the active protein.[45,46] Pancreatic amylase differs in this as in other respects from malted barley beta amylase which was observed to require free sulfhydryl groups but not free primary amino groups for its activity.[47] In this connection, it is of interest to note that Meyer and coworkers[34] have recently reported the absence of sulfur from their crystallized homogeneous pancreatic amylase.

Reacting mixtures of pancreatic amylase and starch exhibit alpha mutarotation.[48,49]

b. *Conditions Which Favor Activity.*—Not only the stability but the activity of pancreatic amylase is dependent upon its chemical environment. The general statement may be made that pancreatic amylase is

(40) H. Nakamura, *J. Soc. Chem. Ind., Japan*, **34**, (supp.) 265 (1931).

(41) Roslyn B. Alfin and Mary L. Caldwell, *J. Am. Chem. Soc.*, **70**, 2534 (1948).

(42) H. C. Sherman and Mary L. Caldwell, *J. Am. Chem. Soc.*, **43**, 2469 (1921); **44**, 2926 (1922).

(43) H. C. Sherman, Mary L. Caldwell and Nellie M. Naylor, *J. Am. Chem. Soc.*, **47**, 1702 (1925).

(44) J. E. Little and Mary L. Caldwell, *J. Biol. Chem.*, **142**, 585 (1942).

(45) J. E. Little and Mary L. Caldwell, *J. Biol. Chem.*, **147**, 229 (1943).

(46) Mary L. Caldwell, C. E. Weill and Ruth S. Weil, *J. Am. Chem. Soc.*, **67**, 1079 (1945).

(47) C. E. Weill and Mary L. Caldwell, *J. Am. Chem. Soc.*, **67**, 212, 214 (1945).

(48) R. Kuhn, *Ber.*, **57**, 1965 (1924).

(49) G. G. Freeman and R. H. Hopkins, *Biochem. J.*, **30**, 451 (1936).

most active in solutions which are close to neutrality but the exact hydrogen ion activity which favors the amylase activity depends upon other factors such as the kind and concentration of electrolyte.[37-39] When measured at 40° in the presence of 0.01 M phosphate and 0.02 M chloride, pancreatic amylase is most active at pH 7.2.[37]

Early evidence showed that electrolyte was needed for the activation of pancreatic amylase.[50] Later evidence showed that certain anions

TABLE I

Comparison of the Influence of Certain Ions upon the Activity of Pancreatic Amylase
(Data of Sherman, Caldwell and Adams[39])

Salt	Concentration of salt[a] (molarity)	Reaction mixture (pH)	Relative activity of amylase
Sodium chloride	0.03	7.1	100
Potassium chloride	0.03	7.1	100
Lithium chloride	0.02–0.05	7.1	80–90
Sodium bromide	0.03	7.1	77
Sodium nitrate	0.10	7.1	40
Sodium chlorate	0.10	6.9	27
Sodium thiocyanate	0.15	6.7	28
Sodium fluoride	0.20	6.7	21
Sodium sulfate	0.01–0.10	5.7–7.7	0
Sodium phosphate	0.01	5.7–7.7	0

[a] Each salt was present in the concentration and at the hydrogen ion activity which had been found to be most favorable to the action of pancreatic amylase in the presence of that salt.

were responsible for this activation.[39] The chloride ion is the most effective anion but it is possible to activate pancreatic amylase with several other anions. A comparison of the relative activation of pancreatic amylase by a number of anions is given in Table I.[39]

Pancreatic amylase is so far the only amylase which has been found to be completely inactive in the absence of certain anions.[39,50] Other amylases of animal origin, such as salivary and blood amylases, show increased activity in the presence of electrolyte[51,52] and, when highly purified, may also be found to require the presence of certain anions for their activation. Kneen and coworkers[53,54] have reported the purifica-

(50) E. C. Kendall and H. C. Sherman, *J. Am. Chem. Soc.*, **32**, 1087 (1910).
(51) A. Hahn and H. Meyer, *Z. Biol.*, **76**, 227 (1922).
(52) K. Myrbäck, *Z. physiol. Chem.*, **159**, 1 (1926).
(53) E. Kneen and R. M. Sandstedt, *J. Am. Chem. Soc.*, **65**, 1247 (1943); *Arch. Biochem.*, **9**, 235 (1946).
(54) W. Militzer, C. Ikeda and E. Kneen, *Arch. Biochem.*, **9**, 309, 321 (1946).

tion of an inhibitor from cereal grains which retards the action of pancreatic amylase.

c. Activity Measurements and Ratios.—In general, three different types of methods are used to measure the activity of alpha amylases and each of these methods measures somewhat different properties of the enzyme. A comparison of the results obtained by means of more than one method for evaluating the activity of the alpha amylases often gives considerable insight into the mode of action of these enzymes and occasionally furnishes evidence for the presence of contaminating carbohydrases.

Alpha amylases cause a rapid fragmentation of starch with an accompanying marked decrease in viscosity of the starch solutions. Therefore, viscosity determinations may be used to follow the early stages in the hydrolysis of starch by these amylases. Although this type of measurement is of considerable importance for certain industrial applications, it has not been used to any great extent in investigations with pancreatic amylase.

A second method that is frequently used to measure the dextrinogenic activity of alpha amylases is based upon the hydrolysis of starch to dextrins of low molecular weight which give a red or a colorless solution with iodine.

The classical Wohlgemuth method[55] was one of the first to make use of this property of the alpha amylases and measures the weight of starch hydrolyzed by a given weight of enzyme to products that give a clear red end point with iodine. Several modifications of this method have been developed and some investigators have chosen as the end point the "achroic stage" at which no color is given with iodine. Hanes and Cattle[56] have introduced the use of spectrophotometric measurements to follow the changes in color. These methods have proved very useful, particularly as evidence of differences in the mode of action of different alpha amylases.

The third type of method measures the saccharogenic activity of the alpha amylases and is based upon a determination of the increase in the reducing sugars in the reaction mixture. There are a variety of methods for measuring the reducing sugars in starch hydrolyzates but for fundamental investigations on the mode of action of the amylases, the iodometric methods are preferable. They are stoichiometric and, in contrast to other methods frequently used in measuring reducing sugars, give a direct measure of the glucosidic linkages of the substrate that have been broken. It has become customary to report the reducing values of the

(55) J. Wohlgemuth, *Biochem. Z.*, **9**, 1 (1908).
(56) C. S. Hanes and M. Cattle, *Proc. Roy. Soc. London*, **B125**, 387 (1938).

products in amylase reaction mixtures as their equivalents of theoretical maltose.

The ratio of the liquefying or dextrinogenic activity of an amylase preparation to the saccharogenic activity of this same enzyme depends upon the substrate used and upon the conditions maintained for each type of measurement. In order to obtain comparable ratios for a given amylase it is essential that activity measurements be made on the same substrate and under uniform conditions. For different amylases, comparisons should also be made on the same substrate but under the optimal conditions for the activity of each enzyme. In this way, remarkably constant values have been found for the ratio of the dextrinogenic to the saccharogenic activity of pancreatic amylase preparations at different stages of its purification.[41,44,45,57] In highly purified preparations of this amylase no appreciable change in this ratio has been observed as the result of partial inactivation of this enzyme by means of heat, by means of unfavorable hydrogen ion activities or by means of nitrous acid.[41,44,45] Such observations have led to the conclusion that pancreatic amylase is not a mixture of amylases. However, it should be borne in mind that measurements of amylase activities are not yet sufficiently precise to eliminate the possibility of traces of a second amylase even when the activity ratios remain reasonably constant after so-called preferential inactivation of an amylase. It should also be borne in mind that comparisons of dextrinogenic and saccharogenic activities are made at relatively early stages of the hydrolysis of starch and that, therefore, such comparisons would not be influenced appreciably by the presence of maltase or of other glucosidases. The influence of such enzymic impurities would become more important in the later stages of the hydrolysis.

The ratio of dextrinogenic to saccharogenic activities for pancreatic amylase, measured with Lintner's soluble potato starch under comparable conditions at 40°, is approximately 2 to 1.[41,44,45,57] The achroic point is reached in the hydrolysis of potato starch by highly purified pancreatic amylase when approximately 20% of the glucose linkages of the starch have been broken.[41]

d. *Extent of Hydrolysis of Starch.*—The data summarized in Figure 2 were obtained[41] when different concentrations of purified pancreatic amylase reacted with Lintner's soluble potato starch under conditions designed to favor the action of the amylase and to protect it from inactivation.[37-39] Concentrations of the amylase were chosen so that the reactions would proceed rapidly and be practically complete before contamination by yeasts and bacteria might be expected appreciably to influence the results. The data show that the extent of the hydrolysis

(57) H. C. Sherman and M. D. Schlesinger, *J. Am. Chem. Soc.*, **35**, 1784 (1913).

of starch by pancreatic amylase depends within wide limits upon the concentration of amylase used. The hydrolysis curves show a change from a rapid to a slow phase of the reaction, typical of many other enzyme reactions, but tend to flatten at higher values as the concentration of amylase is increased. With different concentrations of pancreatic amylase, there is no evidence of a common limit in the hydrolysis of starch, such as is observed with different concentrations of beta amylase (Figure 1) or such as has been reported for pancreatic amylase.[29,58] These results illustrate an important difference between the action of pancreatic amylase and that of beta amylase.[1,3,4,6,19] They also cast doubt upon the rather common practice of assuming a limit in the hydrolysis of starch

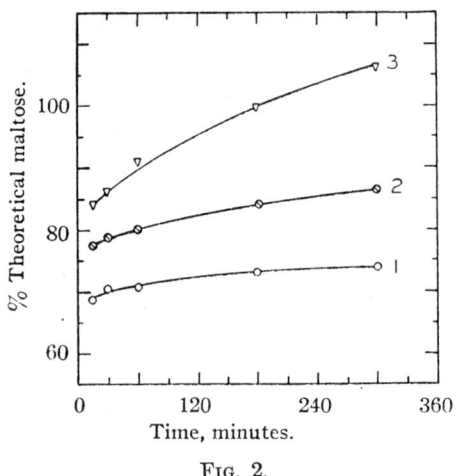

FIG. 2.

by pancreatic amylase at 75% theoretical maltose when solutions of unknown amylase concentrations are being compared and evaluated for amylase activity.[29,58]

It is important to note that no evidence of maltase activity was found in the amylase preparations even when the highest concentrations used in these comparisons were allowed to react for twenty-four hours with one per cent maltose under the conditions for the hydrolysis of starch. Similarly, no evidence was obtained for the presence of any other contaminating or extraneous carbohydrases in the amylase preparations. Partial inactivation of the amylase under a number of different conditions failed to give any evidence of selective inactivation such as might be expected if more than one enzyme were present. The substrates used for measuring the activity of the partially inactivated amylase were starch and starch hydrolyzates that had already been extensively

(58) J. Blom, Agnete Bak and B. Braae, Z. physiol. Chem., **250**, 104 (1937).

hydrolyzed by pancreatic amylase. With the hydrolyzates, low molecular weight saccharides were available for the action of glucosidases if they were present.

The influence of different concentrations of pancreatic amylase was also studied under similar conditions with each of the following substrates: potato starch;[41] exhaustively defatted corn starch;[59] exhaustively defatted waxy maize starch;[60] and corn amylose.[41] The waxy maize starch gave no evidence of the presence of straight chain components either by potentiometric titration[61] or by precipitation.[10] The corn amylose was not appreciably contaminated with branched chain components. It was hydrolyzed completely to fermentable sugar by beta amylase and corresponded to 100% amylose by potentiometric titration.[61] Similar results to those reported in Figure 2 for Lintner's soluble potato starch were obtained with all of the substrates investigated.[41,59,60] Likewise, Bernfeld and Studer-Pécha[62] have recently reported that the extent of the hydrolysis of amylose depends upon the concentration of crystallized pancreatic amylase. The flattening of the reaction curves at different extents of hydrolysis with different concentrations of enzyme appears to be typical of the action of this amylase. It is evident that pancreatic amylase hydrolyzes extensively the branched as well as the straight chain components of starches. Unlike beta amylase,[1,3,4,6,19] pancreatic amylase appears to be able to reach the inner 1,4 α-D-glucopyranosidic linkages between branched positions as well as those in the outer branches of the amylopectin components; it appears to cause the random hydrolysis of its substrates.

e. Stages of Very Slow Rates of Change.—Many investigators have suggested that the slowing down of enzyme reactions such as those observed in Figure 2 may be due to inactivation of the enzyme. However, inactivation of the enzyme does not explain the results reported here with pancreatic amylase. The evidence presented in Figure 3 shows[41] that the slowing down of the reaction occurs with pancreatic amylase even when conditions are used which prevent any appreciable inactivation of the amylase.[37-39] The presence of active amylase in reaction mixtures which had reached stages of very slow rates of change was shown by the hydrolysis of additional substrate. The extent of the hydrolysis after the introduction of additional substrate was practically the same as that reached in comparable reaction mixtures which had

(59) Marie M. Daly, Dissertation, Columbia University, New York, N. Y. (1947).
(60) Florence M. Mindell, A. Louise Agnew and Mary L. Caldwell, *J. Am. Chem. Soc.*, **71**, 1779 (1949).
(61) F. L. Bates, D. French and R. E. Rundle, *J. Am. Chem. Soc.*, **65**, 142 (1943).
(62) P. Bernfeld and H. Studer-Pécha, *Helv. Chim. Acta*, **30**, 1895, 1904 (1947).

contained initially the same relative concentrations of amylase and substrate. The data given in Figure 3 are strictly comparable as the reaction mixtures which had reached stages of very slow rates of change were divided and one portion was treated with additional substrate while the other was continued at 40° as before.

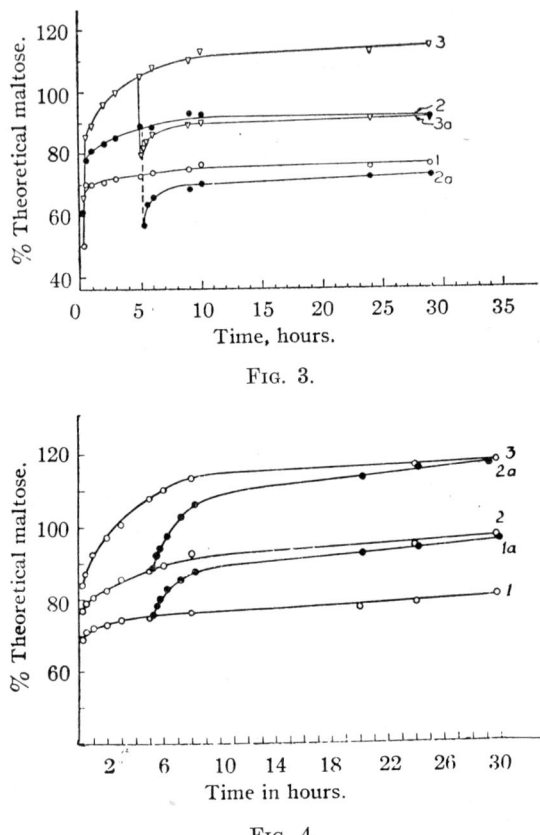

Fig. 3.

Fig. 4.

Similarly, increased hydrolysis was repeatedly found to occur when additional amylase was introduced into reaction mixtures which had reached stages of very slow rates of change. The extent of the increased hydrolysis depended upon the extent of the increase in the concentration of amylase. Typical data[41] are given in Figure 4. It is evident that products capable of further hydrolysis by the enzyme remained in the reaction mixtures which had reached stages of very slow rates of change. Moreover, the data summarized in Figure 4 show that the extent of hydrolysis attained was practically the same whether a given concentration of amylase was added at the start of the reaction or in part after the reaction had reached a stage of slow action.

The slowing down of enzyme reactions has often been attributed to reaction with, or equilibrium between, the enzyme and its substrate or between the enzyme and the products of its action. In order to determine the influence of the products of the action of pancreatic amylase on the extent of the hydrolysis of starch, portions of its hydrolysis mixtures were subjected to efficient dialysis during hydrolysis and the results compared with aliquots of the reaction mixture which had been treated in the same way except for dialysis.[41] The results of such experiments

TABLE II

Influence of Dialysis During Hydrolysis upon the Extent of the Hydrolysis of Lintner's Soluble Potato Starch by Pancreatic Amylase
(Data of Alfin and Caldwell[41])

Reaction time (minutes)	Theoretical maltose			
	Hydrolysis accompanied by dialysis			Hydrolysis without dialysis
	Inside	Outside	Total	
A—300	12.4%	56.3%	68.7%	68.8%
	11.0	57.8	68.8	
B 300	Not measurable	70.8	70.8	71.7
C 300	Not measurable	66.4	66.4	69.7
D 360	Not measurable	74.3	74.3	72.6
E 360	6.5	75.8	82.3	82.9
F 600	2.8	82.2	85.0	83.5

with Lintner's soluble potato starch showed that the extent of the hydrolysis of the dialyzed portions was the same, within experimental error, as that attained in the undialyzed reaction mixtures; with potato starch the extent of the hydrolysis was somewhat lower for dialyzed than for undialyzed reaction mixtures. These results show that the slowing down of the reactions in the hydrolysis of starch by pancreatic amylase cannot be explained by assuming reaction with or equilibrium between the amylase and maltose or glucose or other readily dialyzable products. If this were the case, the removal of such products would be expected to increase the extent of the hydrolysis of the dialyzed reaction mixtures. Typical data[41] for this point are summarized in Table II.

The addition of more pancreatic amylase to reaction mixtures that had been dialyzed during the hydrolysis and that had reached stages of very slow rates of change often resulted in small but measurable increases in their reducing values. Such results show the presence in the dialyzed hydrolysis mixtures of products capable of further hydrolysis by the amylase.[41]

Similarly, the addition of substrate to hydrolysis mixtures that had reached stages of very slow rates of change with dialysis resulted in extensive hydrolysis of the added substrate. Comparisons showed that the new substrate was hydrolyzed to practically the same extent in the same time as the original substrate.[41] These findings show not only that active amylase was present but that no appreciable inactivation of the amylase had taken place in the dialyzing hydrolysis mixtures when the

TABLE III

Influence of Dialysis During the Hydrolysis upon the Products Formed from Lintner's Soluble Potato Starch by Purified Pancreatic Amylase
(Data of Alfin and Caldwell[41])

Hydrolyses	Time	Reducing values as per cent theoretical maltose								
		Total			Unfermented reducing dextrins[a]			Maltose and glucose[a]		
	Hours	Inside	Outside	Total	Inside	Outside	Total	Inside	Outside	Total
Dialyzed	6	1.4%	80.0%	81.4%	1.4%	32.8%	34.2%	0%	47.2%	47.2%
Undialyzed	6			82.9			24.3			58.6

[a] Determined by selective fermentation with washed baker's yeast; dextrins by difference.

dialysis was carried out against buffer solutions of the same electrolyte concentration and pH value as those of the reaction mixtures.[41]

On the other hand, marked inactivation of the amylase (86 to 87%) occurred when solutions of the same purified pancreatic amylase preparations were dialyzed under the same conditions but in the absence of substrate.[41] These results give experimental evidence for the suggestion often advanced that the amylase unites with its substrate, in this case with the larger less readily dialyzable products of the hydrolysis of starch, and thus is protected from appreciable inactivation due to dialysis.

The reducing values of the reaction mixtures considered here were determined by iodometric titration[63] and, therefore, measure directly the number of glucosidic linkages that had been broken. The results given in Table III show that, although the number of glucosidic bonds

(63) Mary L. Caldwell, Susie E. Doebbeling and S. H. Manian, *Ind. Eng. Chem., Anal. Ed.*, **8**, 181 (1936).

broken was approximately the same, whether or not the substrate-enzyme mixture was subjected to dialysis during hydrolysis, the products of the hydrolysis were not the same. When hydrolysis was accompanied by dialysis, higher concentrations of reducing dextrins and lower concentrations of glucose and maltose were present than in a comparable undialyzed reaction mixture that had reached practically the same per cent of theoretical maltose. The presence in the dialyzate of reducing dextrins apparently capable of further hydrolysis gives evidence that pancreatic amylase hydrolyzes dialyzable products of relatively low molecular weights.

At present there seems to be no adequate explanation for the finding that dialysis during hydrolysis influences the products of hydrolysis without influencing the number of glucosidic linkages broken. These results are probably closely related to those already reported in Figures 3 and 4 that show, at the stage of negligible hydrolysis, a constant relationship between the relative concentrations of enzyme to substrate and the number of glucosidic linkages broken whether the entire amount of enzyme or substrate is present from the beginning of the hydrolysis or is added in part at a later stage in the hydrolysis. Further investigations are essential before it will be possible to draw any definite conclusions from these results.

The results of this work as a whole indicate that the slowing down of the hydrolysis of starch by pancreatic amylase is due largely to the replacement of the original substrate by products for which the amylase has relatively low affinities.

f. Products.—Considerable information concerning the mechanism of the enzymic hydrolysis of starch has been obtained from investigations of the action of purified maltase-free pancreatic amylase on a number of different substrates. The substrates studied were ordinary unfractionated but exhaustively defatted[10] potato and corn starches; a branched chain substrate, waxy maize starch; and amylose, the linear component of corn starch.[41,59,60,64] These investigations included comparisons not only of the rates of the hydrolysis of the different substrates but also of the products formed from them.

With the same concentration of pancreatic amylase reacting under comparable conditions, no marked differences were observed in the rate of the hydrolysis of any of the unfractionated ordinary starches studied.[41,59,60,64] On the other hand significant differences were observed in the rate of the hydrolysis of straight and of branched chain substrates. The data[60] in Table IV show that waxy maize starch is hydrolyzed more slowly than unfractionated corn starch and much more slowly than the

(64) Roslyn B. Alfin and Mary L. Caldwell, *J. Am. Chem. Soc.*, **71**, 128 (1949).

linear fraction from corn starch. The data in Table V show that during the early stages of hydrolysis, unfractionated potato starch is hydrolyzed more rapidly than amylose. However, after approximately three per cent of the glucosidic linkages of these substrates had been broken the linear substrate was hydrolyzed more rapidly than the unfractionated starch. These results suggest that there are more points of attack for pancreatic amylase in the branched chain than in the straight chain components of starch but that the reducing dextrins formed from the branched

TABLE IV

A Comparison of the Action of Purified Maltase-free Pancreatic Amylase on Waxy Maize Starch and on Other Substrates
(Data of Mindell, Agnew and Caldwell[60])
(Reducing Values as Percent Theoretical Maltose)

Reaction time (minutes)	Relative concentrations of amylase					
	1			8		
	Waxy maize starch	Corn starch	Linear substrate	Waxy maize starch	Corn starch	Linear substrate
2.5	36%	40%	60%	62%	65%	84%
5	50	56	80	64	66	85
10	55	61	81	65	68	86
20	60	66	86	65	70	85
30	62	68	87	68	70	85
45	63	69	87	69	72	88
60	63	—	87	70	74	89
90	65	70	87	71	74	90
120	65	71	87	73	75	91
180	67	71	87	75	79	93
300	69	72	90	79	82	94

chain components are hydrolyzed less rapidly than are the reducing dextrins formed from the linear substrate.

The data given in Table V show not only that pancreatic amylase hydrolyzes unfractionated starch and a linear substrate at different rates but also that, for equivalent time intervals with the same concentration of pancreatic amylase, the relative concentrations of the products formed from these two substrates differ. In addition, Table VI[60,64] summarizes comparative data for the products of the hydrolyses of potato starch, of corn amylose, and of waxy maize starch when equivalent numbers of glucosidic linkages of these substrates had been broken.

Maltose is liberated by pancreatic amylase in measurable concentrations even during the early stages of the hydrolysis of all of the substrates investigated. It was present in considerably higher concentrations in the hydrolyzates from the linear substrate than in those from unfractionated potato starch or from the branched chain substrate, waxy maize starch. This observation holds whether the comparisons are made for the same time intervals, with equal concentrations of amylase (Table V), or whether they are made for these substrates at equivalent stages in their hydrolyses (Table VI).[61,64]

TABLE V

A Comparison of the Hydrolysis of Potato Starch and of Corn Amylose by Purified Maltase-free Pancreatic Amylase
(Data of Alfin and Caldwell[64])

Reaction time (minutes)	Reducing values as percent theoretical maltose				Glucose[a] as percent of theoretical glucose		Dextrins[a]			
	Total		Maltose[a]				Percent of total products by weight		Average degrees of polymerization[b]	
	Potato starch	Amylose	Potato starch	Amylose	Potato starch	Amylose	Potato starch	Amylose	Potato starch	Amylose
5	2.94%	1.55%	1.57%	1.32%			98.4%	98.7%	144 $\overline{DP}$	860 $\overline{DP}$
10	4.08	3.34	1.92	2.29			98.1	97.7	89	464
15	5.36	5.27	2.12	3.56			97.9	96.4	60	113
20	6.40	6.64	2.53	3.83			97.5	96.2	51	68
25	7.49	8.19	2.94	4.09			97.1	95.9	43	47
30	8.58	9.97	3.46	5.27			96.5	94.7	38	40
45	12.8	14.8	4.50	6.42	0.24%		95.3	93.6	24	22
60	16.6	19.3	6.00	8.00	0.55	0.15%	93.5	91.9	20	17
90	23.2	27.9	7.92	10.5	0.89	0.65	91.2	88.9	13	11

[a] Glucose or maltose by selective fermentation with washed baker's yeast; dextrins by difference.
[b] Average degrees of polymerization of dextrins were calculated from their reducing values as follows:

$$\overline{DP} = \frac{\text{weight of dextrins (mg)}}{\text{reducing value as maltose (mg)}} \times 2$$

Glucose, also, is liberated by maltase-free pancreatic amylase from all of the substrates investigated, although this sugar does not appear in the very early stages of the hydrolysis of these substrates. Even after fifteen per cent of the glucosidic linkages of the substrates had been broken, only about one per cent of glucose was present (Table VI).

Not only does glucose appear somewhat earlier but also it is liberated in higher concentrations in the hydrolyzates from potato starch than in those from amylose. In addition, Table VI presents evidence for the more rapid liberation of glucose from waxy maize starch than from

TABLE VI

Products Formed from Potato Starch, from Waxy Maize Starch and from Corn Amylose by Purified Maltase-free Pancreatic Amylase
(Data by Mindell, Agnew and Caldwell[60] and by Alfin and Caldwell[64])

Reducing values as percent theoretical maltose				Glucose[a] as percent theoretical glucose			Dextrins[a]			Average degrees of polymerization[b]		
	Maltose[a]						Percent of total products by weight					
Total	Waxy maize starch	Potato starch	Amylose	Waxy maize starch	Potato starch	Amylose	Waxy maize starch	Potato starch	Amylose	Waxy maize starch	Potato starch	Amylose
2.9%		1.57%	2.0%					98.4%	98.0%		148$\overline{DP}$	392$\overline{DP}$
5		2.2	2.7					97.8	97.3		70	85
10		3.5	4.5		0.40%			96.5	95.5		30	35
15		4.7	6.5		0.60			94.9	93.5		20	22
20		6.5	8.3		1.0	0.2%		92.9	91.5		15	16
25		8.0	10.5		1.3	0.5		91.0	89.0		12	13
30		9.0	12.5		1.6	0.8		89.7	86.7		9.8	11
35		10.5	14.5	3.0%	1.8	1.2		87.9	84.3		8.3	9.3
40	9.0%	12.5	17.5	4.5	1.8	1.6	88 %	85.7	80.9	7.0$\overline{DP}$	7.2	8.4
50	15.0	16.5	23.0	5.7	2.4	2.3	80.5	81.1	74.7	6.2	5.7	6.7
60	21.0	21.0	29.5	8.9	4.0	3.2	73.3	75.0	67.3	5.3	4.8	5.6
70	29.5	25.5	38.5	13.2	6.6	4.0	61.6	67.9	57.5	5.4	4.3	4.9
80	35.5	31.0		16.0	9.4		51.3	59.6		5.7	3.9	
90	40.0	41.5		17.5	13.0		44.0	45.5		4.9	4.0	
100	45.8	50.5			17.0		36.7	32.5		3.8	4.2	
110		52.0			21.4			26.6			3.5	

[a] Glucose by selective fermentation with a yeast which does not ferment maltose; maltose by selective fermentation with washed baker's yeast; dextrins by difference.
[b] Please see footnote b, Table V.

unfractionated potato starch or from amylose. These results suggest that pancreatic amylase liberates glucose more readily from its branched chain than from its linear substrates.[60,64]

A study of the reducing dextrins (Tables V and VI),[60,64] provides further evidence for differences in the mechanism of the hydrolysis of these three substrates by pancreatic amylase. During the early stages of hydrolyses the high average molecular weight of the reducing dextrins liberated from the linear substrate as compared to the average molecular weight of the dextrins from unfractionated potato starch again suggests that the branched chain components of starch offer more points of attack for pancreatic amylase than do the linear components. Eventually however, the linear dextrins appear to be hydrolyzed more extensively than the branched chain dextrins as indicated by the relative concentrations of dextrin and maltose present in their respective hydrolyzates.

None of the substrates investigated were hydrolyzed completely to maltose and glucose. During the hydrolysis of the linear substrate there appear to accumulate dextrins that are too short for the amylase to hydrolyze readily. The average molecular weights of the dextrins remaining after the extensive hydrolysis of waxy maize starch by pancreatic amylase decrease more slowly than do those of dextrins from potato starch or from the linear substrate. It is probable that this is due to a higher concentration of low molecular weight dextrins that contain 1,6 α-D-glucopyranosidic linkages in addition to dextrins such as maltotriose that contain only 1,4 α-D-glucopyranosidic linkages.

Additional information about the action of pancreatic amylase is obtained from a study of the dextrins formed by this amylase from exhaustively defatted corn starch.[59] The hydrolyzate studied had reached a stage of very slow rate of change at the equivalent of sixty-eight per cent theoretical maltose. Reducing dextrins accounted for sixty-two per cent by weight of the total hydrolysis products and had an average degree of polymerization of 4.6 glucose units. This average degree of polymerization is similar to that observed for the dextrins from amylose and from potato starch and is somewhat lower than that observed for the dextrins from waxy maize starch when these substrates had been hydrolyzed by pancreatic amylase to an equivalent extent (Table VI).[60,64] The hydrolyzate was fractionated with methanol. The fractions were dried and analyzed for glucose, maltose and reducing dextrins.[59] After the fractionation, the average degrees of polymerization of the reducing dextrins in the different fractions ranged from 9.9 to 3.6 glucose units with more than half of the dextrins present in the fraction of lowest molecular weight. This fractionation of dextrins with average degrees of polymerization of 4.6 into fractions with such widely different average molecular weights gives additional evidence for the random hydrolysis of starch by pancreatic amylase.

Studies of the specific rotations of these dextrins, of the action of beta amylase upon them, and comparisons of the reducing values of these dextrins and of the beta-dextrins formed from them by beta amylase, all lead to the conclusion that the slowing down of the hydrolysis of starch by pancreatic amylase is due in part to the presence of "anomalous"[65] 1,6 α-D-glucosidic linkages in the dextrins, especially in those of higher average molecular weights.[59,66,67] The evidence indicates that the more soluble dextrins of lower average molecular weights remaining after the action of pancreatic amylase are made up largely of linear glucosidic chains but too short to be hydrolyzed readily by pancreatic amylase. A large proportion of these latter dextrins was hydrolyzed completely to reducing sugars by beta amylase.[59]

g. Action with Methylated Dextrins.—Experiments with methylated dextrins showed that pancreatic amylase requires free hydroxyl groups in its substrate. Pancreatic amylase caused no measurable hydrolysis of methylated dextrins with average degrees of polymerization of 13 glucose units and methoxyl contents of 30.8 or 43.3 per cent (67 or 94 per cent of theory, respectively). Under the same conditions the same amylase preparation caused the rapid hydrolysis of the unmethylated dextrins.[59]

2. Amylase of Aspergillus oryzae

a. Chemical Nature and Properties.—The mold, *Aspergillus oryzae*, is an important commercial source of amylase. Takamine[68] described a method for the concentration of the amylase which is the chief active constituent of "taka diastase." Taka diastase and similar products have found wide use in medicine and in industry. The amylase has been purified by a number of investigators[3] and the most highly purified preparations so far reported are protein.[69]

Even after extensive purification, the amylase of *Aspergillus oryzae* is relatively stable in aqueous solutions held at ordinary room temperature. Its lability increases with increasing temperatures and becomes very rapid between 50° and 60°. This loss of activity may be retarded by the presence of substrate and by the presence of calcium ions.[40,70]

(65) K. Myrbäck and K. Ahlborg, *Svensk Kem. Tid.*, **49**, 216 (1937).
(66) K. Ahlborg and K. Myrbäck, *Biochem. Z.*, **308**, 187 (1941).
(67) S. Peat, E. Schlüchterer and M. Stacey, *J. Chem. Soc.*, 581 (1939).
(68) J. Takamine, *J. Soc. Chem. Ind. London*, **17**, 118 (1898).
(69) Mary L. Caldwell, Ruth M. Chester, A. H. Doebbeling and Gertrude W. Volz, *J. Biol. Chem.*, **161**, 361 (1945).
(70) Virginia Hanrahan and Mary L. Caldwell (unpublished).

The amylase of *Aspergillus oryzae* is most active in slightly acid solutions. When reacting in the presence of 0.01 M acetate at 40° it is most active[71] at pH 5.0.

The amylase of *Aspergillus oryzae* causes a very rapid decrease in the viscosity of its substrates and a very rapid disappearance from its reaction mixtures of products which give color with iodine. When examined under favorable conditions[71] at 40° with Lintner's soluble potato starch, the achroic point was reached with highly purified maltase-free amylase when approximately 12% of the glucose linkages of the substrate had been ruptured.

When measured at 40° with Lintner's soluble potato starch, the ratio of the dextrinogenic to the saccharogenic activities is approximately 6 to 1 for the amylase of *Aspergillus oryzae*. This value is given by both crude and purified preparations of the amylase if the measurements are carried out under comparable conditions. This constancy in the ratio of these two activities has led to the conclusion that, like pancreatic amylase, the amylase of *Aspergillus oryzae* is not accompanied in nature by beta amylase. Reacting mixtures of the amylase of *Aspergillus oryzae* and starch exhibit alpha mutarotation.[72]

b. Extent of Hydrolysis of Starch.—The data summarized in Figure 5 were obtained[73] when different concentrations of highly purified maltase-free amylase of *Aspergillus oryzae* reacted with Lintner's soluble potato starch under conditions designed to favor the action of the amylase and to protect it from inactivation.[71]

The data show that the extent of the hydrolysis of starch by the amylase of *Aspergillus oryzae* depends within wide limits upon the concentration of amylase used. Like those for pancreatic amylase already discussed (Figure 2), these hydrolysis curves show a change from a rapid to a slow phase of the reaction and tend to flatten at higher values as the concentration of amylase is increased. Again, with different concentrations of the amylase of *Aspergillus oryzae* there is no evidence of a common limit such as is observed with different concentrations of beta amylase (Figure 1).

Although the extent of the hydrolysis of starch by pancreatic amylase or by the amylase of *Aspergillus oryzae* depends within wide limits upon the concentration of amylase used, marked differences have been noted in the relative increases in amylase concentration required for these two alpha amylases to effect an equivalent increase in the extent of the

(71) Mary L. Caldwell and Susie E. Doebbeling, *J. Am. Chem. Soc.*, **59**, 1835 (1937).
(72) R. Kuhn, *Ann.*, **443**, 1 (1925).
(73) Gertrude W. Volz and Mary L. Caldwell, *J. Biol. Chem.*, **171**, 667 (1947).

hydrolysis of a given substrate. This point is illustrated by the comparison given[41,73] in Table VII for the action of purified maltase-free preparations of pancreatic amylase and of the amylase of *Aspergillus oryzae*. Starting with concentrations of these amylases which gave superimposable hydrolysis curves, it was found that a two-hundred-fold increase in the concentration of pancreatic amylase caused a less extensive hydrolysis of starch than a thirty-two-fold increase in the concentration

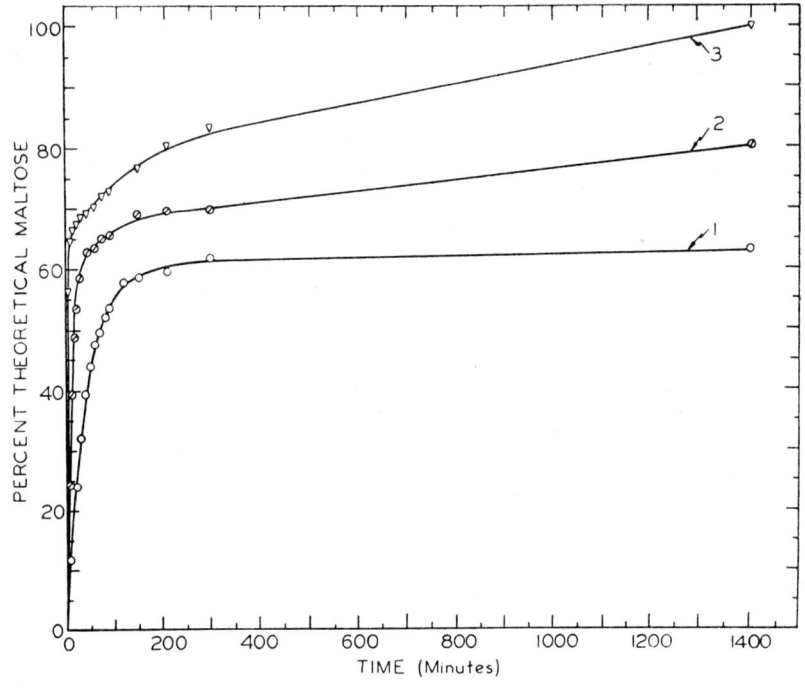

FIG. 5.

of the amylase of *Aspergillus oryzae*. The preparation of the amylase of *Aspergillus oryzae* which, in relatively low concentrations was only about one-half as potent as that of pancreatic amylase, weight for weight, was much more potent at higher concentrations when comparisons were made in the later stages of the hydrolysis of the same substrate. Although no evidence was obtained for the presence of contaminating carbohydrases, this possibility has not been entirely excluded, particularly for the amylase of *Aspergillus oryzae*, and may explain in part the differences observed for the relative activities of the more concentrated solutions of these enzyme preparations.

c. *Stages of Very Slow Rates of Change.*—Results very similar to those described for pancreatic amylase and illustrated in Figure 3, were obtained[41] with highly purified maltase-free amylase of *Aspergillus oryzae*[73] and lead to the conclusion that with this amylase also, inactivation does not explain the slowing down of the reactions at different extents of hydrolysis observed with different concentrations of the amylase.

Similarly, the addition of maltose to reaction mixtures failed to influence the extent of the hydrolysis of Lintner's soluble potato starch

TABLE VII

A Comparison of the Hydrolysis of Soluble Potato Starch by Purified Maltase-free Pancreatic Amylase or Amylase of Aspergillus oryzae
(Extent of Hydrolysis as Percent Theoretical Maltose)
(Data of Alfin and Caldwell[41] *and of Volz and Caldwell*[73])

Reaction time (minutes)	Relative amylase concentrations as mg preparation per 1000 mg starch					
	Aspergillus 0.09 (1)	Pancreatic 0.04 (1)	Aspergillus 0.72 (8)	Pancreatic 1.00 (25)	Aspergillus 2.88 (32)	Pancreatic 8.00 (200)
15	14.5%	14.0%	48.0%	65.5%	66.0%	72.0%
20	18.9	18.3	53.5	67.0	67.5	73.0
30	25.8	25.5	58.5	69.0	68.8	74.0
45	34.0	34.0	62.5	70.0	70.0	74.8
60	41.0	40.8	64.0	71.0	71.2	75.5
90		48.0	66.5	72.0	73.0	77.0
120		52.2	67.5	73.0	74.5	78.3
240			69.8	74.0	81.5	80.0
300			69.8	74.0	83.5	81.0
1400			80.4	76.0	100.0	91.5

and led to the conclusion that the slowing down of the hydrolysis with the amylase of *Aspergillus oryzae* is not due to equilibrium between the enzyme and maltose, one of the products of the reaction.[73]

d. *Products.*—The data summarized in Table VIII[64,73] give average values for the distribution of the products formed at different stages in the hydrolysis of Lintner's soluble potato starch by highly purified maltase-free preparations of pancreatic amylase and of the amylase of *Aspergillus oryzae*. The amylase of *Aspergillus oryzae* causes a rapid breakdown of starch to reducing dextrins of low molecular weight. Glucose and maltose appeared in small amounts when the hydrolysis had reached 20 to 30% theoretical maltose, and accounted for 90% of the reducing value of the reaction mixture at 100% theoretical maltose.

The degree of polymerization of the reducing dextrins dropped rapidly to 10 and then slowly to 4. These findings are in accord with the observations of many investigators that the amylase of *Aspergillus oryzae* causes a very rapid decrease in the viscosities of its substrates and a very rapid disappearance of products which give color with iodine.

Myrbäck[65,74–77] studied products obtained from a number of different starches after the prolonged action of unpurified taka diastase. He

TABLE VIII

Products Formed from Lintner's Soluble Potato Starch by Purified Maltase-free Pancreatic Amylase or Amylase of Aspergillus oryzae
(Data of Alfin and Caldwell[64] and of Volz and Caldwell[73])

Reducing values as percent theoretical maltose		Glucose^a as percent theoretical glucose		Dextrins^a			
				Percent of total products by weight		Average degrees of polymerization^a	
Total	Maltose^a						
	P^b	A^b	P^b	A^b	P^b	A^b	P^b	A^b
10%	1.0%		0.6%		98.4%	100%	25 DP	20 DP
15	2.0		0.6		97.4	100	17	13
20	4.0	0.5%	0.7		95.3	99.5	13	10
25	6.0	1.5	0.8	0.15%	93.2	98.3	11	8.5
30	9.0	2.5	1.0	0.30	90.0	97.2	9.5	7.2
35	11.5	4.0	1.3	0.50	87.2	95.5	8.3	6.4
40	14.5	6.0	1.7	0.80	83.8	93.2	7.6	5.7
50	21.0	13.5	3.0	1.45	76.0	85.0	6.6	5.1
60	27.0	26.0	4.8	2.35	68.2	71.6	5.9	4.9
70	35.0	38.0	6.5	4.2	58.5	57.8	5.3	4.9
80	43.0	44.5	7.9	8.0	49.1	47.5	4.6	4.9
90		54.5		10.2		35.3		4.7
100		68.5		11.0		20.5		4.3

^{a} Please see footnotes of Table V.
^{b} P = Pancreatic Amylase; A = Amylase of *Aspergillus oryzae*.

reports that hexa-saccharides predominated in all cases but that longer and shorter saccharides also were present. He concludes from his work that 1,6 α-D-glucopyranosidic linkages of starches accumulate in the unhydrolyzed products and are responsible in part for the resistance of these products to hydrolysis by taka diastase or by other alpha amylases. He was able to isolate and identify isomaltose from among the products

(74) K. Ahlborg and K. Myrbäck, *Biochem. Z.*, **297**, 172 (1938).
(75) K. Myrbäck, B. Örtenblad and K. Ahlborg, *Biochem. Z.*, **315**, 240 (1943).
(76) K. Mrybäck, B. Örtenblad and K. Ahlborg, *Biochem. Z.*, **316**, 424 (1944).
(77) K. Myrbäck, K. Ahlborg and B. Örtenblad, *Biochem. Z.*, **316**, 444 (1944).

obtained by prolonged action of taka diastase on certain fractions which in turn had been obtained after the extensive action of taka diastase on corn starch.[74]

Myrbäck[76,77] found relatively high concentrations of phosphorus in some of the dextrins of higher molecular weight from potato, wheat, barley and arrowroot starches after prolonged action by taka diastase, and suggests that linkages involving phosphorus also appear to present difficulties to the action of the amylase of *Aspergillus oryzae*.

Taken as a whole, the results indicate that the amylase of *Aspergillus oryzae* causes the rapid random hydrolysis both of the straight and of the branched chain components of starch and that it hydrolyzes very slowly products with average molecular weights of penta- and tetra-saccharides.

A study of the comparison given in Table VIII shows that pancreatic amylase and the amylase of *Aspergillus oryzae* break down starch very differently, especially in the earlier stages of its hydrolysis. The data in this table were obtained with samples of the same starch and, therefore, are comparable. In the earlier stages of the hydrolyses, up to 50% theoretical maltose, there was liberated much more maltose, more glucose, and less reducing dextrins by pancreatic amylase than by the amylase of *Aspergillus oryzae*; the reducing dextrins formed by pancreatic amylase were of higher average molecular weight than those formed by the amylase of *Aspergillus oryzae*. These results are in accord with the more rapid decrease in the viscosities and with the more rapid appearance of the achroic point with the amylase of *Aspergillus oryzae* than with pancreatic amylase.

Upon extensive hydrolysis of starch by either of these enzymes, only small differences were observed in the concentrations of the products or in the average degrees of polymerization of the dextrins. These comparisons were made for equivalent stages of hydrolyses and are not necessarily related to the rates of the hydrolysis of starch by these two amylases.

3. *Alpha Amylase of Malted Barley*

a. Occurrence.—Alpha amylase does not appear to be present in active form in appreciable concentrations in sound ungerminated barley grains. Upon the germination and malting of barley, the available or measurable beta amylase is markedly increased and alpha amylase is formed or set free. Thus, untreated crude extracts of malted barley contain both alpha and beta amylases.[3]

In crude aqueous extracts of malted barley the alpha amylase is usually much more thermostable than the beta amylase. This difference in the stability of the two amylases in crude aqueous extracts was recognized by early investigators and was utilized by Ohlsson to remove beta amylase from its mixtures with alpha amylase. According to Ohlsson,[23] if crude aqueous extracts of malted barley, at their natural acidity, are held at 70° for 15 minutes, the beta amylase will be inactivated completely while much of the alpha amylase activity will remain. This observation of Ohlsson's has been confirmed by many investigators and is the basis for procedures widely used to prepare alpha amylase solutions from malted grains.

On the other hand, in crude aqueous extracts of malted barley the alpha amylase is much more sensitive to high hydrogen ion activities than the beta amylase. This difference between the two amylases also was utilized by Ohlsson[23] to inactivate the alpha amylase and to prepare beta amylase relatively free from alpha amylase.

Examination of the ratios of the dextrinogenic to the saccharogenic activities of malted barley extracts before and after treatment shows that the results of the Ohlsson procedures[23] are not always predictable.[3] The concentration of the amylases in the extracts, and the kinds and concentrations of substances which accompany them, influence the results. The presence or the absence of calcium ions is an important factor. Calcium ions increase the inactivation of beta amylase of malted barley and protect the alpha amylase from inactivation at unfavorable temperatures and also at unfavorable hydrogen ion activities.[28] With purification, both amylases become increasingly thermolabile and increasingly sensitive to unfavorable hydrogen ion activities.[78]

The marked changes in the ratios of dextrinogenic to saccharogenic activities found when crude extracts of malted barley are heated or acidified, led to the recognition of two amylases in these extracts. Conversely, relatively small changes in these ratios in amylase solutions before and after various procedures are often taken as proof that only one amylase is present.

b. Chemical Nature and Properties.—The alpha amylase of malted barley has been purified by a number of investigators[3] and the most highly active preparations so far reported are protein.[77–79] They are very thermolabile. Highly purified preparations were inactivated completely when held in dilute aqueous solution for one minute at 50°.[78] The amylase is also inactivated rapidly in aqueous solutions at high hydrogen ion activities.[78] Calcium ions protect malted barley alpha amylase from inactivation at unfavorable temperatures and also at unfavorable hydrogen ion activities.[28]

The alpha amylase of malted barley is most active in slightly acid

(78) Mary L. Caldwell and Susie E. Doebbeling, *J. Biol. Chem.*, **110**, 739 (1935).
(79) S. Schwimmer and A. K. Balls, *J. Biol. Chem.*, **176**, 465 (1948).

solutions. When reacting in the presence of 0.01 M acetate at 40°, it is most active at pH 4.5.[78]

The alpha amylase of malted barley causes a rapid decrease in the viscosity of its substrates and the rapid disappearance from its reaction mixtures of products which give color with iodine. Myrbäck[80] reports a drop of 50% in the relative viscosity of potato starch when only approximately 0.1% of the glucose linkages of the substrate had been ruptured under the influence of this amylase. The achroic point for reactions of malted barley alpha amylase usually is stated to occur at approximately 30% theoretical maltose.[1,3] In contrast, the achroic point is reached with pancreatic amylase at 40% and with the amylase of *Aspergillus oryzae* at 25%.[41,73]

Very highly purified preparations of alpha amylase of malted barley give a value of approximately 4 to 1 for the ratio of their dextrinogenic to their saccharogenic activities when the measurements are made at 40° with Lintner's soluble potato starch.[81] Under the same conditions approximately the same value is obtained with products precipitated by alcohol from malted barley extracts which had been treated to inactivate beta amylase.[23,81] However, a constant value for these ratios is not proof that beta amylase is entirely absent. There is at present no satisfactory way of making certain that malted barley alpha amylase is not contaminated with traces of beta amylase. The crystallization of alpha amylase from malted barley has been reported since this manuscript was written.[79]

The reacting mixtures of malted barley alpha amylase and starch exhibit alpha mutarotation.[23,72] Myrbäck[82] has presented evidence to show that this mutarotation is exhibited by the dextrins as well as by the sugars formed from starch by this enzyme.

Free aldehyde groups do not appear to be essential to the action of malted barley alpha amylase. Myrbäck[82,83] reports that, like beta amylase,[1,3] this amylase also causes the hydrolysis of dextrins after their aldehyde groups have been oxidized to the corresponding acids.

c. Extent of Hydrolysis of Starch.—The data given in Figure 6 were taken from a report by Myrbäck[84] for the action on arrowroot starch of different concentrations of purified alpha amylase from malted barley. The amylase was prepared from extracts of malted barley which had been

(80) K. Myrbäck and L. G. Gjörling, *Arkiv Kemi, Mineral, Geol.*, **20A**, No. 5 (1945).
(81) Mary L. Caldwell and A. H. Doebbeling (unpublished).
(82) K. Myrbäck, *Biochem. Z.*, **307**, 140 (1941).
(83) B. Örtenblad and K. Myrbäck, *Biochem. Z.*, **307**, 129 (1941).
(84) K. Myrbäck, *Biochem. Z.*, **311**, 227 (1942).

treated according to Ohlsson[23] to remove beta amylase. It was maltase-free and gave no evidence of the presence of contaminating glucosidases. Approximately two per cent arrowroot starch was used in both reaction mixtures while the concentration of amylase in the reaction mixture represented by curve 2 was approximately three times as high as that used in the reaction mixture represented by curve 1.

These data with arrowroot starch are similar to those reported by many previous investigators for the action of different concentrations of malted barley alpha amylase on other substrates.[1,3,19,81] It is evident

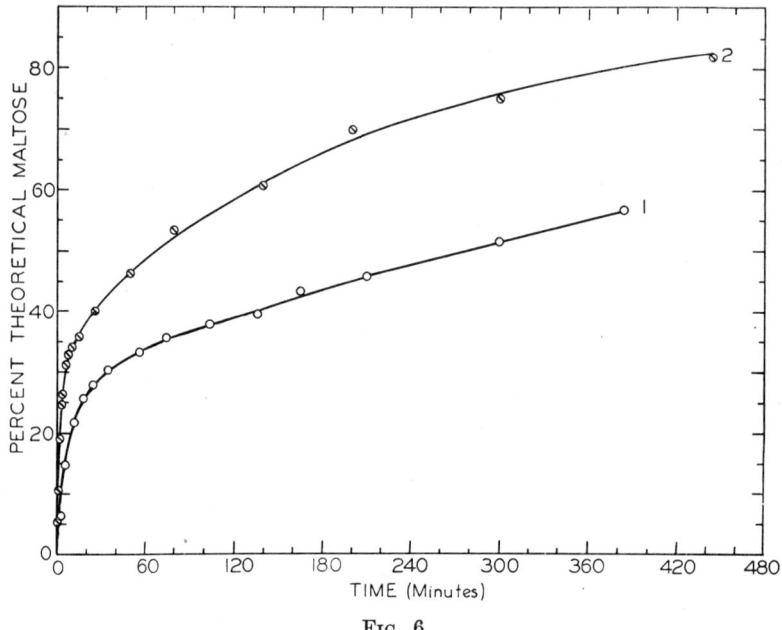

FIG. 6.

that with malted barley alpha amylase also, the extent of the hydrolysis of starch depends within wide limits upon the concentration of amylase used. Again, these data are very different from those reported in Figure 1 for beta amylase.

The data summarized in Figures 2, 5 and 6 show that a flattening of the reaction curves at different extents in the hydrolysis of starch is characteristic of the action of the three alpha amylases discussed here. Both the position of the break in the reaction curves and the extent of the hydrolysis depend within wide limits upon the concentration of amylase used. There is no evidence for the suggestion[85] that the flattening of the reaction curves occurs at a definite per cent of hydrolysis or

(85) K. Myrbäck and W. Thorsell, *Svensk Kem. Tid.*, **54**, 50 (1942).

that a final equilibrium is reached which is characteristic of each amylase. Therefore, in the opinion of the authors, generalizations about the way these amylases act do not seem justified when based upon a reaction curve obtained with one concentration of an amylase.

Studies of the rate of the hydrolysis of dextrins isolated from a reaction mixture after the extensive hydrolysis of starch by maltase-free malted barley alpha amylase, led Myrbäck[11] to conclude that the flattening of the reaction curves with this amylase is not due to equilibrium between the amylase and the products of the hydrolysis. As indicated above, similar conclusions have been reached for pancreatic amylase and for the amylase of *Aspergillus oryzae*.[41,73]

d. *Products.*—Table IX summarizes data reported by Myrbäck[84] for the hydrolysis of arrowroot starch by purified maltase-free malted

TABLE IX

Products Formed from Arrowroot Starch by Purified Maltase-free Alpha Amylase from Malted Barley

(Data of Myrbäck[84])

Time (minutes)	Color with iodine	Theoretical maltose from reducing value[a]	Glucose linkages hydrolyzed	Fermentable sugar		Dextrins	
				Glucose	Maltose	Yield	Average degrees of polymerization[b]
2.8	Blue	6.3%	3.2%	0%	0%	100%	19$\overline{DP}$
6	Violet	14.5	7.3	0	0	100	12.3
12	Red	21.8	11.0	< 2	~ 5	92	10.9
18	Red?	25.6	12.8	< 2	7.6	91	8.1
25	Colorless	27.9	13.9	~ 2	8.5	88	7.8
35	"	30.4	15.2	~ 3	11.1	87	7.5
56	"	33.4	16.7	4	12.5	85	7.2
75	"	35.7	17.8	4	13.3	85	7.1
105	"	37.9	18.9	4	14.7	85	6.9
135	"	39.6	19.8	6	14.7	80	6.7
165	"	43.5	21.7	6	21.3	78	6.3
210	"	46.0	23.0	6	21.3	75	6.4
300	"	51.7	25.8	10	23.1	73	5.7
385	"	56.5	28.3		25.8	67	5.5
520	"	58.2	29.1		29.3	62	5.5
12 hrs.	"	59.2	29.6		38.1	54	5.3
23 "	"	72.0	36.0	12	53.8	39	5.1

[a] Arrow root starch, approximately 2%; pH 5.3; amylase prepared from malted barley extract according to Ohlsson[23] and purified by precipitation with alcohol.

[b] Average degrees of polymerization calculated from results of acid hydrolysis of dried precipitated dextrins that remained after removal of fermentable sugars.

barley alpha amylase. No measurable amounts of maltose or of glucose were present in the very early stages of the hydrolysis of this starch. After the hydrolysis had reached approximately 20% theoretical maltose, both glucose and maltose were present and increased steadily in concentration throughout the hydrolysis. In the early stages of the hydrolysis, the average degree of polymerization of the dextrins decreased very rapidly and, when the hydrolysis had reached 26% theoretical maltose, had already dropped to 8. From then on, the average degree of polymerization of the dextrins decreased slowly and at 72% theoretical maltose was 5.

Although the data are not strictly comparable, a comparison of these results with the data for the hydrolysis of potato starch by pancreatic amylase (Table VI) suggests that the hydrolysis of starch by these two maltase-free enzymes follows quite a different course. Malt alpha amylase, except for the very early stages of hydrolysis, appears to cause the liberation of more maltose and glucose than pancreatic amylase; it also seems to cause a more rapid decrease in the concentration of dextrins and to liberate dextrins of lower average molecular weights. These results are in accord with the differences already noted in the ratios of the dextrinogenic to the saccharogenic activities of these two enzymes, 2:1 for that of pancreatic amylase and 4:1 for that of malted barley alpha amylase. In the very late stages of the hydrolysis by pancreatic amylase, the residual dextrins were of slightly lower average molecular weight than those resulting from hydrolysis by malt alpha amylase. Further data obtained under strictly comparable conditions should add considerable information concerning the properties of the different alpha amylases and should provide further information concerning the substrates which they hydrolyze.

Table X[84] summarizes similar data for the hydrolysis by maltase-free malt alpha amylase of beta dextrins obtained from arrowroot starch by the action of beta amylase. The beta dextrins were precipitated with alcohol from the reaction mixture of arrowroot starch after it had reached a limit in the hydrolysis at 60% theoretical maltose. The beta dextrins were hydrolyzed extensively by malt alpha amylase. Glucose was liberated in very small amounts even in the later stages of the hydrolysis of these beta dextrins; maltose was liberated in appreciable amounts and, at equivalent hydrolyses, appeared to be formed somewhat more rapidly from the beta dextrins (Table X) than from the untreated starch (Table IX). Upon hydrolysis with malt alpha amylase the molecular weights of the beta dextrins dropped appreciably but not as extensively as when arrowroot starch was hydrolyzed directly by malt alpha amylase.

The hydrolysis of amylose by malt alpha amylase has been investigated by Meyer[86] and by Myrbäck.[85] According to Meyer, malt alpha amylase causes a random hydrolysis of the straight chain substrate, corn amylose, and a relatively slow hydrolysis of the short fragments such as maltotriose. A reaction mixture which had reached 98.5% theoretical maltose contained 6.1% glucose, 72.7% maltose and 21.1% dextrins with an average degree of polymerization of 3.1.

TABLE X

Products Formed by Malted Barley Alpha Amylase from Beta Dextrins Formed from Arrowroot Starch by Beta Amylase
(Data of Myrbäck[84])

Time (minutes)	Color with iodine	Theoretical maltose from reducing values[a]	Glucose linkages hydrolyzed	Fermentable sugar		Dextrins	
				Glucose	Maltose	Yield	Average degrees of polymerization[b]
2	Red-violet	2.2%	1.1%	0%	< 2%	98%	22DP
3	Red-violet	—	—	—	—	—	—
5	Red	6.0	3.0	0	< 3	98	15.5
10	Red-brown	10.7	5.4	0	~ 5	96	12.7
15	Brown	14.5	7.3	0	~ 6	95	12.3
24	Brown?	18.7	9.4	0	~ 6	95	10.9
36	Colorless?	23.0	11.5	0	8	94	10.2
50	Colorless	23.0	11.5	0	8	90	10.1
70	Colorless	23.0	11.5	0	8	90	9.6
100	Colorless	24.2	12.1	1	9	90	9.5
145	Colorless	28.8	14.4	2	11	88	9.2
200	Colorless	28.8	14.4	2	12	84	8.7
45 hrs.	Colorless	68.8	34.4	5	52	45	5.7

[a] Arrowroot starch, approximately 2%; pH 5.3; maltase-free amylase prepared according to Ohlsson[23] from malted barley extract and purified by precipitation with alcohol.

[b] Average degrees of polymerization were calculated from results of acid hydrolysis of dried precipitated dextrins that remained after removal of fermentable sugars.

Tables XI and XII summarize data obtained by Myrbäck[85] for the hydrolysis of amylose by purified maltase-free malted barley alpha amylase. The hydrolysis curve with this linear substrate is much the same as those obtained with unfractionated starches, and also is similar to the curves representing the hydrolysis of amylose by pancreatic amylase.[41] The flattening of the hydrolysis curves during the later stages

(86) K. H. Meyer and P. Bernfeld, *Helv. Chim. Acta*, **24**, 359E (1941).

of the hydrolysis of this linear substrate indicates that the slowing down of the reaction is due, to a considerable extent, to the presence of dextrins of such low molecular weights that they are attacked very slowly by the amylase. Presumably in this case the influence of "anomalous"[65] linkages has been ruled out by the use of the linear fraction of starch.

TABLE XI

A Comparison of the Action of Purified Maltase-free Alpha Amylase from Malted Barley upon Corn Amylose, Corn Starch and upon Beta Dextrins Formed from Corn Starch by the Action of Beta Amylase
(Data of Myrbäck and Thorsell[85])

Time	Corn amylose				Corn starch		Beta dextrins	
	Experiment No. 1	Experiment No. 2			Theoretical maltose from reducing values	Fermentable sugar	Theoretical maltose from reducing values	Fermentable sugar
	Theoretical maltose from reducing values	Theoretical maltose from reducing values	Fermentable sugar					
			Maltose	Glucose				
2 min.	—	10.8%	3.3%	< 1%	—	—	—	—
3 "	—	—	—	—	—	—	1.3%	0%
6 "	22.2%	—	—	—	4.6%	0%	3.1	—
9 "	26.2	30.0	6.0	< 1	6.4	0	—	—
15 "	35.0	37.0	—	—	10.1	3.1	6.2	2
30 "	44.3	44.2	13.6	< 1	17.4	5.7	10.2	4.4
60 "	—	45.9	17.2	—	22.9	14.6	14.2	6.1
120 "	—	48.4	—	—	29.3	18.9	18.2	8.2
12 hrs.	—	63.4	33.4	3.5	—	—	—	—
24 hrs.	80.0	—	—	—	52.7	43.0	20.8	—
4 days	—	91.0	82.0	—	71.0	—	22.6	13.0
8 days	93.0	96.8	—	—	79.0	—	23.0	14.0
32 days	—	107	91.8	8.6	87	—	—	—

Both maltose and glucose were present in the reaction mixtures with amylose,[85] (Table XI). Therefore, glucose is liberated in addition to maltose from this straight-chain substrate as well as from unfractionated starch by maltase-free malted barley alpha amylase (Table IX).[84] It appears from the results reported by Myrbäck (Table XI)[85] that amylose can be hydrolyzed completely to fermentable sugar by malted barley alpha amylase but only after a prolonged period of hydrolysis (32 days). A more recent report by Myrbäck[87] confirms and extends this con-

(87) K. Myrbäck, *Arch. Biochem.*, **14**, 53 (1947).

clusion. The substrate used in this work was a dextrin fraction made up chiefly of hexa-saccharides.[88] These dextrins were hydrolyzed completely to fermentable sugar by beta amylase and, therefore, were considered to contain only 1,4 α-D-glucopyranosidic linkages.[89] These dextrins were hydrolyzed completely to glucose, maltose and maltotriose after prolonged action (20+ days) by maltase-free malted barley alpha amylase.[87] When all of the products were calculated to their equivalents of glucose, the action of malted barley alpha amylase on the dextrins (chiefly hexa-saccharides) gave 17% glucose, 60% maltose and 23% maltotriose. Myrbäck[87] concludes that malted barley alpha amylase

TABLE XII
A Fractionation of the Dextrins Formed from Corn Amylose by Purified Maltase-free Malted Barley Alpha Amylase
(Data of Myrbäck and Thorsell[85])

Fraction[a] (number)	Yield in percent of amylose	Reducing value as percent glucose	Average molecular weights	Average degrees of polymerization
1	0.4%			
2	2.2	10.9%	1650	10
3	18.5	15.2	1180	7
4	26.7	17.8	1010	6
5	22.0	26.1	690	4
Total	69.8			

[a] Corn amylose was hydrolyzed for 1 hour to 45% theoretical maltose. Sugars were removed by fermentation and remaining dextrins precipitated with alcohol.

does not attack maltotriose. This observation is in accord with the conclusion by Meyer[86] referred to above.

An interesting comparison is also given in Table XI[85] of the action of malted barley alpha amylase upon corn starch, beta dextrins obtained from corn starch by the action of beta amylase, and upon corn amylose prepared according to the method of Meyer.[6,90] These data show that, with equivalent amounts of enzyme, amylose was hydrolyzed much more rapidly and more extensively than unfractionated corn starch. Although the beta dextrins were hydrolyzed by this amylase, they were hydrolyzed much more slowly and less extensively than either of the other two substrates. These results are similar to those already considered for pancreatic amylase.[41,64]

(88) B. Örtenblad and K. Myrbäck, *Biochem. Z.*, **307**, 123 (1941).
(89) K. Myrbäck, *Biochem. Z.*, **307**, 132 (1941).
(90) K. H. Meyer, W. Brentano and P. Bernfeld, *Helv. Chim. Acta*, **23**, 845 (1940).

Myrbäck has carried out numerous investigations of the properties of dextrins fractionated from hydrolysis mixtures obtained by the action of malted barley alpha amylase upon a variety of starches.[88,89,91–96] With all of the starches thus investigated, the fractions isolated consisted of dextrins of various chain lengths. In several cases the fractions of lower average molecular weights were hydrolyzed completely to fermentable sugar by beta amylase.[89,93,96] The fractions containing alpha dextrins of higher average molecular weights were hydrolyzed slowly and incompletely by beta amylase.[89,93,96] This behavior toward beta amylase and also the optical rotations of the isolated fractions[88] gave evidence of the presence of 1,6 α-D-glucopyranosidic linkages in the less readily hydrolyzed alpha dextrins of higher average molecular weights. Phosphorus, when present in the starch, was also found to be concentrated in the less readily hydrolyzed dextrins.[65,88]

Myrbäck[87] compares the action of malted barley beta and alpha amylases. The beta amylase is believed to attach itself to free non-reducing end groups of normally formed glucosidic chains and to break off maltose until the substrate deviates from the simple 1,4 α-D-glucopyranosidic linkage pattern. According to Myrbäck[87], malted barley alpha amylase "is not only independent of free end groups, but its action is successively hindered by the proximity of end groups. The enzyme has the capacity of attacking and rupturing any maltose linkage in a chain molecule of the starch type, but the velocity is greater for linkages at a distance from end groups."[97] Myrbäck[87] considers that branching points or other "anomalies" act as end groups. Like Meyer[90] and other investigators,[4] Myrbäck[87] believes that the more rapid action of malted barley alpha amylase on linkages far from the end group is due to a higher affinity of the amylase for longer than for shorter normal glucosidic chains. Myrbäck has repeatedly reported that malt alpha amylase hydrolyzes only slowly dextrins with average degrees of polymerization of approximately eight or less.[87]

In general, the results with malted barley alpha amylase show that this amylase hydrolyzes readily the 1,4 α-D-glucopyranosidic linkages of the straight chain components and of the outer branches of the branched chain components of starch; it also hydrolyzes the inner 1,4 α-D-glucopyranosidic linkages between the branches of the amylopectin com-

(91) K. Myrbäck and G. Stenlid, *Svensk Kem. Tid.*, **54**, 103 (1942).
(92) B. Örtenblad and K. Myrbäck, *Biochem. Z.*, **315**, 233 (1943).
(93) K. Myrbäck and B. Martelins, *Biochem. Z.*, **316**, 414 (1943).
(94) K. Myrbäck and B. Örtenblad, *Biochem. Z.*, **316**, 429 (1943).
(95) L. G. Sillén, and K. Myrbäck, *Svensk. Kem. Tid.*, **55**, 294 (1943).
(96) K. Myrbäck, G. Stenlid and G. Nycander, *Biochem. Z.*, **316**, 433 (1944).
(97) L. G. Sillén and K. Myrbäck, *Svensk Kem. Tid.*, **56**, 42 (1944).

ponents but does this less readily and the resulting dextrins are of higher molecular weights. The presence of dextrins containing 1,6 α-D-glucopyranosidic linkages and of dextrins containing phosphorus as well as the presence of short chain low molecular dextrins are responsible, in part at least, for the slowing down of the reaction of malt alpha amylase.

4. Other Alpha Amylases

a. Salivary Amylase.—This enzyme has recently been crystallized by Meyer and his coworkers.[98] It is a protein. Studies of the action of highly purified salivary amylase, similar to those reported here for other amylases, are not available for comparison and this amylase will not be discussed further.

b. Bacterial Amylases.—Various bacterial amylases are known. These amylases are becoming of increasing importance in industry. They have the general properties of alpha amylases. They cause a very rapid decrease in the viscosities of their substrates and the rapid disappearance of products that give color with iodine. Several of these amylases are very thermostable. The amylase of *Bacillus subtilus*, for example, exerts marked liquefying activity at temperatures as high as 95°. In the presence of substrate, it is possible to boil reaction mixtures of this amylase for a short time without causing the complete destruction of the liquefying activity.[99] Meyer and his coworkers[100] have recently reported the crystallization of a bacterial alpha amylase obtained from "biolase". Bernfeld and Studer-Pécha[101] have compared the action of highly purified preparations of a bacterial amylase, of pancreatic amylase and of malt alpha amylase upon amylose and upon amylopectin. They conclude that the action of these three alpha amylases is qualitatively similar but quantitatively very different. The authors attribute the differences in the action of these amylases to differences in their affinities for their substrates.

Kneen[53,54] reports that inhibitors obtained from cereal grains have different retarding effects upon the activity of bacterial and pancreatic amylases when examined under comparable conditions. These findings give further evidence for differences in the action of these alpha amylases.

Unfortunately, much of the work with the bacterial amylases has been carried out with impure products of undefined origin so that it is difficult to compare and evaluate the results. For this reason, with the

(98) K. H. Meyer, Ed. H. Fischer, P. Bernfeld and A. Staub, *Experientia*, **3**, 455 (1947).
(99) L. Wallerstein, *Ind. Eng. Chem.*, **31**, 1218 (1939).
(100) K. H. Meyer, Maria Fuld and P. Bernfeld, *Experientia*, **3**, 411 (1947).
(101) P. Bernfeld and H. Studer-Pécha, *Helv. Chim. Acta*, **30**, 1895, 1904 (1947).

exception of a brief reference to the amylase of *Bacillus macerans*, the bacterial amylases will not be discussed more fully here.

Bacillus macerans forms an amylase which has distinctive action on starches. It hydrolyzes starches rapidly to a mixture of water-soluble non-reducing dextrins from which two characteristic crystalline compounds may be isolated with ease. These compounds are known as the alpha and beta Schardinger dextrins.[102,103] They are closed-ring compounds composed of glucose residues united by 1,4 α-D-glucopyranosidic linkages.[104-106] The alpha dextrin is made up of six glucose residues while the beta dextrin is made up of seven glucose residues.[107] Yields of crystalline dextrins amounting to approximately 40% of a starch substrate[108] and to approximately 70% of a crystalline amylose substrate have been obtained.[109] There is no evidence that these rings are preformed in starches or amyloses in any such concentrations. Therefore, the alpha and beta Schardinger dextrins offer examples of hydrolysis products which were not present as such in the original starches.[109-111]

IV. Discussion and Summary

The alpha amylase of malted barley, the amylase of *Aspergillus oryzae* and pancreatic amylase all are thermolabile proteins that rapidly lose their amylase activities upon exposure to unfavorable temperatures, to unfavorable hydrogen ion activities, or to other unfavorable chemical environments. The loss of amylase activity in aqueous solutions increases with increasing temperatures and is exceedingly rapid for each of these amylases at 50°. The inactivation of each of these amylases at unfavorable temperatures or at unfavorable hydrogen ion activities may be retarded by the presence of suitable concentrations of calcium ions.

Each of these amylases causes a rapid decrease in the viscosity of starch pastes and the rapid disappearance from its reaction mixtures of products that give color with iodine. During the early stages of the hydrolysis of starch, the relative decrease in the viscosity of the substrates for hydrolyzates of equivalent reducing value is most marked

(102) F. Schardinger, *Zentr. Bakt. Parasitenk.*, Abt. II, **22**, 98 (1908-9).
(103) F. Schardinger, *Zentr. Bakt. Parasitenk.*, Abt. II, **29**, 188 (1911).
(104) K. Freudenberg and R. Jacobi, *Ann.*, **518**, 102 (1935).
(105) K. Freudenberg, G. Blomqvist, Lisa Ewald and K. Soff, *Ber.*, **69**, 1258 (1936).
(106) K. Freudenberg and W. Rapp, *Ber.*, **69**, 2041 (1936).
(107) D. French and R. E. Rundle, *J. Am. Chem. Soc.*, **64**, 1651 (1942).
(108) Evelyn B. Tilden and C. S. Hudson, *J. Am. Chem. Soc.*, **61**, 2900 (1939).
(109) R. W. Kerr, *J. Am. Chem. Soc.*, **64**, 3044 (1942).
(110) R. W. Kerr and G. M. Severson, *J. Am. Chem. Soc.*, **65**, 193 (1943).
(111) R. W. Kerr, *J. Am. Chem. Soc.*, **65**, 188 (1943).

with the amylase of *Aspergillus oryzae*, less so with the alpha amylase of malted barley and still less marked with pancreatic amylase.

These amylases all cause a rapid and apparently a random breakdown of starches to reducing dextrins of low molecular weights. After extensive hydrolysis, the residual dextrins have an average chain length of 4–5 glucose units. Maltose and glucose are liberated by all three of the amylases although the rate at which these sugars are liberated and their concentrations depend upon the enzyme and the substrate used. The liberation of glucose by these amylases cannot be attributed to the presence of maltase because the highest concentrations of amylase preparations used in these investigations failed to give any evidence of maltase activity when they reacted with maltose under comparable conditions. However, when excessive concentrations of highly purified pancreatic amylase reacted with maltose in the proportion of 500 mg amylase preparation to 1000 mg maltose, a considerable increase in the reducing value of the reaction mixture was observed.[59] It is impossible to state, with our present information, whether the hydrolysis of maltose by excessive concentrations of pancreatic amylase preparations indicates the presence of a trace of maltase or whether, in accord with Weidenhagen's theory,[112] amylase, as an alpha glucosidase, will also hydrolyze maltose. According to Myrbäck,[87] malt alpha amylase attacks very slowly chains with eight glucose residues or less. It is possible, therefore, that alpha amylases will hydrolyze any 1,4 α-D-glucopyranosidic linkage but that these enzymes have very little affinity for the low molecular weight glucosides and, therefore, that the hydrolysis of maltose itself is exceedingly slow.

The hydrolysis curves for the three alpha amylases considered in this review are all similar in shape. They show a rapid hydrolysis in the early stages with a slow secondary stage of hydrolysis which cannot be explained as due either to inactivation of the enzymes or to the influence of products formed during the hydrolysis. The extent of the hydrolysis and the position of the break in the hydrolysis curve depend upon the concentration of enzyme; the break in the hydrolysis curve does not appear to be a fixed point characteristic of any one enzyme. With each of the alpha amylases discussed here the slowing down of the hydrolysis appears to be due largely to the replacement of the original substrate by products for which the amylase has relatively low affinity.

Differences in the structure of the substrates are reflected in the products formed and in the rate and extent of the hydrolysis. Under comparable conditions, a straight chain substrate is hydrolyzed more rapidly than unfractionated starches that contain mixtures of straight

(112) R. Weidenhagen, *Ergeb. Enzymforsch.*, **1**, 168 (1932).

and branched chain components and these, in turn, are hydrolyzed more rapidly than a branched chain substrate such as waxy maize starch or a substrate containing a high concentration of branched chain components such as beta dextrins. Although there appears to be a tendency for these alpha amylases to liberate more glucose from the branched chain components than from the straight chain components of starches, the data on this point are not entirely consistent and more data under strictly comparable conditions are needed.

The three alpha amylases considered in detail here all hydrolyze readily the 1,4 α-D-glucopyranosidic linkages of the straight chain components and of the outer branches of the amylopectin components of starch; they also hydrolyze the 1,4 α-D-glucopyranosidic linkages between the branches of the amylopectin components. There is considerable evidence that the 1,6 α-D-glucopyranosidic linkages of starches are concentrated in the dextrins that remain in the hydrolysis mixtures of these amylases at stages of slow rates of reaction. Phosphorus has been found also to accumulate in the residual dextrins resulting from the hydrolysis of potato starch by malt alpha amylase. Evidence has been presented to show that short straight chain dextrins are also hydrolyzed very slowly by these amylases.

Although the mechanism of the hydrolysis of starch by beta amylase has been well established, the hydrolysis of starch by the alpha amylases has proven much more complicated. The data already available show that alpha amylases from different sources hydrolyze starches very differently and that these differences are more marked in the early than in the late stages of the hydrolysis of starch. Unfortunately, sufficient strictly comparable data are not available at present to make possible clear cut statements as to the similarities and differences in the mode of action of the three amylases discussed here. However, it is evident that further work with these and other amylases will be amply repaid as it adds to our exact information and increases our understanding of these important catalysts and of the substrates upon which they act.

XYLAN

By Roy L. Whistler

Department of Agricultural Chemistry, Purdue University, Lafayette, Indiana

Contents

I. Introduction..... 269
II. Occurrence..... 270
III. Pretreatment of Plant Material for Polysaccharide Isolation..... 272
IV. Removal of Lignin..... 274
V. Extractive Isolation of Xylan..... 274
VI. Purification..... 276
VII. Composition and Structure..... 278
VIII. Oxidation..... 284
IX. Degree of Polymerization..... 285
X. Derivatives..... 286
    1. Xylan Diacetate..... 286
    2. Other Derivatives..... 287
XI. Biological Decomposition of Xylan..... 288
XII. Industrial Uses..... 288
XIII. Addendum..... 289

I. Introduction

Xylan, a polysaccharide occurring in nearly all plants, is composed, for the most part, of a chain of anhydro-D-xylose units to which may be attached a single L-arabinose unit, and, at least in some instances, a D-glucuronic acid unit. It has not been proven whether more than one type of xylan occurs in nature. At present, a weight of evidence indicates that not less than two similar types of xylan polysaccharides occur, one with and one without glycosidically-bound D-glucuronic acid. The name xylan was assigned to these substances by early workers[1] who believed xylose to be the sole product of hydrolysis.

Xylan has the general properties of insolubility in water, solubility in alkaline solutions, ease of acid hydrolysis, high negative optical rotation, and non-reducing action toward Fehling's solution. It can be placed in three general polysaccharide classes: (1) pentosan, (2) glycan, and (3) hemicellulose. It is classed as a pentosan because it is principally a polymer of a pentose. It is by far the most abundant pentosan.

(1) See E. W. Allen and B. Tollens, *Ann.*, **260**, 289 (1890); E. S. Schulze, *Ber.*, **24**, 2277 (1891).

Araban, another member of this group, is much less abundant and is generally found as an associate of pectic substances.[2] The pentosans lyxan and riban have not been observed in nature. Xylan is classified as a glycan because it is largely, if not entirely, a polymer of an unmodified sugar or sugars. Alternatively, the word cellulosan,[3] is sometimes used to designate this group of compounds, but the term is not completely satisfactory.[4] Xylan is classified most frequently as a hemicellulose because it is removed by hemicellulose extraction procedures and is often the principal component of hemicelluloses. Purified hemicellulose-A is, in many instances, identical to xylan.

Determination of xylan is frequently made by estimation of furfural production. Data so obtained, even when appropriately corrected for furfural that arises from uronic acid, may be high if araban is present in the polysaccharide preparation. In the absence of interfering carbohydrates, furfural estimation may lead to accurate xylan values.

Since the terms xylan, pentosan, hemicellulose and hemicellulose-A often refer to the same substance or to substances which differ only slightly, the literature is sometimes confusing. In this review the term xylan will be used to signify preparations which are believed to be reasonably pure, the term pentosan will be used to designate material the composition of which is based on furfural estimation, and the term hemicellulose will be used to denote polysaccharide preparations of uncertain composition.

II. Occurrence

Xylan occurs in practically all land plants and is said to be present in some marine algae.[5] In both wide botanical distribution and abundance in nature it closely follows cellulose and starch. It is most abundant in annual crops, particularly in agricultural residues such as corn cobs, corn stalks, grain hulls and stems. Here it occurs in amounts ranging from 15 to 30%. Hard woods contain 20 to 25% xylan while soft woods contain 7 to 12%. Spring wood has more pentosan than summer wood.[6,7] Low strength vegetable fibers of commerce such as jute, sisal, Manila

(2) See E. L. Hirst and J. K. N. Jones, *Advances in Carbohydrate Chem.*, **2**, 235 (1946).

(3) L. F. Hawley and A. G. Norman, *Ind. Eng. Chem.*, **24**, 1190 (1932).

(4) The word cellulosan has the disadvantages that (a) it employs an ending ordinarily used for designating simple anhydro-sugars, and (b) that the name implies a relation with cellulose, whereas in fact no relationship is intended and polysaccharides placed in this class are quite different from cellulose in both composition and properties. Their connection with cellulose seems to be only that they occur with cellulose as constituents of the plant cell wall.

(5) V. C. Barry and T. Dillon, *Nature*, **146**, 620 (1940).

(6) G. J. Ritter and L. C. Fleck, *Ind. Eng. Chem.*, **18**, 608 (1926).

(7) R. D. Preston and A. Allsopp, *Biodynamica*, No. **53**, 8 pp. (1939).

hemp and coir may contain 5 to 20% xylan. High strength fibers such as ramie, flax and cotton are devoid or almost devoid of xylan. The approximate distribution of xylan in various plant products is shown in Table I.

Xylan is found principally as a constituent of the plant cell wall although analysis of microdissected middle lamellae of Douglas fir shows the presence of 14% pentosan in the intercellular layer.[8] Just how xylan

TABLE I
Pentosan Content of Various Natural Products

Substance analyzed	Pentosan content, per cent	Investigator
CASCARA	23.6	Kurth[9]
PACIFIC DOGWOOD	23.0	
DOUGLAS FIR	10.1	
SIERRA JUNIPER	16.0	
PACIFIC MADRONE	19.4	
CALIFORNIA LAUREL	22.4	
PACIFIC YEW	11.7	
PONDEROSA PINE	7.4	Ritter and Fleck[10]
TANBARK OAK	19.6	
ASPEN	17.6	Heuser and Broetz[11]
BEECHWOOD	26–28.7	Müller[12]
NETTLE STALK	14.1	Routala and Sevon[13]
TREE BARK	9–18	Richter[14]
WESTERN RED CEDAR BARK	10.3–10.9	Cram and coworkers[15]
WHITE SPRUCE	10.7	Sherrard and Blanco[16]
CORN COBS	28.1	Dunning and Lathrop[17]
OAT HULLS	29.5	
RYE STRAW	26.4	Panasyuk[18]
CORN COBS	34.9	
CORN STALK	27.6	Wills and Steller[19]

(8) A. J. Bailey, *Paper Ind.*, **18**, 379 (1936); *Ind. Eng. Chem., Anal. Ed.*, **8**, 52, 389 (1936).
(9) E. F. Kurth, *Paper Trade J.*, **126**, No. 6, 56 (1948).
(10) G. J. Ritter and L. C. Fleck, *Ind. Eng. Chem.*, **14**, 1050 (1922).
(11) E. Heuser and A. Broetz, *Papier-Fabr.*, **23**, 69 (1925).
(12) O. Müller, *Ann.*, **558**, 81 (1947).
(13) O. Routala and J. Sevon, *Cellulosechem.*, **8**, 16 (1927).
(14) G. A. Richter, *Ind. Eng. Chem.*, **33**, 75 (1941).
(15) K. H. Cram, J. A. Eastwood, F. W. King and H. Schwartz, *Dominion (Canada) Forest Serv. Circ.*, No. 62, 12 pp. (1947).
(16) E. C. Sherrard and G. W. Blanco, *Ind. Eng. Chem.*, **15**, 611 (1923).
(17) J. W. Dunning and E. C. Lathrop, *Ind. Eng. Chem.*, **37**, 24 (1945).
(18) V. G. Panasyuk, *Chem. Zentr.*, II, 1506 (1940).
(19) S. D. Wells and R. L. Steller, *Paper Trade J.*, **116**, No. 15, 45 (1943).

is bound in the fabric of the cell wall is not known. Certain experimental results based on the solubility characteristics of native xylan are interpretated as evidence that a chemical combination occurs between cellulose and xylan. So far, no such interconnecting bonds have been proven to exist. While a few such bonds might reasonably become established in cell wall formation, the presence of covalent bonds is not required to explain why xylan or portions of it are difficult to extract from plant tissue. Since the xylan chain is similar to the cellulose chain except that it lacks a projecting carbinol group on each ring unit, the molecules presumably could pack into dense, highly associated groups or could fit snugly into the cellulose matrix and substitute for cellulose molecules. In such a relationship, a large number of secondary forces could exist to bind strongly xylan molecules to xylan molecules or xylan molecules to cellulose molecules. This may explain why xylan is not easily separated from the cellulose matrix by solvent extraction. Additional reasons for the observed difficulty of separating these polysaccharides may be derived from the possibility that some xylan molecules are entangled or overlaid with protective cellulose molecules. That xylan molecules are not constituents of the crystalline micellar regions of the cell wall is suggested by the observation that extractive removal of xylan from the tissue does not greatly alter the crystalline X-ray pattern.[20,21]

The origin and function of xylan in the cell wall are also not explained. Postulations that it is a plasticizer or is a reserve food are not fully substantiated. Its derivation from cellulose through the decarboxylation of an intermediary polyglucuronic acid seems very unlikely. There is evidence from a number of sources to indicate that the xylan polysaccharide is deposited along with cellulose in cell wall elaboration.

III. Pretreatment of Plant Material for Polysaccharide Isolation

The extractability of the polysaccharides of plants can be profoundly affected by the chemical and physical changes which occur between the end of active vegetative growth and the time the plant parts are subjected to chemical extraction.

As a plant tissue ages, the solubility of the polysaccharides decreases to some extent. Decreased solubility is most pronounced when the tissue material is dried. Insolubilization is not inherent in the drying process itself because it is possible to dry plant material in such a way as to prevent loss of solubility or chemical reactivity. Optimum drying

(20) W. T. Astbury, R. D. Preston and A. G. Norman, *Nature*, **136**, 391 (1935).
(21) R. D. Preston and A. Allsopp, *Biodynamica*, No. 53, 8 pp. (1939).

appears to require a rapid withdrawal of water at relatively low temperatures (ca. 0–40°). Good drying may be accomplished by desiccation in a vacuum oven at low temperature, by lyophilization, or by thorough agitation of the finely ground material with successive fresh portions of alcohol and removal of the alcohol in a vacuum desiccator over calcium chloride. Presumably when highly hydrated polysaccharide material is slowly air dried, the gradual removal of water molecules allows neighboring polysaccharide molecules or chain segments to come gradually into contact and thereby establish strong secondary unions which later restrain the separation of the molecules and, hence, the penetration of water or solvent molecules.

O'Dwyer[22] finds that increased severity in the conditions of drying of English oak decreases the yield of hemicellulose extracted by 4% sodium hydroxide solution. Likewise, Meller[23] notes that there occurs an increased amount of pentosan resistant to extraction with 7% cold alkali whenever wood pulps are dried at high temperatures. Similar changes, associated principally with the furfural-yielding complexes, can be produced in both hard and soft woods.[24]

In some instances, it is desirable to remove lipids and other so-called extractives before proceeding with polysaccharide separation. This removal is generally accomplished through Soxhlet extraction with an azeotropic mixture of ethanol and benzene.[25,26,27]

Pectin and pectic substances are sometimes removed prior to xylan extraction. This is especially desirable when dealing with plant material containing large amounts of pectic substances such as the cambium layer of wood[28] or the leaves and stems of plants.[29,30] Extraction for one or more separate periods with 0.5% solutions of ammonium oxalate at 90–100° removes water-insoluble pectin substances[31,32,33] that are not mechanically trapped or chemically bound in the plant struc-

(22) M. H. O'Dwyer, *Biochem. J.*, **28**, 2116 (1934); *J. Soc. Chem. Ind.*, **51**, 968 (1932).
(23) A. Meller, *Paper Trade J.*, **125**, No. 11, 57 (1947).
(24) W. G. Campbell and J. Booth, *Biochem. J.*, **24**, 641 (1930); **25**, 756 (1931).
(25) E. F. Kurth, *Ind. Eng. Chem., Anal. Ed.*, **11**, 203 (1939).
(26) R. L. Whistler, D. R. Bowman, J. Bachrach, *Arch. Biochem.*, **19**, 25 (1948).
(27) TAPPI, Method No. T6m-45.
(28) E. Anderson, *J. Biol. Chem.*, **112**, 531 (1936).
(29) H. D. Weihe and M. Phillips, *J. Agr. Research*, **60**, 781 (1940); **64**, 401 (1942).
(30) M. Phillips and B. L. Davis, *J. Agr. Research*, **60**, 775 (1940); B. L. Davis and M. Phillips, *ibid.*, **63**, 241 (1941).
(31) M. Lüdtke and H. Felser, *Ann.*, **549**, 1 (1941).
(32) S. T. Henderson, *J. Chem. Soc.*, 2117 (1928).
(33) M. H. Branfoot, "A Critical and Historical Study of the Pectic Substances of Plants," His Majesty's Stationery Office, London, 1929.

ture. Ammonium oxalate may, however, remain to some extent as a contaminant.

IV. Removal of Lignin

The ubiquity of lignin in plant tissue presents an obstacle to the removal and purification of xylan. Lignin retards or prevents the complete solution of xylan either because of mechanical obstruction or perhaps by reason of attachment through as yet unidentified covalent bonds. Furthermore, lignin is partially soluble in the various aqueous alkaline solutions used for dissolving xylan and, consequently, poses a purification problem in various subsequent steps designed to isolate the pure polysaccharide.

For xylan preparation, some workers prefer to use a specially prepared, highly delignified pulp called holocellulose.[34] Holocellulose represents the total water-insoluble carbohydrate portion of the plant. It is usually prepared by alternate treatment of plant material with water, chlorine, ethanol and a 3% solution of ethanolamine in ethanol[35] or by continuous extraction for several hours with water containing chlorine dioxide.[26,36–40] Evidence from a number of sources indicates that the polysaccharide mixture is damaged very little during the delignification.[41–46] In some instances, however, it is noted that a rather considerable loss of pentosan may occur.[47] In all cases, marked loss of carbohydrate is observed if the delignification treatment is continued beyond the time necessary to reduce the lignin value to about 0.5 to 2.0%.

V. Extractive Isolation of Xylan

Classically, the procedure for the isolation of xylan consists in the extraction of plant material with alkaline solutions. Because of the

(34) G. J. Ritter and E. F. Kurth, *Ind. Eng. Chem.*, **25**, 1250 (1933).
(35) W. G. Van Beckum and G. J. Ritter, *Paper Trade J.*, **104**, No. 19, 49 (1937); **105**, No. 18, 127 (1937); **108**, No. 7, 27 (1939); **109**, No. 22, 107 (1939); *Tech. Assoc. Papers*, **21**, 431 (1938); **22**, 619 (1939); **23**, 652 (1940).
(36) G. Jayme, *Cellulosechem.*, **20**, 43 (1942).
(37) L. E. Wise, *Ind. Eng. Chem., Anal. Ed.*, **17**, 63 (1945).
(38) L. E. Wise, M. Murphy, and A. A. D'Addieco, *Paper Trade J.*, **122**, No. 2, 35 (1946).
(39) G. A. Adams and A. E. Castagne, *Can. J. Research*, **B26**, 325 (1948).
(40) E. Bennett, *Anal. Chem.*, **19**, 215 (1947).
(41) H. Wenzl, *Papier-Fabr.*, **39**, 177 (1941).
(42) J. F. White and G. P. Vincent, *Paper Trade J.*, **111**, No. 12, 39 (1940).
(43) G. Jayme and S. Mo, *Papier-Fabr.*, **39**, 193 (1941).
(44) A. W. Sohn and F. Reiff, *Papier-Fabr.*, **40**, 105 (1942).
(45) H. Staudinger and J. Jurisch, *Papier-Fabr.*, **35**, 462 (1937).
(46) W. B. Van Beckum and G. J. Ritter, *Paper Trade J.*, **108**, No. 7, 27 (1939).
(47) C. V. Holmberg and E. C. Jahn, *Paper Trade J.*, **111**, No. 1, 33 (1940).

solubility of xylan in alkali as contrasted to the insolubility of cellulose, this procedure is widely used today albeit with numerous modifications.

Aqueous alkaline extraction of wood was employed by Poumarede and Figuier[48] in 1846 for the removal of a substance called wood gum. Similar very crude xylan or hemicellulose preparations were made by other workers of the early period. Extractions were made not only from wood sawdust, but from annual plant materials such as wheat straw, corn cob, etc.[49-56]

Solutions of 2 to 4% alkali concentration have ordinarily been used for extractions. For extensive xylan removal, however, it would seem that higher alkali concentrations are necessary. For example, Wise and coworkers[57] show that the solubility of hemicelluloses of slash pine holocellulose increases as the potassium hydroxide concentration of the extracting solution rises from 2 to about 16% and that the solubility of aspen holocelluloses increases as the potassium hydroxide concentration increases from zero to about 10% but thereafter changes little for increasing concentrations up to 30%. Whistler, Bowman, and Bachrach[26] find similar solubility characteristics for the hemicelluloses of corn cobs (Figure 1). These solubility characteristics seem to be quite general irrespective of plant type or species. However, because some xylan may be extracted with dilute alkaline solutions while the remainder is removed only with strong solutions, Schmidt and coworkers[58] subdivide xylan into "easily soluble" and "difficultly soluble" fractions.

Usually it is considered that xylan has been effectively removed if the extracted residue compares in composition to α-cellulose. Yet, even α-cellulose contains small amounts of xylan as well as other hemicelluloses such as mannan.[57,59] Beta and γ-cellulose are mixtures which contain pentosans along with other alkali-soluble extractives.[16,60]

(48) J. A. Poumarede and L. Figuier, *Compt. rend.*, **23**, 918 (1846).
(49) J. Thomsen, *J. prakt. Chem.*, [2] **19**, 146 (1879).
(50) H. J. Wheeler and B. Tollens, *Ber.*, **22**, 1046 (1889); *Ann.*, **254**, 304 (1889).
(51) E. W. Allen and B. Tollens, *Ber.*, **23**, 137 (1890); *Ann.*, **260**, 289 (1891).
(52) E. Schulze, *Ber.*, **24**, 2277 (1891).
(53) E. Schulze, *Z. physiol. Chem.*, **16**, 387 (1892).
(54) E. Winterstein, *Z. physiol. Chem.*, **17**, 381 (1893).
(55) S. W. Johnson, *J. Am. Chem. Soc.*, **18**, 214 (1897).
(56) E. Schulze and N. Castoro, *Z. physiol. Chem.*, **39**, 318 (1903).
(57) L. E. Wise and E. K. Ratliff, *Anal. Chem.*, **19**, 459 (1947); see also reference 37.
(58) E. Schmidt, K. Meinel, K. Nevros, and W. Jandebeur, *Cellulosechem.*, **11**, 49 (1930); see also R. Runkel and G. Lange, *ibid.*, **12**, 185 (1931).
(59) R. E. Dörr, *Reichsamt Wirtschaftsausbau*, **Prüf.-Nr. 54(PB52006)**, 29 (1940); *Chem. Abstracts*, **41**, 5301 (1942).
(60) E. C. Sherrard and G. W. Blanco, *Ind. Eng. Chem.*, **15**, 1166 (1923).

For the most part alkaline extractions are conducted at room temperature because hot alkaline solutions possess a marked degradative effect on pentose constituents.[61] Previously, little or no attempt was made to exclude oxygen although, at present, extractions are often performed in an atmosphere of nitrogen. In view of the ease with which

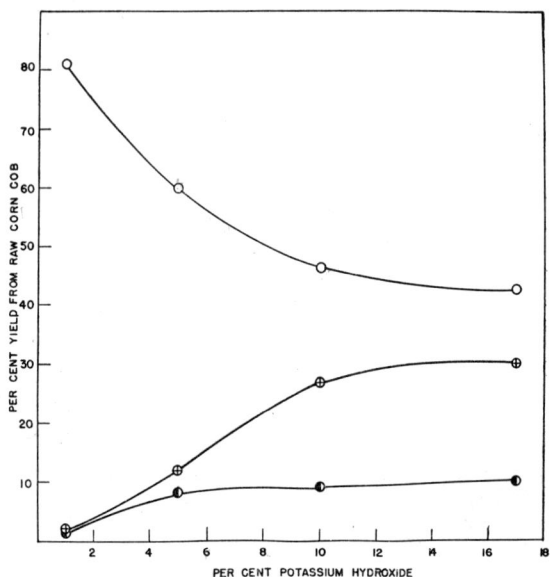

FIG. 1.—Yield of hemicellulose-A and -B and residue obtained by alkali extraction of raw corn cob. ○ Residue, ⊕ A-fraction, ◐ B-fraction.

carbohydrates are oxidized in alkaline solutions, it would seem desirable to conduct all alkaline extractions in the absence of oxygen.

VI. Purification

Many substances besides xylan may be extracted from plant material and, hence, if this polysaccharide alone is desired one or preferably more purification steps must be employed.

Prior to xylan removal it is possible to extract from the plant material those hemicelluloses soluble in dilute alkali. Soluble hemicelluloses of this type are the principal impurities extracted with xylan and consist for the most part of low molecular weight polysaccharides and polyuronides. Corn cob holocellulose may be freed of these compounds (called hemicellulose-B fraction, see Figure 2) by extraction with potas-

(61) I. A. Preece, *Biochem. J.*, **34**, 251 (1940).

sium hydroxide solutions of 0.2 to 0.5% concentration.[26] Subsequent extraction of the plant material with stronger alkaline solution removes a relatively pure xylan. Similar extractive pretreatments with dilute alkaline solutions or with solutions of lower pH such as those obtained with sodium carbonate have been employed by other workers.

Another general procedure for effecting initial subdivision of the crude mixture of polysaccharides obtained on alkaline extraction is that of Schulze[53] and O'Dwyer[62] which takes advantage of the fact that neutralization of the solution causes precipitation of the high molecular weight

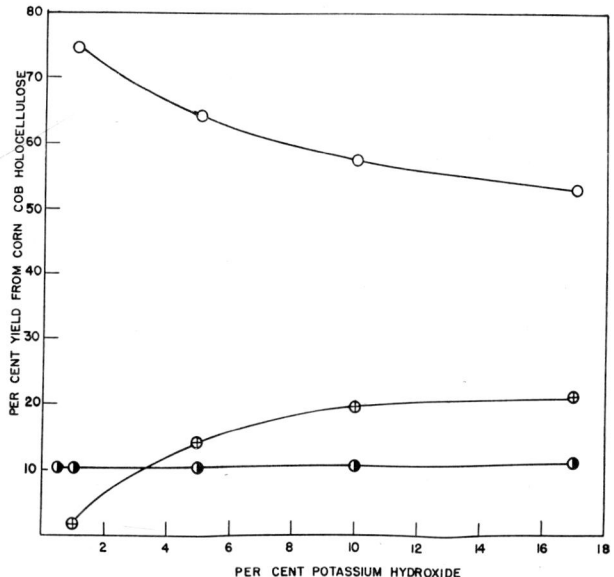

FIG. 2.—Yield of hemicellulose-A and -B and residue obtained by alkali extraction of corn cob holocellulose. ○ Residue, ⊕ A-fraction, ◐ B-fraction.

polysaccharides and leaves in solution polyuronides and molecules of low molecular weight. According to Angell and Norris[63] the pH of the precipitation is critical and to produce maximum yield, at least in the case of corn cob extract, the optimum value is pH 3.7 to 4.2. The precipitated fraction called hemicellulose-A, following the nomenclature of O'Dwyer,[62] consists of fairly pure xylan. It will be contaminated by mannan or high molecular weight glycans, if present, and by small amounts of other material which may be occluded in the precipitate.

For further purification, the xylan can be dissolved in dilute alkali

(62) M. H. O'Dwyer, *Biochem. J.*, **20**, 656 (1926).
(63) S. Angell and F. W. Norris, *Biochem. J.*, **30**, 2155 (1936).

and reprecipitated as a gelatinous complex by the addition of Fehling's solution at room temperature. Many investigators employ this purification technique. It was introduced by Salkowski[64] who at first considered it a means of separating xylan from araban but later found that some araban also precipitates, particularly if an excess of Fehling's solution is added. Some workers[52] substitute for Fehling's solution a mixture of glycerine and copper sulfate which is equally effective in bringing about the promotion of a copper–polysaccharide precipitate.

Yundt[64a] obtains a crystalline xylan from the xylan-rich fraction of straw or birchwood hemicellulose. The fraction is hydrolyzed with 0.2% oxalic acid solution for five hours at 100°. Upon autoclaving the insoluble residue approximately 0.2% dissolves and precipitates as hexagonal platelets when the filtrate is cooled to 60–70°. Very likely this crystalline material is of low molecular weight.

VII. Composition and Structure

Wheeler and Tollens[65] demonstrated that the sugar D-xylose[66] which was obtained on hydrolysis of xylan had the empirical composition $C_5H_{10}O_5$ while the parent xylan had the expected composition $(C_5H_8O_4)_n$. Although controversy arose over this latter formula, later workers have shown it to be correct.[67–70] Analytical difficulties were encountered because xylan, like other polysaccharides, is strongly hygroscopic at low moisture levels and is therefore difficult to obtain in an undegraded anhydrous condition. Under ordinary atmospheric conditions, xylan contains 8 to 11% moisture.

Hydrolysis of xylan produces principally D-xylose.[71] By hydrolysis of straw xylan in 3% nitric acid, Heuser and Jayme[72] obtained crystalline D-xylose in 85% yield. Using a similar procedure, Hampton, Haworth and Hirst[70] obtained a yield of 93% crystalline D-xylose from esparto xylan. A practically identical yield was indicated by reducing sugar determination made on the hydrolysis products of corn seedling xylan.[68]

(64) E. Salkowski, *Z. physiol. Chem.*, **34**, 162 (1901); **35**, 240 (1902).
(64a) A. P. Yundt, *J. Am. Chem. Soc.*, **71**, 757 (1949).
(65) H. J. Wheeler and B. Tollens, *Ann.*, **254**, 304 (1889).
(66) Xylose was originally isolated by F. Koch, *Pharm. Z. Russland*, **25**, 619, (1886).
(67) S. W. Johnson, *J. Am. Chem. Soc.*, **18**, 214 (1896).
(68) K. P. Link, *J. Am. Chem. Soc.*, **51**, 2506 (1929).
(69) K. P. Link, *J. Am. Chem. Soc.*, **52**, 2091 (1930).
(70) H. A. Hampton, W. N. Haworth and E. L. Hirst, *J. Chem. Soc.*, 1739 (1929).
(71) In the early nomenclature this sugar was termed L-xylose.
(72) E. Heuser and G. Jayme, *J. prakt. Chem.*, **105**, 232 (1923).

Relatively pure xylan isolated from the holocellulose of aspen (*Populus*) wood is said to contain 85% of xylose residues.[73] One of the characteristic properties of xylan is its ease of hydrolysis. Because it hydrolyzes much more readily than cellulose, mild acid treatment may be employed to bring about preferential hydrolysis of xylan from plant material. Xylose is ordinarily prepared in the laboratory by direct sulfuric acid hydrolysis of the native xylan in ground corn cobs.[74] Hydrolysis in hydrochloric acid proceeds rapidly, but decomposition to furfural also occurs to some extent.[75] A commercial method for the production of D-xylose from cottonseed hulls[76] and straw[77] and from corn cobs[17,78] has been described.

The native xylan molecule may contain in addition to D-xylose a small amount of L-arabinose and, in some instances, a small amount of a glycuronic acid, possibly D-glucuronic acid. Hydrolysis of esparto[79,80] xylan and corn cob[81] xylan with solutions of 0.2% oxalic acid or 0.02N nitric acid removes L-arabinose with only slight depolymerization of the xylan molecule. L-Arabinose is identified by conversion to the α-benzyl-α-phenylhydrazone. L-Arabinose is also found in a yield of 2.7% when xylan of wheat straw or beechwood is hydrolyzed completely with 3% nitric acid.[82] If a purified hemicellulose (probably crude xylan) of apple wood[83] is hydrolyzed, xylose and arabinose are found in the ratio of 7:1. One hemicellulose[84] of sugar cane bagasse is said to contain 13%

(73) B. B. Thomas, *Paper Ind. and Paper World*, **27**, 374 (1945).
(74) C. S. Hudson and T. S. Harding, *J. Am. Chem. Soc.*, **40**, 1601 (1918); F. B. LaForge and C. S. Hudson, *Ind. Eng. Chem.*, **10**, 925 (1918). See also "Polarimetry, Saccharimetry and the Sugars," by F. J. Bates and Associates. Circular C440, Natl. Bur. Standards, U. S. Dept. Commerce, p. 481 (1942).
(75) E. Heuser and L. Brunner, *J. prakt. Chem.*, **104**, 264 (1922).
(76) W. T. Schreiber, N. V. Geib, B. Wingfield and S. F. Acree, *Ind. Eng. Chem.*, **22**, 497 (1930).
(77) R. E. Keller, *J. Applied Chem. (U.S.S.R.)*, **10**, 2041 (1937); *Chem. Abstracts*, **32**, 5203 (1938).
(78) See also Fiat Final Report No. 499, 14 November 1945, "Production of Wood Sugar in Germany and its Conversion to Yeast and Alcohol," E. G. Locke, S. F. Saeman, and G. K. Dickerman. Field Information Agency, Technical.
(79) W. N. Haworth, E. L. Hirst and E. Oliver, *J. Chem. Soc.*, 1917 (1934).
(80) R. A. S. Bywater, W. N. Haworth, E. L. Hirst and S. Peat, *J. Chem. Soc.*, 1983 (1937).
(81) J. Bachrach, *Ph. D. Dissertation, Purdue University, Lafayette, Indiana* (1948).
(82) G. Jayme and M. Sätre, *Ber.*, **75**, 1840 (1942).
(83) F. Gerhardt, *Plant Physiol.*, **4**, 373 (1929).
(84) Y. Hachihama, *J. Soc. Chem. Ind., Japan*, **36**, 634B (1933); *Chem. Abstracts*, **28**, 926 (1934).

anhydroarabinose and 87% anhydroxylose and another[85] to contain 7 to 8% anhydroarabinose and 92 to 93% anhydroxylose.

Indication of the presence of a glycuronic acid unit in some xylan molecules has been obtained. It is suggested that the acid unit is a monomethyl derivative. Hemicellulose-A of flax is believed to contain a chain of about ten D-xylopyranose residues united by β-1,4-glycosidic linkages and terminated by a uronic acid unit.[86] By acid hydrolysis of the hemicellulose-A of both sapwood and heartwood of English oak there is obtained an aldobiuronic acid consisting of D-xylose and a monomethylhexuronic acid. In the heartwood, approximately eleven D-xylose units occur for each methylhexuronic acid unit.[87] Hemicellulose from cottonseed hulls possesses ten to sixteen D-xylose units for each D-glucuronic acid unit.[86,88] Xylan from corn cobs is found by one group of workers[63] to contain 95% xylan, 5% anhydroglycuronic acid and 0.5% methoxyl, and by another group[89] to contain 94% xylan and 3% anhydroglycuronic acid. It is stated that esparto xylan contains no methoxyl groups. Glycuronic acid analyses have not been reported for this polysaccharide.[70] Corn stalk hemicellulose contains D-glucuronic acid, L-arabinose, and D-xylose in the approximate ratio[90] of 2:7:19. Hemicellulose of wheat straw and alfalfa hay contains these monosaccharides in the ratio[91] of 1:0.9:23.

Substantial evidence regarding the linkage between xylose units has recently been obtained by Bachrach and Whistler.[91a] They find a 10 per cent yield of xylobiose on stopping the hydrolysis of xylan at a reducing value corresponding to two-thirds of that for complete hydrolysis. The xylobiose can be easily converted to a crystalline hexaacetate as well as a crystalline methyl xylobioside pentaacetate, a methyl xylobioside and a methyl pentamethylxylobioside. These are the first crystalline derivatives of a pentose disaccharide. The presence of a 1,4-glycosidic linkage is indicated by performing the following reactions and identification of the products as crystalline derivatives.

(85) S. Miyaki, E. Hamaguti and H. Kurasaw, *J. Soc. Trop. Agr., Taihoku Imp. Univ.*, **11**, 215 (1939); *Chem. Abstracts*, **34**, 6472 (1940).
(86) R. J. McIlroy, G. S. Holmes and R. P. Mauger, *J. Chem. Soc.*, **796** (1945).
(87) M. H. O'Dwyer, *Biochem. J.*, **33**, 713 (1939).
(88) E. Anderson, J. Hechtman and M. Seeley, *J. Biol. Chem.*, **126**, 175 (1938).
(89) R. L. Whistler, D. R. Bowman, and J. Bachrach, unpublished results.
(90) H. D. Weihe and M. Phillips, *J. Agr. Research*, **64**, 401 (1942).
(91) H. D. Weihe and M. Phillips, *J. Agr. Research*, **60**, 781 (1940).
(91a) J. Bachrach and R. L. Whistler, Paper presented before the Division of Sugar Chemistry, 116th Meeting of the American Chemical Society, Atlantic City, 1949.

Alkaline solutions of xylan are highly laevorotatory as shown by the values in Table II.

Natural xylan may occur partially esterified with acetic acid. This possibility arises from the observation that most holocellulose preparations contain a small amount of acetic acid. So far, however, it has not been established which polysaccharides are esterified. Treatment of the holocellulose with the alkaline solutions necessary for xylan extraction brings about complete deacetylation.

Most structural work on xylan has been done on that from esparto grass and the principal attack made by way of the methyl ether. Xylan can be methylated by heating with methyl iodide and silver oxide,[92,93] but complete etherification is difficult and considerable degradation probably occurs. On the other hand, complete etherification is attained by methylation in two operations with potassium hydroxide and dimethyl sulfate to give a dimethylxylan in almost quantitative yield[70] showing $[\alpha]^{22}_D - 92°$ in chloroform. Methylation with potassium hydroxide appears to proceed more readily than with sodium hydroxide.[70,92]

Dimethylxylan, like methylated starch and cellulose, is soluble in cold water and insoluble in hot water. This behavior has been explained as being due to a weakly bonded hydrate which causes solubility at low temperature but which is broken up on heating.

Methanolysis of dimethylxylan in the presence of hydrogen chloride leads to the isolation of methyl 2,3-dimethyl-D-xylopyranoside in 90% yield,[70] methyl 2,3,5-trimethyl-L-arabinofuranoside in 6% yield, and

(92) E. Heuser and W. Ruppel, *Ber.*, **55**, 2084 (1922).
(93) S. Komatsu, T. Inoue and R. Nakai, *Mem. Coll. Sci. Kyoto Univ.*, **7**, 25 (1923); *Chem. Abstracts*, **18**, 666 (1924).

Table II
Optical Rotations of Xylan in Sodium Hydroxide Solutions

Xylan source	$[\alpha]_D$	Conc. of NaOH solution, percent	Temperature, °C.	Workers
Esparto	−109.5°	2.5	22	Hampton, Haworth, and Hirst[70]
Wheat Straw, Beechwood	−107.3	8.0	20	Jayme and Sätre[82]
Wheat Straw, Beechwood	−87	6.0	—	Husemann[94]
Apple Wood (Hemicellulose)	−104	1.0	25	Gerhardt[83]
Corn Cob (Hemicellulose)	−96 to −108	4.0	20	Preece[95]
Corn Cob	−106	4.0	20	Whistler, Heyne and Bachrach[97]
Corn Seedling	−80 to −83.9	1.0	20	Link[68]
Oat Hulls (Pentosan)	−78.2	5.0	20	Krznarish[96]

methyl monomethyl-D-xylopyranoside in 5% yield.[79] Isolation of the 2,3-dimethylxylose is also reported by Japanese workers.[93] Proof of a pyranose ring structure[98] for the xylose units is obtained through acetolysis of dimethylxylan to produce an unidentified disaccharide derivative which is then saponified, oxidized with bromine in aqueous barium

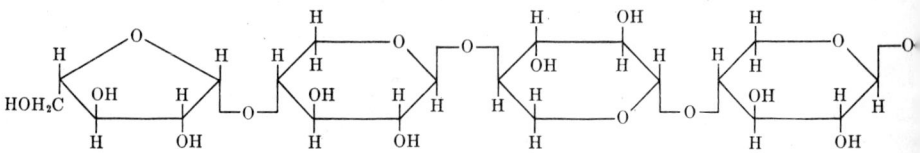

FIG. 3.—Section of xylan chain with non-reducing end of L-arabinofuranoside.

benzoate suspension, methylated and subjected to hydrolysis. The separated products, which are 2,3,4-trimethyl-D-xylose in 80% yield and 2,3,5-trimethyl-γ-D-xylonolactone in 73% yield, are evidence of the 1,4-glycosidic linkage and pyranose ring in the xylan chain. These observations, in concert with the lack of reducing action of xylan and its

(94) E. Husemann, *J. prakt. Chem.*, **155**, 13 (1940).
(95) I. A. Preece, *Biochem. J.*, **24**, 973 (1930).
(96) P. W. Krznarish, *Cereal Chem.*, **17**, 457 (1940).
(97) R. L. Whistler, E. Heyne and J. Bachrach, *J. Am. Chem. Soc.*, **71**, 1476 (1949).
(98) W. N. Haworth and E. G. V. Percival, *J. Chem. Soc.*, 2850 (**1931**).

marked increase in optical rotation on acid hydrolysis, suggest a branched β-glycosidically-linked chain structure with one branch on the average for every eighteen to twenty D-xylose units, the branches being terminated by an L-arabinofuranose unit. Such a structure would yield derivatives which would show on analysis a ratio of two substituents for each anhydrosugar unit. If branching occurs through a glycosidic bond, there would be available at most only one potential reducing group per molecule; hence, the polysaccharide might exhibit very little reducing action. Absence of reducing action can be explained in several ways. Schmidt's[99] evidence that xylan contains a carboxyl group is not confirmed by Haworth, Hirst and Oliver[79] for esparto xylan.

Additional evidence that the non-reducing terminal group is an arabinofuranoside unit is obtained through hydrolytic removal of it in 0.2% aqueous oxalic acid at 100° with minimum cleavage of the xylan chain.[80] Methylation of the acid-treated xylan, with subsequent methanolysis and separation of products, gives 2,3-dimethyl-D-xylose in 85 to 90% yield, 2-monomethyl-D-xylose in 6 to 10% yield, and 2,3,4-trimethyl-D-xylose in 4% yield. Approximately 2% of 2,3,5-trimethyl-L-arabinose is also found. The yield of trimethylxylose corresponds to one end group for each eighteen to nineteen sugar units. This experiment also indicates that the penultimate unit in the xylan chain, in other words the last D-xylose unit, does not constitute a branch point.

While methylation evidence suggests a branched structure, no complete proof of branching exists. In fact, some characteristic properties of xylan may be taken to indicate that if branching does occur, the branches must be very few or very short. These deductions have credence because xylan derivatives, particularly the diacetate, can form strong, pliable, cellophane-like films.[100,101] Such films can be elongated under stress and thereupon become anisotropic to polarized light and exhibit other evidence of alignment of the chains by becoming increasingly strong and by ultimately fracturing in lines parallel to the direction of elongation. Although polysaccharide molecules with side chains of only one unit can be formed into pliable films,[102] it is observed that molecules which possess long branches form only very brittle films.[103]

Methylation of a purified alkaline-soluble polysaccharide from New Zealand flax is accomplished by subjecting it to five successive methyla-

(99) E. Schmidt, *Cellulosechem.*, **13**, 129 (1932).
(100) C. L. Smart and R. L. Whistler, *Science*, **110**, 713 (1949).
(101) N. Y. Solechnik, *J. Applied Chem. (U.S.S.R.)*, **7**, 1029 (1934); *Chem. Abstracts*, **29**, 5418 (1935).
(102) C. L. Smart and R. L. Whistler, *J. Polymer Sci.*, **4**, 87 (1949).
(103) See for example R. L. Whistler and G. E. Hilbert, *Ind. Eng. Chem.*, **36**, 796 (1944).

tions with dimethyl sulfate and sodium hydroxide and finally with methyl iodide and silver oxide.[86] Fractional precipitation leads to the inference that the compound is homogeneous. After methanolysis, 2,3-dimethyl-D-xylose can be isolated in large yield, 2,3,4-trimethyl-D-xylose in about 11% yield, monomethyl-D-xylose in very small amount and there is present a fragment from which can be isolated a dimethyl-D-xylose and 2,3-dimethyl-D-glucuronic acid. Optical rotation changes lead one to infer a preponderance of β-glycosidic linkage. It is also believed that the original polysaccharide contains a chain of about ten D-xylopyranose residues united through 1,4-glycosidic linkages and terminated at the reducing end by a D-glucuronic acid combination. The investigators are of the opinion that the uronic acid portion is branched but do not feel that the xylose portion of the molecule exists in a branched structure.

VIII. Oxidation

On treatment with periodate ion, straw and beechwood xylan oxidize rapidly.[104] As with other polysaccharides, the oxidation comes to a more definite end point in solutions which are buffered to pH 4–5. While somewhat more than the theoretical amount of periodate ion is consumed, the reaction apparently proceeds uniformly with oxidative cleavage of the 2,3 carbon bond to produce the structure indicated in Figure 4. In the course of the reaction the xylan passes into solution. The optical rotation of the oxidized product is surprisingly high (*ca.* 100°).

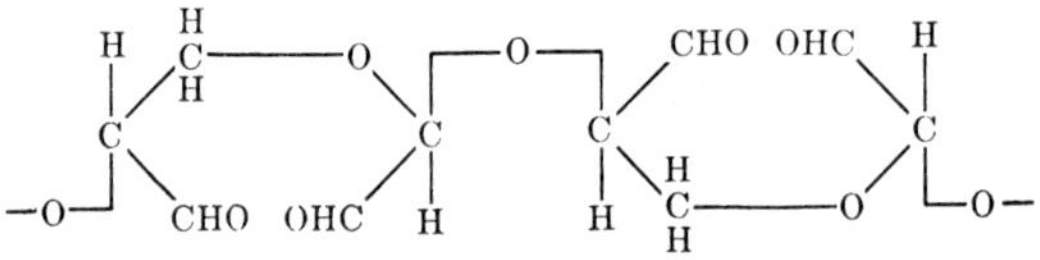

Fig. 4.—The oxidation of xylan by periodate ion.

The extent of oxidation, when determined by the reaction of the product with phenylhydrazine, is 85%. Hydrolysis of the oxidized xylan should produce approximately equimolar quantities of D-glyceraldehyde and glyoxal. Experimental determination of glyceraldehyde indicates 67% of the theoretical when the oxidized xylan is distilled with sulfuric acid and the evolved methylglyoxal measured as the phenylosazone. Glyoxal is isolated in 63% yield, when separated as the phenylosazone or as the dioxime. Aldehyde groups in the oxidized xylan may be further

(104) (a) G. Jayme and M. Sätre, *Ber.*, **75**, 1840 (1942); (b) *Ber.*, **77**, 242 (1944); (c) *Ber.*, **77**, 248 (1944).

oxidized to carboxyl by means of hypobromite or they may be reduced to the corresponding alcohol groups by hydrogenation over Raney nickel activated by platinum.

These results may be explained on the basis that xylan possesses a linear chain of 1,4-linked anhydroxylose units. However, it should be pointed out that on hydrolysis of the oxidized xylan, some D-xylose is found in addition to the expected glyceraldehyde and glyoxal.[104(b)] While it is possible that the oxidation reaction is not complete, another explanation is that a branched xylan chain may be present. If branching occurs on a pentose chain unit, only one free hydroxyl would remain on the unit; hence, it would not be oxidized or degraded during the course of the periodate reaction.

Recently a quantitative method for the determination of carbonyl groups in oxidized xylan has been developed.[105] The method involves the reaction of the carbonyl group with O-methylhydroxylamine hydrochloride to form the corresponding oxime with the liberation of one mole of titratable acid.

Heating xylan at 45° with nitric acid for sixteen hours produces xylotrihydroxyglutaric acid in 22% yield.[72]

IX. Degree of Polymerization

Molecular weight studies have not been made extensively on xylan. From measurement of the viscosity of dimethylxylan in m-cresol, a degree of polymerization (D. P.) of 75 to 80 is calculated.[79] Husemann[94] reports viscosity and osmotic pressure measurements on xylan and xylan derivatives from straw and beechwood. Because of the insolubility of the homogeneous derivatives, mixed derivatives such as the methylacetyl- and benzylacetylxylan are also investigated. Osmotic pressure determinations[106] indicate a D. P. of 113 to 150 as compared to a minimum value of 1500 for beechwood cellulose. The viscosity of xylan in cuprammonium solution is such as to indicate a Staudinger constant of 5×10^{-4}, which is the same as the constant for cellulose, which suggests that the xylan molecule is of linear nature. Fractionation of beechwood xylan indicates that 95% of the molecules are of roughly the same chain length. From viscosity measurements on fractionated holocellulose nitrates of Western hemlock, Mitchell[107] concludes that two groups of molecules occur; namely, the hemicelluloses with a D. P. of about 70 and α-cellulose with a D. P. of 2000 to 2500.

(105) B. Meesook and C. B. Purves, *Paper Trade J.*, **123**, No. 18, 35 (1946).
(106) Method of G. V. Schulz, *Z. physik. Chem.*, **A176**, 317 (1936).
(107) R. L. Mitchell, *Ind. Eng. Chem.*, **38**, 843 (1946).

Acetyl values from fully acetylated xylan preparations indicate a D. P. of about 100 or greater. The theoretical acetyl value for a fully acetylated xylan molecule of infinite size is 39.8%. As the molecule decreases in length, its acetyl content increases because of the increasing proportion of triacetyl-bearing end units. Assuming an accuracy of ±0.1% for the acetyl determination, a noticeable deviation from theory would be apparent in acetylated xylan of less than D. P. 100.

The film-forming property of xylan derivatives also leads to the inference of an average D. P. greater than about 50. This deduction arises from the general observation that pliable polysaccharide acetate films are only formed from molecules with a D. P. of 50 or more.

X. Derivatives

1. *Xylan Diacetate*

Bader[108] prepared a xylan diacetate by treatment of the crude polysaccharide or the monoacetate with acetic anhydride at 140–150°. A similar diacetate[109] can be prepared by esterification in the presence of pyridine at 70° or by the action of acetyl chloride at 50°. Esterification occurs when xylan is treated at 70° for seven hours with acetic anhydride containing nitric acid in 0.25% concentration.[97,110] A small amount of nitrate equivalent to about 0.2% nitrogen is introduced but otherwise the acetylation is complete. All of these acetates may be considered more or less degraded because of the harsh acetylation conditions which are needed to bring the preparations into reaction. Some polysaccharide preparations are particularly unreactive. For example, O'Dwyer reports[111] that xylan of oak acetylates with difficulty even with Barnett's[112] mixture, which is composed of acetic acid, acetic anhydride, and a catalyst consisting of equal parts of sulfur dioxide and chlorine. A similar unreactivity is observed when very dry xylan is treated with acetic anhydride containing sulfuric acid.[101]

Esterification is not difficult if xylan is first carefully precipitated from solution and dried to a fluffy, non-horny powder, or if the xylan can be highly swollen so as to make the polysaccharide molecules easily accessible to the esterifying reagents. Solechnik[101] finds that swollen undried xylan, when treated with acetic anhydride containing sulfuric

(108) R. Bader, *Chem. Ztl.*, **19**, 55, 78 (1895).
(109) E. Heuser and P. Schlosser, *Ber.*, **56**, 392 (1923).
(110) C. D. Hurd and N. R. Currie, *J. Am. Chem. Soc.*, **55**, 1521 (1933).
(111) M. H. O'Dwyer, *Biochem. J.*, **22**, 381 (1928).
(112) W. L. Barnett, *J. Soc. Chem. Ind.*, **40**, 8T (1921).

acid, acetylates smoothly to form the diacetate. Carson and Maclay[113] esterify hemicelluloses by first swelling them in formamide. The powerful swelling action of formamide and acetamide on such compounds was earlier demonstrated.[114] When swollen by these agents, xylan can be easily acetylated with acetic anhydride and pyridine.[97]

High quality xylan diacetate is insoluble in most reagents although the acetates of degraded xylan become progressively more soluble as the molecular weight decreases. Solubility in pyridine first appears, while with further decrease in molecular size, solubility in chloroform occurs. Because of their insolubility, the acetates of undegraded xylan cause filtration difficulties when present in commercial cellulose acetates.

2. Other Derivatives

Nitration occurs when xylan is treated with nitric acid or mixtures of nitric acid and sulfuric or phosphoric acids.[94,108,115,116] The dinitrate prepared by esterification of xylan with a mixture of nitric and phosphoric acids[117] is insoluble in most solvents.[94]

Sulfuric acid esters of xylan can be prepared[118] by the method of Karrer.

Other esters such as the benzoyl,[108] stearoyl,[101] oleoyl,[101] can be prepared by treating xylan with the appropriate chloride in the presence of sodium hydroxide, potassium hydroxide, pyridine or quinoline. These derivatives have not been characterized, perhaps because of their insoluble nature.

Xylan xanthate can be made by the action of carbon disulfide on xylan in sodium hydroxide solution. Analysis of the products indicates substitution below the diester stage, but the reaction appears similar to that occurring with cellulose.[119,120] However, neither Heuser and Schorsch[119] nor Dörr[121] find that xylan gives a viscose-like solution.

(113) J. F. Carson and W. D. Maclay, *J. Am. Chem. Soc.*, **70**, 293 (1948); **68**, 1015 (1946).

(114) J. Reilly, P. P. Donovan and K. Burns, *Proc. Royal Irish Acad.*, **39B**, 505 (1930).

(115) W. Will and F. Lenze, *Ber.*, **31**, 68 (1898).

(116) N. Y. Solechnik, *J. Applied Chem. (U.S.S.R.)*, **6**, 93 (1933); *Chem. Abstracts*, **27**, 5964 (1933).

(117) See H. Staudinger, R. Mohr, H. Haas and K. Feuerstein, *Ber.*, **70**, 2296 (1937).

(118) E. Husemann, K. N. v. Kaulla and R. Kappesser, *Z. Naturforsch.*, **1**, 584 (1946).

(119) E. Heuser and G. Schorsch, *Cellulosechem.*, **9**, 93 (1928).

(120) T. Liser and A. Hackl, *Ann.*, **511**, 128 (1934).

(121) R. E. Dörr, *Angew. Chem.*, **53**, 292 (1940).

Heuser and Schorsch[119] discuss the possibility of compound formation between xylan and sodium hydroxide, thus suggesting a further analogy with cellulose.[122]

XI. Biological Decomposition of Xylan

Little information is available on the enzymatic hydrolysis of xylan. The kinetics of breakdown of xylan are reported.[123] Hemicellulose from linden wood is said to be extensively hydrolyzed[124] by the digestive juices of *Helix pomatia*. Xylanase[125] of *Aspergillus niger*, while active on xylan between pH 2–8, has its optimum at pH 4.0.

A number of microorganisms decompose xylan. This decomposition was probably first studied by Hoppe-Seyler.[126] While fungi and bacteria both act on xylan,[127] fungi are primarily active in nature. The presence of xylan in a cellulose apparently assists organisms in their attack on the cellulose.[128,129]

XII. Industrial Uses

Pure xylan is not employed in industry but crude xylan or pentosans are of industrial importance. Xylan has been proposed as a textile size but is not employed as yet for this purpose.[130] Perhaps the largest use of pentosans is in their conversion to furfural, which has many applications and serves as the source of other furan derivatives. At the present time, large quantities of furfural are used in the extractive purification of petroleum products, and recently a large plant has been constructed to convert furfural by a series of reactions to adipic acid and hexamethylene-diamine, basic ingredients in the synthesis of nylon. In commercial furfural manufacture, rough ground corn cobs are subjected to steam distillation in the presence of hydrochloric acid. As mentioned above, direct preferential hydrolysis of the pentosan in cobs or other pentosan-bearing products could be used for the commercial manufacture of D-xylose.

(122) See also S. P. Saric and R. K. Schofield, *Proc. Roy. Soc. (London)*, **A185**, 431 (1946).
(123) W. Voss and G. Butter, *Ann.*, **534**, 161 (1938).
(124) T. Ploetz, *Ber.*, **72**, 1885 (1939); **73**, 57 (1940).
(125) W. Grassmann, R. Stadler and R. Bender, *Ann.*, **502**, 20 (1933).
(126) F. Hoppe-Seyler, *Z. physiol. Chem.*, **13**, 66 (1889).
(127) Y. Ziemiecka, *Polish Agr. Forest Ann.*, **25**, 313 (1931); *Chem. Abstracts*, **27**, 1438 (1933).
(128) W. H. Fuller and A. G. Norman, *J. Bact.*, **46**, 281 (1943).
(129) A. G. Norman, *Iowa Agr. Expt. Sta. Ann. Rpt.*, **77**, (1939); **92**, (1940).
(130) A. R. Fuller, U. S. Pat. 2,403,515, (July 9, 1946).

Hemicelluloses and pentosans are said to be of value when present in small quantities in paper pulps. These polysaccharides decrease the beating time and give the finished paper a higher breaking strength.[131-136] The presence of large amounts of hemicellulose may give the paper an undesirable glassine appearance. However, this latter effect is believed to be due more to the polyuronides than to xylan.

XIII. Addendum

Sometime after the above review was written the author learned of further unpublished work on xylan through private communications with the research workers concerned. These new findings bring to light important new concepts regarding the structure of xylan and are presented here in brief summary. Happily, part of this work has just been published.

Xylan obtained from esparto holocellulose[137] can be purified by repeated precipitation as the copper complex.[138] The L-arabinose content is successively reduced with each purification step and is absent in the final product, as evidenced by analysis and by chromatographic examination of the hydrolysis mixture. Earlier evidence[80] of L-arabinose in esparto xylan likely arose from the presence of araban as a contaminant. In conformity with earlier work, the purified esparto xylan is found to contain no uronic acid. Methanolysis of the fully methylated xylan and separation of the hydrolysis products on a cellulose column[139] yields the methyl glycosides of 2,3,4-trimethyl-D-xylopyranose, 2,3-dimethyl-D-xylopyranose and 2-monomethyl-D-xylopyranose. Isolation of the latter compound indicates that it represented the branch point and that branching occurred through the C3 position of a xylose residue in the 1,4' linked chain. The molar ratio of the methylated products coupled with molecular weight estimations of the polysaccharide suggest that xylan consists of a singly branched molecule containing 70 to 80

(131) G. Jayme, *Papier-Fabr. Wochbl. Papierfabr.*, 295 (1944); *Chem. Abstracts*, **40**, 3895 (1946).
(132) F. J. Greenane, *Proc. Tech. Sect. Paper Makers' Assoc. Gt. Brit. and Ireland*, **25**, 220 (1944).
(133) H. E. Obermanns, *Paper Trade J.*, **103**, No. 7, 83 (1936).
(134) H. H. Houtz and E. F. Kurth, *Paper Trade J.*, **109**, No. 24, 38 (1939).
(135) J. M. Limerick and A. J. Corey, *Pulp Paper Mag. Canada*, **47**, No. 4, 62 (1946).
(136) L. E. Wise, *Paper Ind. and Paper World*, **29**, 825 (1947).
(137) S. K. Chanda, E. L. Hirst, J. K. N. Jones and E. G. V. Percival, *J. Chem. Soc.*, 1289 (1950).
(138) See method of reference 64.
(139) L. Hough, J. K. N. Jones and W. H. Wadman, *J. Chem. Soc.*, 2511 (1949).

D-xylopyranose units. This structural concept is substantiated by estimation of the formic acid obtained when the xylan is oxidized by periodate ions. On hydrolysis of the fully oxidized xylan there is obtained a small amount of D-xylose which presumably occupied the branch points in the polysaccharide and consequently was protected from periodate oxidation by possessing no adjacent free hydroxyl groups.

Other workers[140] have also examined the products derived by hydrolysis of periodate-oxidized xylan. Both wheat straw and corn cob xylan, after oxidation and hydrolysis, yield small amounts of L-arabinose and D-xylose. These sugars are obtained even after the xylans have been subjected to extended periods of oxidation. It is concluded that the D-xylose constituted branch points in the xylan. Likewise, the L-arabinose molecules must not have been terminal units in a xylan chain but must have been either interior units in the xylan molecule or have constituted an araban-like polysaccharide which is in combination or admixture with the xylan polysaccharide.

(140) J. Ehrenthal and F. Smith. Unpublished work. Private communication from Dr. F. Smith.

AUTHOR INDEX*

A

Abderhalden, E., 80, 82 (4)
Abitz, W., 106, 108 (12)
Acree, S. F., 279
Adams, B. A., 138
Adams, G. A., 274
Adams, M. H., 90, 162, 164 (83, 84)
Adams, Mildred, 66, 68, 69 (45), 70 (45), 230, 231 (3), 232 (3), 233 (3), 234 (3), 236 (30), 237 (37, 38, 39), 240 (3), 241 (3, 37, 38, 39), 250 (3), 255 (3), 256 (3), 257 (3), 258 (3)
Adkins, H., 2, 3, 7
Agnew, A. Louise, 241, 245 (60), 246, 248, 249 (60)
Ahlborg, K., 250, 254 (65), 255 (74, 76, 77), 256 (77), 262 (65), 264 (65)
Akiya, S., 162, 163 (80)
Alfin, Roslyn B., 236, 239 (41), 241 (41), 242 (41), 243 (41), 244 (41), 245 (41), 246 (64), 247 (64), 248, 249 (64), 252 (41), 253 (41, 64), 257 (41), 259 (41), 263 (41, 64)
Allen, E. W., 269, 275
Allen, M. C., 89
Allsopp, A., 270, 272
Anderson, C. C., 160, 162 (63), 163 (63), 219
Anderson, C. G., 152, 154, 155 (23), 185 (23), 186 (23), 190 (23)
Anderson, E., 273, 280
Angell, S., 277, 280 (63)
von Antropoff, A., 53
Appel, H., 160, 162 (65), 163 (65)
Armstrong, E. F., 54, 105
Armstrong, K. F., 105
Asahina, Y., 5, 15, 198, 202 (18), 226 (16, 17, 18)
Assaf, A. G., 123, 124
Astbury, W. T., 272

Atkins, D. C., 223
Avinieri-Shapiro, S., 47

B

Bachrach, J., 273, 274 (26), 275 (26), 276 (26), 279, 280, 286 (97), 287 (97)
Bader, R., 286, 287 (108)
Badgley, W., 106, 112 (9), 122 (9)
Badollet, M. S., 139
Baehren, F., 58
Bailey, A. J., 271
Bak, Agnete, 232, 240
Baker, G. L., 81, 93, 96 (52), 101
Baker, H. R., 223
Ballou, G. A., 82, 93, 94 (49), 96 (49), 97 (49)
Balls, A. K., 234, 256, 257 (79)
Bamann, E., 80, 82 (4)
Barker, C. C., 160, 180, 181 (151), 182 (151)
Barker, H. A., 32, 34, 38 (12), 39 (27), 41 (28, 29), 43 (30), 46, 47, 59, 71, 72 (58a)
Barnard, T. H., 132
Barnett, W. L., 286
Barry, V. C., 180, 270
Bashford, V. G., 193, 195, 196, 216, 226 (12), 227 (12)
Bates, F. J., 279
Bates, F. L., 241, 247 (61)
Bauer, L., 79, 80 (2)
Bauer, W., 83, 99 (17)
Bear, R. S., 105
Becker, Johanna, 158, 161 (52), 173 (52)
Beckmann, C. O., 232
Behr, A., 128
Bell, D. J., 150, 155, 156 (45), 159, 161 (56), 165, 166 (17, 19, 92), 168, 169 (95), 170, 171 (98, 100)
Bell, F. K., 211

* The numbers in parentheses are reference numbers.

Bell, R. P., 52
Bender, R., 288
Bennett, E., 274
Berger, L., 73
Bergmann, M., 55, 187, 198, 200 (21a)
Berkman, S., 52
Bernfeld, P., 232, 234 (18), 235, 236 (33, 34), 241, 261, 263 (86), 265
Bernhart, F. W., 223
Berthelot, M., 203
Billica, H. R., 3
Black, I. M. A., 177, 178 (125)
Blanco, G. W., 271, 275 (16)
Blåch, H., 223
Blom, J., 232, 240
Blomqvist, G., 266
Bloor, W. R., 222
Bock, H., 80, 81, 82 (4)
Bodansky, O., 89
Bollenback, G. N., 7, 26 (22)
Bolliger, H. R., 20, 27 (60), 152, 155 (31)
Bomhard, H. v., 189, 190 (188)
Bonner, R. E., 223
Booher, Lela E., 235
Booth, J., 273
Boppel, H., 160, 163 (69)
Bouchardat, G., 192, 226 (2)
Bougault, J., 1, 202
Bourquelot, E., 57, 67
Boutron-Charlard, 60
Bowman, D. R., 273, 274 (26), 275 (26), 276 (26), 280
Braae, B., 232, 240
Brandner, J. D., 223, 228 (92)
Branfoot, M. H. (neé Carré), 99, 273
Brauns, D. H., 69
Brauns, F., 153, 161, 163 (73), 173 (73)
Bredereck, H., 160, 163 (67)
Bredig, G., 53
Brentano, W., 263
Brewster, M. D., 223, 228 (92)
Bridel, M., 67
Bridgett, R. C., 172, 175 (102)
Brigl, P., 149, 150 (5, 9), 151 (5), 183, 193, 204, 209, 214, 226 (7, 37)
Brink, N. G., 4, 12, 13, 19, 26 (14, 38), 27 (14, 57)
Brockway, L. O., 73
Broetz, A., 271

Brown, K. R., 223, 224
Browne, C. A., 87
Bruce, T. A., 136
Brunner, L., 279
Buckland, Irene K., 161, 163 (73), 173 (73)
Burkard, J., 33
Burn, J. H., 225
Burns, K., 287
Burrill, M. W., 90
Butter, G., 288
Bywater, R. A. S., 279, 283 (80)

C

Caldwell, Mary L., 230, 231 (3), 232 (3), 233 (3, 20), 234 (3, 20), 235, 236 (30), 237 (37, 38, 39), 239 (37, 38, 39, 41, 44, 45), 240 (3), 241 (3, 37, 38, 39, 41), 242 (41), 243 (41), 244 (41), 245 (41, 60), 246 (64), 247 (64), 248, 249 (60, 64), 250 (3), 251, 252 (73) (41), 253 (41, 64, 73), 254, 255 (3), 256 (3), 257 (3, 41, 73, 78), 258 (3, 41, 73, 81), 263 (41, 64)
Camargo, P. F., 4
Cameron, A., 186, 188 (167)
Campbell, W. G., 273
Cantor, S. M., 140, 141
Caputto, R., 50
Carney, D. M., 154, 155 (42), 178 (42)
Carpenter, D. C., 86, 102 (29)
Carr, C. J., 5, 17 (17), 26 (17), 202, 211, 225, 227 (99)
Carré, M. H., 87
Carré, P., 204, 228
Carrington, H. C., 178 (143), 179, 190 (143)
Carson, J. F., 221, 225 (74), 287
Castagne, A. E., 274
Castan, P., 148, 182
Castoro, N., 275
Cattelain, E., 1, 202
Cattle, M., 238
Centola, G., 123
Chabrier, P., 1, 202
Challinor, S. W., 174, 175 (119)
Chamberlain, K. A., 150
Chance, B., 53
Chanda, S. K., 289

Charlton, W., 152, 154, 155 (23), 173, 174 (106, 114), 175 (114), 179, 185 (23), 186 (23), 188 (142), 189 (142), 190 (23, 142)
Charney, J., 223
Chaudun, A., 81
Chester, Ruth M., 250
Chodat, R., 198
Clark, E. D., 235
Coleman, G. H., 152, 160, 163 (60), 174, 175 (118), 178 (118), 181 (118), 182 (28), 183 (28), 184 (28, 118), 188 (118)
Coles, H. W., 185
Colin, H., 81
Colowick, S. P., 73
Compton, J., 105
Conley, Maryalice, 5, 17, 194, 202, 226 (11)
Conrad, C. C., 109, 110, 112 (20), 113 (20), 114 (20), 122
Conrad, C. M., 120, 125 (29)
Cooper, C. J. A., 177, 179 (126)
Corey, A. J., 289
Cori, C. F., 34, 73
Cori, Gerty T., 75
Courtois, J., 31
Covert, L. W., 3
Cox, E. G., 160
Cox, E. H., 182, 183 (157), 184 (157)
Cram, K. H., 271
Cramer, M., 182, 183 (157), 184 (157)
Criegee, R., 199
Currie, N. R., 286

D

D'Addieco, A. A., 274
Dale, J. K., 136
Daly, Marie M., 241, 245 (59), 249 (59), 250 (59)
D'Arcy Hart, P., 223
Dauphine, A., 81
Davidson, G. F., 117, 119, 120
Davis, B. D., 223
Davis, B. L., 273
Davis, H. A., 175, 176 (122)
Dean, G. R., 141
Denham, W. S., 177, 178 (135)

Deuel, H., 83, 86, 88, 89 (25), 90, 93 (25), 99 (17), 101 (25), 102 (25)
Dewalt, C. W., 19, 27 (58)
Dewar, J., 152, 155 (30), 163, 164 (86, 87), 167, 168 (30, 87), 180 (86), 181 (86)
Dick, J. S., 172, 174 (104), 175 (104)
Dickerman, G. K., 279
Dickson, A. D., 109
Diehl, H. W., 25, 28 (71)
Dienes, Margaret, 201
Dikanowa, A., 30
Dillon, R. T., 154
Dillon, T., 270
Dimler, R. J., 175, 176 (122)
Doebbeling, A. H., 250, 257, 258 (81)
Doebbeling, Susie E., 232, 233 (20), 244, 251, 256
Dörr, R. E., 275, 287
Donovan, P. P., 287
Dore, W. H., 35, 39 (27), 41 (29)
Doser, A., 153
Doudoroff, M., 32, 33, 34, 35, 36, 37, 38 (12), 39 (27), 41 (28, 29), 43 (30), 46, 47, 48, 57, 58 (22), 59, 69 (22), 70, 71, 72 (58a)
Drew, H. D. K., 190
Dubos, R. J., 223
Dürr, W., 154, 155 (39)
Duff, R. B., 169
Dumas, J. B., 103
Dunning, J. W., 271
Dustman, R. B., 86, 93 (28), 94 (28), 98 (28), 102 (28)

E

Eastwood, J. A., 271
Ebert, C., 132
Eddy, C. R., 81, 98, 101
Egloff, G., 52
Ehrenthal, J., 290
Ehrlich, F., 80, 82 (4), 84 (4), 85, 88, 102 (19)
Eisfeld, K., 23
Ellis, F. W., 225
Elöd, E., 122
Eriksson, E., 234
Erlbach, H., 197
Erlenmeyer, H., 223

v. Euler, H., 54
Evans, T. H., 160, 163 (64)
Ewald, Lisa, 266

F

Fairhead, E. C., 161, 173 (74)
Fajans, K., 53
Farley, F. F., 154, 155 (42), 178 (42)
Fauconnier, A., 206, 227 (44), 228 (44)
Fellers, C. R., 87
Felser, H., 273
Feuerstein, K., 287
Figuier, L., 275
Fischer, E., 8, 10, 51, 189, 195, 197, 226 (13)
Fischer, Ed. H., 235, 236 (33, 34), 265
Fischer, H. O. L., 8
Fish, V. B., 86, 93 (28), 94 (28), 98 (28), 102 (28)
Fisher, S. F., 99
Fiske, P. S., 53
Fleck, L. C., 270, 271
Fletcher, H. G., Jr., 2, 14, 15, 17 (7), 18, 23 (43), 27 (7, 42, 43, 44, 55), 28 (69) 193, 203, 204, 209, 212, 213, 221, 222, 225 (75, 76, 77), 226 (6, 35), 227 (54), 228 (6)
Fleury, P., 31
Flynn, E. H., 4, 12, 13, 19, 26 (14, 38), 27 (14, 57)
Folkers, K., 1, 3 (3), 4, 12, 13, 19, 26 (14, 38), 27 (14, 57)
Forman, S. E., 225
Forrest, H. S., 223
Fort, G., 152, 155 (30), 163, 164 (86, 87), 167, 168 (30, 87), 180 (86), 181 (86)
Fowler, Frances L., 161, 163 (73), 173 (73)
Frankenburger, W., 51
Freeman, G. G., 236
Freeman, S., 90
French, D., 231, 241, 247 (61), 266
Freudenberg, K., 152 (22), 153, 154 (39, 41), 155 (39, 41), 159, 160 (57), 161 (57), 162 (63), 163 (41, 63, 69), 166 (57), 171 (41), 172 (41), 175 (41), 178 (41), 180, 181 (41, 152), 182 (152), 188 (41), 219, 226 (10), 266
Freudenberg, W., 5 (19), 17, 185, 194, 196, 199, 203, 204

Frey-Wyssling, A., 106
Fried, J., 15, 27 (47), 203
Frilette, V. J., 106, 112 (9), 122 (9)
Frush, Harriet L., 102
Fuld, Maria, 232, 265
Fuller, A. R., 288
Fuller, W. H., 288
Fyfe, A. W., 189, 190 (186)

G

Geddes, W. F., 230
Geib, N. V., 279
Georg, A., 70
Georges, L. W., 178 (146), 179
Gerecs, A., 30
Gerhardt, F., 279, 282
Gerngross, O., 106, 108 (12)
Gilchrist, Helen S., 148
Gille, R., 23
Gilmour, R., 188 (178), 189
Gjorling, L. G., 257
Gnüchtel, A., 65
Goebel, W. F., 162, 164 (83, 84)
Goepp, R. M., Jr., 6, 193, 194, 196, 204, 209, 212, 213, 226 (6, 10, 39), 227 (54), 228 (6)
Goldfinger, G., 121
Goldschmid, O., 114, 115 (23)
Goodhue, L. D., 185
Goodyear, E. H., 190
Gottschalk, A., 39, 60, 74, 75
Grandel, F., 220, 221
Granichstädten, H., 177, 181 (139), 182 (139), 183
Grassmann, W., 288
Greenane, F. J., 289
Gregory, Hilda, 224, 227 (95, 96)
Griffen, W. C., 223
Griffin, Jean H., 83, 99 (17)
Griffith, R. H., 52
Grüner, H., 193, 204, 209, 214, 226 (7, 37)
Grünler, S., 65
Günther, E., 158, 159 (53), 161 (53)
Gut, M., 21, 28 (62)

H

Haas, H., 287
Haas, R. H., 123, 124

AUTHOR INDEX

Habrle, J. A., 109, 117 (19), 119, 120, 125 (27)
Hachihama, Y., 279
Hackl, A., 287
Hahn, A., 237
Hain, G. M., 223
Hall, C. C., 52
Hamaguti, E., 280
Hampton, H. A., 278, 280 (70), 281 (70), 282 (70)
Hanes, C. S., 34, 230, 231 (1), 232(1), 233 (1), 234 (1), 238, 240 (1), 241 (1), 257 (1), 258 (1)
Hann, R. M., 8, 10 (27), 14, 25, 28 (71), 40, 43
Hanrahan, Virginia, 250
Hansen, A., 85, 86, 89 (25), 93 (25), 101 (25), 102 (25)
Hardegger, E., 24
Harding, A. T., 137
Harding, T. S., 279
Harley, C. P., 99
Harris, M., 109
Harris, S. A., 1, 3 (3)
Hart, C. E., 34
Haskins, W. T., 8, 14, 43
Hassid, W. Z., 32, 33, 34, 35, 38 (12), 39 (27), 41 (28, 29), 43 (30), 46, 47, 59, 70, 71, 72 (58a), 119, 232
Hauenstein, H., 22
Hauptmann, H., 4
Hawkins, W. L., 160, 163 (64), 173 (64)
Hawley, L. F., 270
Haworth, W. N., 11, 31, 45 (8), 105, 149, 150 (6), 151 (6), 152, 154, 155 (23), 160, 165, 170, 171 (97), 172, 173, 174 (106, 107, 114), 175 (103, 112, 114, 115, 119), 177, 178 (143, 147), 179 (97, 126, 127, 129), 180, 181 (97), 182, 183, 184 (154), 185 (23, 147), 186 (23, 147), 187, 188 (129, 142), 189 (142), 190 (23, 142, 143), 212, 217, 224, 225, 227 (94, 96), 278, 279, 280 (70), 281 (70), 282 (70, 79), 283 (80), 285 (79)
Haynes, Dorothy, 87
Hechtman, J., 280
Heddle, W. J., 154, 155 (38)
Hehre, E. J., 47

Helferich, B., 55, 61, 62 (29), 63 (29), 64 (19, 29), 65 (29), 66 (29), 67 (29), 154, 155 (40), 156 (40), 158, 159 (53), 160, 161 (52, 53, 58, 59), 162 (65), 163 (65), 173 (52), 187, 201
Henderson, S. T., 273
Henglein, F. A., 79, 80 (2), 81
Herbert, R. W., 173, 174 (114), 175 (114)
Hermans, P. H., 121
Herrmann, K., 106, 108 (12)
Hess, K., 177, 178 (144, 145, 148), 179, 189, 190 (148, 187)
Hesse, A. R. F., 235, 240 (29)
Hestrin, S., 47
Heumann, K. E., 178 (148), 179, 190 (148)
Heuser, E., 126, 271, 278, 279, 281, 285 (72, 79), 286, 287, 288
Heyne, E., 282, 286 (97), 287 (97)
Hibbert, H., 160, 161, 163 (64, 73), 173 (64, 73, 74)
Hickinbottom, W. J., 149, 150 (3), 151 (3), 173, 174 (106)
Hilbert, G. E., 175, 176 (122), 283
Hilditch, T. P., 52
Hill, K., 63
Hills, C. H., 85, 93, 96 (52, 53), 98, 101
Himmen, E., 160, 161 (59), 201
Hinshelwood, C. N., 52
Hirst, E. L., 79, 80 (2), 149, 150 (6), 151 (6), 160, 173, 174 (107), 175 (115, 119), 177, 178 (130, 143) 179 (127, 130), 180, 181 (151), 182 (151), 183, 184 (154), 187, 188, 189 (170, 173), 190 (143), 270, 278, 279, 280 (70), 281 (70), 282 (70, 79), 283 (80), 289
Hixon, R. M., 150, 152, 185
Hochstetter, H. v., 154, 155 (39)
Hockett, R. C., 5, 17, 193, 194, 201, 202, 204, 209, 212, 217, 226 (6, 11, 39), 227 (54), 228 (6, 66)
Hoffman, U., 211
Hofmann, A. W., 60
Hofmann, E., 62, 63 (30)
Hogg, T. P., 152, 153 (21), 155 (21), 189, 190 (186)
Holden, Margaret, 93, 95, (54), 96 (54), 98 (54)
Hollenbeck, C. M., 234, 256 (28)
Holmberg, C. V., 274

Holmes, E. L., 138
Holmes, G. S., 280, 284
Hooper, I. R., 13, 26 (39)
Hopkins, F. G., 54
Hopkins, R. H., 74, 230, 233 (4), 236, 240 (4), 241 (4), 264 (4)
Hoppe-Seyler, F., 288
Hough, L., 289
Houtz, H. H., 289
Hudson, C. S., 2, 5, 7, 8, 9, 10 (27), 11, 14, 15, 16, 17 (7, 17), 18, 23 (43), 25, 26 (17, 23, 28), 27 (7, 42, 43, 44, 50, 55), 28 (71), 32, 33, 39 (10), 40, 43, 47 (40), 56, 66, 68, 69 (45), 70 (45), 75, 121, 136, 202, 203 (103), 204, 221, 222, 225 (75, 76, 77), 226 (35, 35a), 266, 279
Hüll, G., 159, 160 (57), 161 (57), 166 (57), 175 (57)
Humphreys, R. W., 188 (182), 189, 190 (182)
Hunter, M. J., 161, 173 (74)
Hunter, R. H., 223, 228 (92)
Hurd, C. D., 286
Hurwitz, O., 153
Husemann, E., 282, 285, 287 (94)
Hynd, A., 188 (181), 189

I

Ikeda, C., 237, 265 (54)
Inoue, T., 281
Ipatieff, V. N., 52
Irvine, J. C., 30, 148, 152, 153 (21), 155 (20, 21), 160, 161, 162, 163 (61), 172, 173, 174 (104, 109, 111), 175 (104, 109, 111), 176, 177 (71), 178 (125, 130, 141), 179 (130), 184, 185 (20), 186, 188 (109, 167, 168, 178, 181, 183, 184), 189, 190 (186)
Isbell, H. S., 39, 75, 102
Ivy, A. C., 90

J

Jackson, E. L., 40, 121
Jackson, R. F., 130, 168
Jacobi, R., 266
Jahn, E. C., 274
Jandebeur, W., 275

Jang, Rosie, 85, 86 (19a), 89, 102 (39)
Jansen, E. F., 85, 86, 89 (27), 90 (21), 91 (27), 96 (21), 97 (21), 101, 102 (39)
Jayme, G., 274, 278, 279, 282, 284, 285 (72, 104b), 289
Jeanloz, R., 15, 19, 20, 27 (44, 59), 162, 222, 225 (77)
Jeger, O., 23, 28 (67)
Jirak, L., 88
Johnson, S. W., 275, 278
Johnston, N. F., 223
Jones, D. T., 223
Jones, J. K. N., 80, 180, 181 (151), 182 (151), 270, 289
Jones, W. G. M., 217
Jones, W. J. G., 165
Joseph, G. H., 81
Josephson, K., 54, 57
Joslyn, M. A., 99, 100 (61)
Jurisch, J., 274

K

Kagan, B. O., 70
Kaplan, N., 33, 70
Kappesser, R., 287
Karabinos, J. V., 2, 3, 5, 22, 26 (5), 28 (65)
Karrer, P., 153, 177
Kaulla, K. N. v., 287
Keilin, D., 53
Keller, R. E., 279
Kendall, E. C., 235, 237
Kenner, G. W., 4
Keresztesy, J. C., 1
Kerly, M., 75
Kerr, R. W., 266
Kertesz, Z. I., 80, 81, 82 (3), 83 (3), 84 (3), 85 (3, 16), 86 (3, 14), 89 (25), 90 (3), 91, 92 (11, 16), 93 (3, 25), 94 (50), 95, 96 (3, 20, 44, 55), 97 (26, 44), 98 (44), 99 (16, 17, 48a), 100 (20, 44, 48c), 102
King, F. W., 271
Kirschoff, G. S. C., 128
Klages, F., 30
Kleiderer, E. C., 2
Klein, W., 160, 161 (58)
Klemm, L. H., 13, 26 (39)
Knauf, A. E., 8, 10 (27)

Kneen, E., 230, 234, 237, 256 (28), 265
Kobel, M., 93, 96 (51), 101 (51)
Koch, F., 278
Koehler, Leonore H., 18, 27 (55)
Komatsu, S., 281
Kornfeld, E. C., 2
Kraemer, E. O., 106
Krantz, J. C., Jr., 211, 225, 227 (99)
Kratky, O., 106
Kreider, L. C., 80
Kriukova, N., 33, 36 (18)
Kropf, R. T., 120
Krznarish, P. W., 282
Kuehl, F. A., Jr., 4, 12, 13, 19, 26 (14, 38), 27 (14, 57)
Kuhn, R., 68, 236, 251, 257 (72)
Kuna, M., 80
Kunitz, M., 73, 75 (63b)
Kurasaw, H., 280
Kursanov, A., 30, 33, 36 (18)
Kurth, E. F., 271, 273, 274, 289

L

LaForge, F. B., 279
Lake, W. H. G., 11, 180, 181 (153), 182 (153)
Landis, Q., 232
Lang, G., 275
Lang, O., 154, 155 (40), 156 (40)
Langenbeck, W., 51, 58
Lathrop, E. C., 271
Law, J., 31
Learner, A., 173, 174 (107)
Lebedew, A., 30
Leissner, Elsa, 231, 259 (11)
Leitch, Grace C., 174, 177, 179 (129), 188 (129)
Lemieux, R. U., 11, 12, 19, 26 (34, 36), 27 (58)
Lenze, F., 287
Leo, H. T., 95
Letzig, E., 102
Levene, P. A., 80, 149, 151, 153, 154, 155, 156 (44), 159, 160, 161 (47), 162 (62), 169, 172, 185 (62, 105, 165), 186
Levi, I., 33, 160, 163 (64), 173 (64)
Levin, R. H., 3, 22
Lew, B. W., 6

Lewis, W. L., 152, 155 (25), 188 (180), 189
Li, T. W., 90
Liebig, J., 60
Lieser, T., 149, 150 (4), 151 (4)
Lillelund, H., 63
Limerick, J. M., 289
Lineweaver, H., 82, 85, 86 (19a), 90 (21), 93, 94 (49), 96 (21, 49), 97 (21, 49), 101
Link, K. P., 79, 80 (2), 109, 278, 282
Liser, T., 287
Little, J. E., 236, 239 (44, 45)
Llewellyn, F. J., 160
Locke, E. G., 279
Lockwood, A. R., 233
Loconti, J. D., 92, 99 (48a), 102
Loder, D. J., 152, 155 (25)
Lohse, H. W., 52
Lorber, J., 170, 171 (100)
Lotzkar, H., 88
Lovell, E. L., 114, 115 (23)
Lüdtke, M., 273
Lüers, H., 234
Luh, B. S., 92
Lutwak-Mann, C., 54
Lyatker, S. N., 70
Lythgoe, B., 4

M

McArthur, N., 163, 164 (87), 167, 168 (87)
Macbeth, A. K., 177, 179 (133), 187
McCloskey, C. M., 160, 163 (60), 174, 175 (118), 178 (118), 181 (118), 182 (28), 183 (28), 184 (28, 118), 188 (118)
McCloskey, D. E., 152
McColloch, R. J., 82, 85 (16), 86 (14), 90, 92 (16), 93, 94 (44), 95, 96 (44, 55), 97 (44, 45), 98 (44, 45), 99 (16), 100 (44, 48c)
McCready, R. M., 34, 101, 232
McDonald, Emma J., 168
Macdonald, J., 160, 177 (71)
Macdonald, J. L. A., 185
McDonald, M. R., 73, 75 (63b)
MacDonnell, L. R., 85, 86, 89 (27), 90 (21), 91 (27), 94 (44), 96 (21), 97 (21), 102 (39)

McGlynn, R. P., 178 (141), 179
McIlroy, R. J., 280, 284
McIntosh, A. V., Jr., 3, 22
Mackay, J., 177, 179 (133)
Maclay, W. D., 8, 10 (27), 40, 101, 221, 225 (74), 287
McOmie, J. F. W., 2, 3
Maehly, A. C., 20, 28 (61)
Mahn, H., 198
Malkomes, T., 187
Malsch, L., 99
Manian, S. H., 244
Mann, T., 53
Maquenne, L., 203
Mark, H., 79, 80 (2), 106, 107, 108, 112 (9), 120, 121 (11), 122 (9)
Marsh, G., 99
Martelins, B., 264
Martin, A. R., 109
Mason, R. I., 194, 226 (11)
Masure, M. P., 99
Mathers, D. S., 162, 163 (82), 170, 171 (99)
Matus, J., 86, 88, 89 (25), 90 (39a), 91 (39a), 93 (25), 101 (25), 102 (25)
Mauclère, P., 228
Mauger, R. P., 280, 284
Maurer, K., 16, 198
Mayer, W., 9
Meagher, W. A., 35
Meesook, B., 285
Mehlitz, A., 86, 88, 89 (25), 93 (25), 101 (25), 102 (25)
Meinel, K., 275
Meller, A., 273
Melville, D. B., 1
Menten, M. L., 53
Merker, R. L., 223
Merrill, R. C., 88
Mertzweiller, J. K., 154, 155 (42), 178 (42)
Messier, R. L., 86, 102 (30)
Meyer, G. M., 80, 149, 151, 153, 160, 162 (62), 172, 185 (62, 105, 165), 186
Meyer, H., 79, 80 (2), 237
Meyer, K. H., 106, 121 (11), 231, 232 (6), 233 (6), 234 (6, 18), 235, 236 (33, 34), 240 (6), 241 (6), 261, 263, 265
Meyerhof, O., 51, 73
Michaelis, L., 53
Michaelis, M., 122
Micheel, F., 178 (144, 145), 179, 189, 190 (187)
Miekeley, A., 187
Militzer, W., 237, 265 (54)
Miller, E. J., 173, 174 (107), 188, 190
Minaeff, M., 53
Mindell, Florence M., 241, 245 (60), 246, 248, 249 (60)
Mitchell, R. L., 285
Mittelman, N., 50
Mitts, Eleanor, 150
Miyaki, S., 280
Mo, S., 274
Mohr, R., 287
Møller, C., 63
Montavon, R. M., 24
Montgomery, R., 197, 209, 211, 212, 213, 216, 225, 226 (14, 47, 51, 56), 227 (50, 51, 56, 101), 228 (47, 101)
Moodie, Agnes M., 188 (184), 189
Moog, K., 177
Morell, J. S., 79, 80 (2)
Morgan, E. J., 54
Morrell, J. C., 52
Morrell R. S., 10
Morton, F. A., 223
Moscowitz, M., 132
Mottern, H. H., 81, 85, 93, 96 (52)
Moyer, J. C., 95, 96 (55)
Mozingo, R., 1, 3, 4
Müller, A., 205, 226, (40)
Müller, O., 271
Mukherjee, S., 22, 28 (63)
Muller, J., 211
Mumford, N. V. S., 129
Munro, J., 156 (49), 157
Murphy, C. M., 223
Murphy, M., 274
Murray, R. H., 233
Muskat, I. E., 152, 155 (26, 27)
Myrbäck, K., 80, 82 (4), 91, 231, 237, 250, 254 (65), 255 (74), 256 (77), 257, 258, 259, 260 (84), 261, 262 (65, 84), 263, 264 (65, 88, 89), 267

N

Nagy, E., 206, 226 (43)
Nakai, R., 281

Nakamura, H., 236, 250 (40)
Naylor, Nellie M., 236
Nelson, J. M., 90
Nelson, Mary L., 115, 116 (24), 120, 125 (29)
Neshida, K., 177
Ness, A. T., 25
Ness, R. K., 18
Neuberg, C., 33, 50, 63, 93, 96 (51), 101 (51)
Nevros, K., 275
Newkirk, W. B., 129, 130, 132, 133, 135, 136
Newth, F. H., 219
Nickerson, R. F., 109, 110 (16), 111, 112 (21), 113 (22), 117 (19), 119, 120, 125 (27)
Niederle, M., 88
Niemann, R., 30
Nord, F. F., 51, 58
Norman, A. G., 270, 272, 288
Norris, F. W., 277, 280 (63)
Norymberski, J., 23, 28 (67)
Nycander, G., 91, 264

O

Obermanns, H. E., 289
O'Dwyer, M. H., 273, 277, 280, 286
Ohle, H., 156, 158, 159, 161 (48, 54), 185 (48), 186 (48), 197
Ohlsson, E., 234, 256, 257 (23), 258, 261
Oldham, J. W. H., 30, 149, 150 (7), 151 (7), 155 (15), 160, 162 (68), 163 (15, 68, 76), 165, 166 (7), 173, 174 (109, 110, 111), 175 (109, 110, 111), 176, 178 (15, 68), 180 (18), 181 (15, 18), 188 (109)
Oldham, Mary A., 150, 180 (18), 181 (18)
Olin, S. M., 11, 26 (34, 36)
Oliver, E., 279, 282 (79), 283, 285 (79)
Olsen, A. G., 81
O'Neal, R., 48
Oparin, A., 30
Oppenheimer, C., 50
Örtenblad, B., 254, 255 (77), 256 (77), 257, 263, 264 (88)
Ossipov, F. M., 86
Østrup, G., 63
Otterson, H., 109

Otto, R., 89
Overend, W. G., 211, 219, 227 (50)
Owen, L. N., 165, 176 (90), 212
Owens, H. S., 88, 101

P

Pacsu, E., 149, 150, 154, 156 (8), 158 (8)
Paine, H. S., 139
Pallmann, H., 86, 88, 89 (25), 93 (25), 101 (25), 102 (25)
Panasyuk, V. G., 271
Panizzon, L., 180, 182, 183, 184 (154)
Papadimitriou, Irene, 5, 200
Passmore, F., 8, 10 (27)
Pasternack, R., 103
Patterson, Jocelyn, 185
Pauling, L., 56, 65, 73
Pavlic, A. A., 3, 7
Payen, A., 103
Peat, S., 11, 153, 155 (32), 165, 171 (32), 173, 174 (116), 175 (116), 176 (90), 177, 178 (123, 147), 179 (126), 180, 181 (153), 182 (153), 183 (32), 184 (32), 185 (147), 186 (147), 188 (142), 189 (142), 190 (142), 192, 197 (1), 219, 250, 279, 283 (80)
Percival, E. G. V., 154, 155 (38), 156 (49), 157, 158 (51), 169, 177, 282, 289
Percival, Elizabeth E., 157, 158 (51)
Perkin, A. G., 203
Perlin, A. S., 122
Peterson, M., 88
Phaff, H. J., 86, 88 (24), 92
Philipp, H. J., 115, 116 (24)
Phillips, M., 273, 280
Picard, R., 198
Pictet, A., 30, 148, 182
Pigman, W. W., 39, 55, 61 (20), 62 (20), 64, 65 (20), 66 (42), 67 (42), 75, 77
Piloty, O., 10
Plankenhorn, E., 154, 155 (41), 163 (41), 171 (41), 172 (41), 175 (41), 179 (41), 180, 181 (41, 152), 182 (152), 188 (41)
Plötner, K., 16
Ploetz, T., 288
Polglase, W. J., 13, 19, 26 (39), 27 (58)
Poore, H. D., 87
Porst, C. E. J., 129

Porter, C. R., 190
Potter, A. L., 35, 43 (30), 71
Poumarede, J. A., 275
Powell, G., 171, 172
Preece, I. A., 276, 282
Prelog, V., 23, 28 (67)
Press, J., 232
Preston, R. D., 270, 272
Pringsheim, H., 188 (179), 189
Prins, D. A., 19, 20, 21, 22, 27 (59, 60), 28 (61, 62), 152, 155 (31)
Proskuriakov, N. J., 86
Pryde, J., 188 (182, 183), 189, 190 (182)
Purdie, T., 172, 175 (102), 186, 188 (167)
Purves, C. B., 14, 33, 68, 69, 70, 105, 123, 124, 187, 189 (170), 285
Puskás, T., 206, 226 (42, 43)
Putman, E. W., 47
Pyenson, L., 223

R

Ramsden, H. E., 201
Raney, M., 3
Rapp, W., 266
Ratliff, E. K., 275
Rauchalles, G., 32, 67
Raybin, H. W., 40
Raymond, A. L., 149, 151, 155, 156 (44), 159, 161 (47), 169
Rees, D. E., 152, 174, 175 (118), 178 (118), 181 (118), 182 (28), 183 (28), 184 (28, 118), 188 (118)
Reeve, W., 2
Reeves, R. E., 162, 164 (83, 84), 165, 166
Regna, P. P., 102
Reichstein, T., 19, 20, 21, 22, 27 (59), 28 (61, 62)
Reiff, F., 274
Reilly, J., 287
Renner, A., 67
Rice, C. C., 87
Richter, G. A., 271
Richtmyer, N. K., 5, 7, 9, 15, 16, 17 (17), 25, 26 (17, 23, 28), 27 (50), 28 (71), 64, 66, 68 (45), 70 (45), 180, 181 (150), 202, 203 (103), 204, 226 (35a)
Rideal, E. K., 53
Ritter, G. J., 270, 271, 274
Robertson, A., 163, 173 (85), 175 (85)

Robertson, G. J., 162, 163 (75, 76, 82) 170, 171 (99), 175 (75), 177 (75), 187, 189 (173)
Robertson, W. R., 83, 99 (17)
Robiquet, 60
Rogers, E. F., 17, 199, 203
Ropes, M. W., 83, 99 (17)
Rosenfeld, D. A., 17, 204, 226 (35a)
Roth, R. W., 223
Routala, O., 271
Roux, E., 219
Ruell, D. A., 173, 175 (115)
Rundle, R. E., 231, 241, 247 (61), 266
Runkel, R., 275
Ruppel, W., 281
Rutherford, Jean K., 149, 150 (7), 151 (7), 160, 162 (68), 163 (68), 165, 166 (7), 174, 176, 178 (68)

S

Saeman, J. F., 116, 117 (25), 279
Sätre, M., 279, 282, 284, 285 (104b)
Säverborn, S., 80, 88 (7)
Salkowski, E., 278
Salmon, M. R., 171, 172
Sandstedt, R. M., 234, 237, 256 (28), 265 (53)
Saric, S. P., 288
Sato, D., 198, 200 (21)
Sato, S., 198, 200 (21)
Schäfer, W., 160, 161 (58)
Schardinger, F., 266
Scheiber, H., 55
Scheuer, M., 88
Schiefer, H. F., 120
Schinle, R., 149, 150 (5, 9, 10), 151 (5), 154, 156 (43)
Schlesinger, M. D., 239
Schlosser, P., 286
Schlubach, H. H., 32, 67, 177, 189, 190 (188)
Schlüchterer, Elsa, 161, 163 (72), 173, 174 (116), 175 (116), 250
Schmid-Bielenberg, H., 122
Schmidt, E., 275, 283
Schmidt, O. T., 152, 153, 155 (24)
Schneider, G., 79 80 (2)
Schneider, W., 23
Schoch, T. J., 231, 241 (10)

AUTHOR INDEX

Schofield, R. K., 288
Schorsch, G., 287, 288
Schreiber, W. T., 279
Schulz, G. V., 285
Schulze, E., 275, 277, 278 (52)
Schulze, E. S., 269
Schwartz, H., 271
Schwimmer, S., 256, 257 (79)
Scott, J. P., 152, 153, 160, 161, 163 (61), 184, 185 (20)
Scroggie, A. G., 109, 110, 112 (20), 113 (20), 114 (20), 122
Sedgwick, W. G., 170, 171 (97), 179 (97), 181 (97)
Sedky, A., 99, 100 (61)
Seeley, M., 280
Sellner, E., 234
Severson, G. M., 266
Sevon, J., 271
Sharp, T. E., 223
Sheehan, J. T., 5, 199
Sheffield, Elizabeth, 193, 204, 209, 212, 217, 226 (6), 227 (54), 228 (6, 66)
Shephard, A. D., 101
Sherman, H. C., 235, 236 (30, 36), 237 (37, 38, 39), 239 (37, 38, 39), 241 (37, 38, 39)
Sherrard, E. C., 271, 275 (16)
Shinoda, J., 198, 200 (21)
Siggia, S., 121
Sillén, L. G., 264
Silsbee, Clara G., 130
Simon, A., 152, 153, 155 (24)
Singh, B., 141
Sisson, W. A., 105
Siwoloboff, A., 209, 217
Sjöberg, K., 234
Skinner, A. F., 30
Slein, M. W., 73, 75
Smart, C. L., 283
Smith, F., 162, 163 (81), 165, 173, 174 (108), 175 (108, 120), 176 (90, 121), 212, 290
Smith, R. K., 52
Soff, K., 266
Sohn, A. W., 274
Solechnik, N. Y., 283, 286 (101), 287 (101)
Soltzberg, S., 193, 194, 196, 204, 209, 226 (6, 10), 228 (6)

Sowden, J. C., 8, 35
Soxhlet, F., 129
Speiser, R., 81, 98, 101
Spencer, C., 4
Spencker, K., 156
Spero, G. B , 3, 22
Sponable, M., 226
Spurlin, H. M., 122
Stacey, M., 161, 163 (72), 173, 174 (108, 116), 175 (108, 116), 217, 250
Stadler, R., 288
Staub, A., 265
Staudinger, H., 274, 287
Stavely, H. E., 15
Stearn, A. E., 59
Steingroever, A., 188 (179), 189
Steller, R. L., 271
Stenlid, G., 91, 264
Stephenson, R. P., 225
Stern, K. G., 53
Stern, P. L., 132
Stevens, J. W., 100
Studer-Pécha, H., 241, 265
Sugg, J. Y., 47
Sundberg, R. L., 152, 174, 175 (118), 178 (118), 181 (118), 182 (28), 183 (28), 184 (28, 118), 188 (118)
Suter, E., 223
Synge, R. L. M., 150, 155, 156 (45), 165, 166 (17), 170, 171 (98)
Szpilfogel, S., 23, 28 (67)

T

Takamine, J., 250
Takimoto, H., 5, 198, 202 (18)
Taylor, C. C., 95
Teece, Ethel G., 149, 150 (6), 151 (6)
Tessmar, K., 158, 159, 161 (54)
Thatcher, R. W., 99
Thomas, B. B., 279
Thompson, A., 7
Thompson, R. R., 234
Thomsen, J., 275
Thorsell, W., 258, 262, 263
Tilden, Evelyn B., 266
Todd, A. R., 4, 22, 28 (63)
Toepffer, H., 160, 162 (63), 163 (63), 219

Tollens, B., 9, 269, 275, 278
Tomarelli, R. M., 223
Tóth, G., 180, 181 (155)
Tressler, D. K., 99
Trucco, R. E., 50
Tsvasman, E. M., 70
Turkevich, J., 52

U

Underkofler, L. A., 7, 26 (22)
Uyeda, Y., 203

V

Vajda, A. M., 185
Valentin, F., 193, 226 (5)
Van Beckum, W. G., 274
van der Burg, J. H. N., 192
van Klinkenberg, G. A., 232, 233 (19), 234, 240 (19), 241 (19), 258 (19)
van Romburgh, P., 192
Vargha, L. v., 152, 155 (29), 156, 157, 158 (50), 159, 161 (48, 54), 172, 185 (48, 50), 186 (48), 205, 206, 219, 226 (40, 41, 42, 43)
Veibel, S., 63
Verdon, E., 57
Vieweg, F., 60
Vigneaud, V., du, 1
Vignon, L., 192
Vincent, G. P., 274
Vogel, H., 30
Volz, Gertrude W., 250, 251 (73), 252, 253 (73), 254, 257 (73), 258 (73)
Voss, W., 288
Votoček, E., 9

W

Wadman, W. H., 289
Wagner, T., 129
Waine, A. C., 190
Waksman, S. A., 89
Walden, M. K., 234
Waldschmidt-Leitz, E., 235, 240 (29)
Walker, J., 223
Wallerstein, L., 265
Walsh, W. F., 86, 102 (29)

Walz, Doris E., 15, 27 (47), 203
Wangel, J., 63
Waters, E. T., 188 (182), 189, 190 (182)
Waters, R. B., 163, 173 (85), 175 (85)
Webb, J. I., 177
Weber, F., 86, 88, 89 (25), 90, 93 (25), 101 (25), 102 (25)
Weber, R., 99
Weidenhagen, R., 32, 39 (11), 51, 58, 67, 267
Weihe, H. D., 273, 280
Weil, Ruth S., 236
Weill, C. E., 236
Weisblat, D. I., 11, 26 (36)
Weitnauer, G., 90, 91 (46)
Wells, S. D., 271
Weltzien, W., 177
Wenzl, H., 274
Wheeler, H. J., 275, 278
Whetstone, J., 178 (147), 179, 185 (147), 186 (147)
Whistler, R. L., 109, 273, 274 (26), 275 (26), 276 (26), 280, 282, 283, 286 (97), 287 (97)
White, J. F., 274
White, J. W., 101
Wichelhaus, H., 128
Widmer, F., 177
Wiggins, L. F., 14, 153, 155 (32), 171 (32), 177, 178 (123), 183 (32), 184 (32), 193, 195, 196, 197, 208, 209, 211, 212, 213, 216, 217, 219, 224, 225, 226 (12, 14, 47, 51, 56, 69), 227 (12, 45, 50, 51, 56, 94, 95, 96, 101), 228 (47, 101)
Will, W., 287
Willaman, J. J., 86, 89 (25), 93 (25), 96 (52, 53), 101 (25), 102 (25)
Willstätter, R., 235, 240 (29)
Wilson, P. I. 173, 174 (108)
Wingfeld, B., 279
Winkler, G., 89
Winterstein, E., 275
Wise, L. E., 274, 275, 289
Wladislaw, B., 4
Wöhler, F., 60
Wohlgemuth, J., 238
Wolf, D. E., 1, 3 (3)

Wolfrom, M. L., 2, 5, 6, 7, 11, 12, 13, 19, 22, 26 (5, 34, 36, 39), 27 (58), 28 (65), 178 (146), 179, 188 (180), 189
Wolochow, H., 47
Woodhouse, Hilda, 177, 178 (135)
Woolvin, C. S., 183
Wylam, B., 173, 175 (112)

Y

Yanovsky, E., 136
Young, G. T., 160
Yundt, A. P., 278
Yusen, M., 194, 226 (11)

Z

Zach, K., 195, 197, 226 (13)
Zechmeister, L., 180, 181 (155)
Zemplén, G., 30, 173
Zerban, F. W., 87
Zervas, L., 5, 15, 17 (46), 198, 200 (21a), 203, 204, 221
Zief, M., 204, 226, (39)
Ziemiecka, Y., 288
Ziifle, Hilda M., 115, 116 (24)
Zimmermann, W., 99
Zisman, W. A., 223
Zissis, E., 9, 26 (28)

Subject Index

A

Acetic acid, glacial, for laboratory crystallization of β-dextrose, 136
Acetobacter suboxydans, 7
β-(N-Acetyl)-glucosaminidase, 61
Acetylation, of xylan, 286
—, rates for various cellulosic fibers, 122, 123
Activated carbon, use in manufacture of dextrose, 140, 143
Adenine, 9-β-D-ribofuranosyl-, 74
Adenosine, 74
Adenosine-5'-triphosphate (ATP), 49, 50, 73, 74
Adipic acid, 288
"Adipo-cellulose," 104
(*dextro*)-Alanine, N-acetyl-, 11
—, configuration, 11
Alcohols, hexahydric, ethylene oxide derivatives of, 218
Aldobiuronic acid, in hemicellulose of English oak, 280
Alfalfa, 93, 95, 96
Algae, marine, presence of xylan in, 270
Alkali cellulose, 113
D-Allitol, 2-desoxy-, 6
—, 1,4:3,6-dianhydro-, 215
D-Allose, 2-desoxy-, 20
α-D-Alloside, methyl 2,3-anhydro-4,6-benzylidene-, 20
—, methyl 2-desoxy-, 20, 27
—, methyl 2-desoxy-3-methyl-, 20, 27
D-Altritol, 1,4:3,6-dianhydro-, 215
α-D-Altroside, methyl 2-methyl-3-methylthio-4,6-benzylidene-, 27
—, methyl 2-methylthio-4,6-benzylidene-, 20, 27
—, methyl 3-methylthio-4,6-benzylidene-, 27
α-D-Altroside, methyl 2-methylthio-3-methyl-4,6-benzylidene-, 20, 27
Amygdalin, 60

Amylase, crystalline beta, of sweet potato, 234
—, crystalline pancreatic alpha, 235, 236, 241
—, crystalline salivary alpha, 265
Amylases, alpha, 229–268
  purification by differential inactivation, 234, 256
—, beta, 231–235, 268
  purification by differential inactivation, 234, 256
Amylopectin, 106, 231, 245, 246
Amylose, 231, 241, 245, 246, 263
Anhydrides, of the pentitols and hexitols, 191–228
Anhydro-. For anhydro- *derivatives see inverted entries under the name of the parent compound, e.g.,* D-sorbitol, 1,5-anhydro-.
Apple wood, L-arabinose and D-xylose in hemicellulose of, 279
  optical rotation of xylan from, 282
Apples, pectin-esterases in, 85, 99
L-Arabinofuranoside, methyl 2,3,5-trimethyl-, 281, 283
L-Arabinopyranose, 3-α-D-glucopyranosyl-, 43, 46, 71, 72
  enzymatic synthesis, 43
  methylation, 44
  phenylosotriazole, 43
α-D-Arabinopyranoside, 2'-naphthyl 1-thio-, triacetate, 14, 27
β-L-Arabinopyranoside, phenyl, 66, 67
α-D-Arabinopyranoside, phenyl 1-thio-, triacetate, 14, 27
D-Arabinose, 8
L-Arabinose, α-benzyl-α-phenylhydrazone, 279
D-Arabinose, diethyl thioacetal, tetraacetate, 7, 26
L-Arabinose, diethyl thioacetal, tetraacetate, 7, 26

304

SUBJECT INDEX

—, enzymatic reaction with α-D-glucose-1-phosphate, 43
—, 2,4-dimethyl-, 44
α-L-Arabinoside. See α-L-Arabinopyranoside.
β-D-Arabinosyl bromide, 2,3,4-triacetyl-, 23
D-Arabitol, configuration, 8
—, 1,5-anhydro-, 14, 23, 27, 221, 222, 225
  2,3,4-triacetate, 225
  2,3,4-tribenzoate, 225
—, 1-desoxy-, tetraacetate, 7, 26
L-Arabitol, 1-desoxy-, tetraacetate, 7, 26
L-Araboketose, enzymatic reaction with α-D-glucose-1-phosphate, 41
L-Araboketoside, α-D-glucopyranosyl-, 35, 41
L-Arabonic acid, 2,4-dimethyl-, 45
  lactone, 45
Arlitan, 194, 226
Ascorbic acid, use of D-galacturonic acid in synthesis of, 102
Aspen, pentosan content, 27
Aspergillus niger, emulsins, 63
—, xylanase enzyme from, 288
Aspergillus oryzae, amylase of, 250–255, 265
ATP. See Adenosine-5′-triphosphate.

B

Bacillus delbrucki, emulsins, 63
Bacillus macerans, amylase of, 266
Bacillus subtilis, 48
  amylase of, 265
Bacterial amylases, 265
Bagasse, sugar cane, L-arabinose and D-xylose in hemicellulose of, 279
Bark, tree, pentosan content, 271
Barley, alpha amylase of malted, 255–265
—, beta amylase of ungerminated, 231–234
Barley shoots, sucrose formation in, 34
"Bastose," 104
Beechwood, L-arabinose in xylan from, 279
—, pentosan content of, 271
Beechwood xylan, 285
Beets, hexose phosphates in, 33
  pectinic acids of, 83

Bentonite, use in manufacture of dextrose, 139, 142
Benzaldehyde, enzymatic formation from amygdalin, 60
—, dibenzyl thioacetal, 4
Bibenzyl, from benzaldehyde dibenzyl thioacetal, 4
Biolase, source of a crystalline alpha amylase, 265

C

Camphor, optically active, formed from inactive (racemic) camphor carboxylic acid in the presence of quinine, quinidine or nicotine, 53
Camphor carboxylic acid. See Camphor, 53
Carbohydrases, 59
Carbohydrate derivatives, formation from thio-carbohydrates through reductive desulfurization by Raney nickel, 1–28
Carbohydrates, enzyme specificity in the domain of, 49–78
—, union of enzyme and substrate, 54, 56
  chemisorption, 56
  forces of attraction, 55
Cascara, pentosan content, 271
Catalase, theory of action on ethyl hydroperoxide, 53
Catalysis, enzyme-substrate and intermediate compound theory in homo- and heterogeneous, 51
Catalyst, Raney nickel for reductive desulfurization, 3
Cedar bark, Western Red, pentosan content of, 271
Cellobial, 2-hydroxy-, heptaacetate, 16
Cellobiitol, 1,5-anhydro-, heptaacetate, 2, 27
Cellobiose, hydrolysis by β-glucosidase, 61
β-Cellobioside, phenyl 1-thio-, heptaacetate, 2, 27
Cellulosan, 270
Cellulose, α, β and γ types of, 275
—, combination with formaldehyde, 126
—, crystalline, X-ray diffractions, 106
—, crystallites and non-crystalline, 107

—, decomposition by *Vibrio perimastrix*, 122
—, deuterium exchange applied to, 122
—, mercerized, 110, 113, 116, 119, 126
—, methods for estimation of crystalline and non crystalline portions, 109–124
—, methyl ethers, 123
—, origin of name, 103
Celluloses, acid hydrolysis, 109–120
—, commercial regenerated, degree of polymerization, 106
—, crystalline and non-crystalline regions of, 109–126
—, esterification and etherification, 122–124
—, oxidation by periodate, 121
—, relative crystallinity of, 103–126
—, swelling and density of, 107, 120–121
Cellulose acetate, accessibility of saponified, 114
—, crystallinity of saponified, 116
Cellulose acetates, commercial, 106, 122
—, commercial, degree of polymerization, 106
Cellulosic fibers, density of, 121
Celtrobiose, β-glucosidase action on, 61
Chaulmoograte, ethyl, 223
Chemisorption, between enzyme and substrate, 56
Chlorine dioxide, use in delignification, 274
Citrus nobilis, emulsins of, 63
Coir, xylan percentage in, 271
Copper, removal in manufacture of dextrose, 139
Cori ester. *See* D-Glucose-1-phosphate.
Corn cobs, optical rotation of xylan from, 282
  as source of D-xylose, 279
  pentosan content of, 271
Corn seedlings, optical rotation of xylan from, 282
Corn stalk, pentosan content of, 271
Cotton (*see also* Cellulose *and* Cotton linters), 104–126
  low xylan content, 271
Cotton linters, 106–124
Cottonseed hulls, as source of D-xylose, 279

Crystalline dextrose, commercial production, 127–143
Crystallinity, relative, of celluloses, 103–126

D

Density of cellulosic fibers, 121
Depolymerase, action on pectic acids, 82, 92
Desulfurization, reductive, by Raney nickel in the carbohydrate field, 1–28
Dextran, produced by *Leuconostoc* organisms, 47
—, 2,3-dimethyl-D-glucose from methylated, 161
Dextrins, from action of amylases, 229–268
  Schardinger crystalline, 266
Dextrose. *See also* D-Glucose
  brewer's sugar, 128
  chip sugar, 128
  "70" sugar, 128
  "80" sugar, 128
Dextrose, alpha anhydrous, 135
  photograph of crystals, 135
Dextrose, alpha monohydrate, 131
  photographs of crystals, 133, 134, 141
Dextrose, beta, 136
  photograph of crystals, 137
Dextrose, commercial production of crystalline, 127–143
  diagram of solubility in water, 130
  process of ion-exchange refining, 137–143
  sodium chloride addition compound, 132
Dialdosyl disulfide, octaacetate, 5
β,β-Di-D-glucopyranosyl disulfide, octaacetate, 5, 26, 202
Diglycolic acid, D,L-hydroxymethyl-, strontium salt, 221
Disaccharides, enzymatic syntheses of, 29–48
Dogwood, pentosan content of Pacific, 271
Douglas fir, pentosan content, 271
DP. Definition as degree of polymerization, 106

Dual affinity, in theory of enzyme action, 54
Dulcitol, 1,4:3,6-dianhydro-, 215
D-Dulcitol, 1,5-anhydro-, 17, 203, 226
  tetraacetate, 17, 27, 204, 226
—, 3,6-anhydro-, 203, 204, 226
Dulcitols, anhydro-, 203–204

E

Eggplant, pectin-methylesterase (PM) in, 93
Emulsin, 60, 66
Enzyme, principles underlying specificity in the domain of carbohydrates, 49–78
—, diagram of postulated union with substrate, 56
Enzyme value, definition, 62
  comparative values from rates of hydrolysis of phenyl hexosides and pentosides, 62
Enzymes, action on pectic substances, 79–102
—, amylolytic. See Amylases.
—, carbohydrates and their specific, 76
—, kinetics and mechanism of activation of, 59
—, pectic, 79–102
—, synthesis of sucrose and other disaccharides by, 29–48
Enzyme-substrate combinations, 53–57, 60
  diagram illustrating sucrose-saccharase combination, 56
1,2-Epoxides, hydrogenation, 22
Esparto xylan, optical rotation, 282
Ethanolamine, use in delignification, 274
Ethyl hydroperoxide, decomposition by catalase, 53
Ethyl sulfuric acid, catalytic formation, 52

F

Fehling's solution, use in purification of xylan, 278, 289
. Fermentation, of only monosaccharides by *Torula monosa*, 38

Flax, 104
  low xylan content, 271
  speed of acetylation, 122
Flax, New Zealand, polysaccharide of, 283
Forces of attraction between enzyme and substrate groupings, 56
Formaldehyde, combination with cellulose, 126
D-Fructofuranose, 31
β-D-Fructofuranose, 42, 58, 59, 74, 75
β-D-Fructofuranosidase, (Invertase, Saccharase), 32, 33, 55, 67–69
  diagram of postulated union with sucrose, 56
α-D-Fructofuranoside, benzyl, 69
—, β-D-glucopyranosyl-, (Isosucrose), 69
  octaacetate, 30
β-D-Fructofuranoside, α-D-glucopyranosyl-, (Sucrose), 29–39, 67–71
  octaacetate, 38
  octamethyl-, 31
α-D-Fructofuranoside, methyl, 69
β-D-Fructofuranoside, methyl, 32, 67
D-Fructose, enzymatic syntheses with, 29–39, 46–48
—, diethyl thioacetal, pentaacetate, 6, 26
—, fermentation of furanose form, 74, 75
—, from styracitol, 200
—, mutarotation, 32, 75
—, 1,3,4,6-tetramethyl-, 31
Fruit juices, pectin-methylesterase (PM) in, 93
Fruit juices, removal of pectin from, 101
L-Fucitol, pentaacetate, 6
"α-Fucohexitol," 9
"Fucohexonic" acids, 9
"α-Fucohexose," 9
L-Fucose, 9
Furan, 2-ethyl-tetrahydro-, 192
—, 2-vinyl-dihydro-, 192
Furfural, commercial production from pentosans, 288

G

D-Galactal, 2-hydroxy-, tetraacetate, 17
D-Galactitol, 1,5-anhydro-, 17, 203, 226
  tetraacetate, 17, 27, 204, 226
—, 3,6-anhydro-, 203, 204, 226

tetraacetate, 226
—, 1-desoxy-, 26
pentaacetate, 6, 26
D-Gala-L-*manno*-heptitol, 1-desoxy-, 10
L-Gala-D-*manno*-heptitol, 7 desoxy-, 9
L-Gala-D-*manno*-heptonic acid, 7-desoxy-, 9
D-Gala-L-*manno*-heptose, 10
Galactokinase, 49, 50
α-D-Galactopyranosidase, 61, 66
α-D-Galactopyranoside, methyl, 66, 67
β-D-Galactopyranoside, 2′-naphthyl 1-thio-, 203
tetraacetate, 17, 27
α-D-Galactopyranoside, phenyl, enzymatic hydrolysis of, 66, 67
β-D-Galactopyranoside, phenyl, enzymatic hydrolysis of, 62, 63
D-Galactose, transphosphorylation between adenosine triphosphate and, 49, 50
L-Galactose, 6-desoxy-, 9
D-Galactose, diethyl thioacetal, 26
pentaacetate, 6, 26
α-D-Galactose, 1-phosphate, 71
α-Galactosidase. See α-D-Galactopyranosidase.
D-Galacturonic acid, production from pectic substances, 102
double salts of, 102
poly-α-pyranose anomer in pectins, 80
use in ascorbic acid synthesis, 102
Gentianose, hydrolysis by saccharase, 68
Gentiobial, 2-hydroxy-, heptaacetate, 16
Gentiobiose, β-glucosidase action on, 61
β-Gentiobioside, phenyl 1-thio, heptaacetate, 16, 17, 27
D-Glucal, 2-hydroxy-, tetraacetate, 15, 198
D-Glucitol. See D-Sorbitol.
D-Glucofuranose, tetramethyl-, 194
α-D-Glucofuranoside, methyl 2,3,5,6-tetramethyl-, 190
β-D-Glucofuranoside, methyl 2,3,5,6-tetramethyl-, 190
—, methyl 2,3,6-trimethyl-, 178
5-benzoate, 178
5-tosylate, 178
α-D-Glucofuranoside, methyl 3,5,6-trimethyl-, 185

β-D-Glucofuranoside, methyl 3,5,6-trimethyl-, 185
β-D-Glucofururonoside, methyl 2,5-dimethyl-, 165
amide, 165
γ-lactone, 165
Glucon, 61
D-Gluconamide, 2,4,6-trimethyl-, 181
D-Gluconate, methyl thiol-, pentaacetate, 23, 28
D-Gluconic acid, 2,6-dimethyl-, phenylhydrazide, 166
—, 2-methyl-, 150
amide, 150
γ-lactone, 150
—, 2-methyl-, 3,4,5,6-di-isopropylidene-, methyl ester, 150
—, 3-methyl-, 155
phenylhydrazide, 155
sodium salt, 155
—, 2,3,4,6-tetramethyl-, 188
amide, 188
δ-lactone, 188
phenylhydrazide, 188
—, 2,3,5,6-tetramethyl-, 190
amide, 190
γ-lactone, 190
phenylhydrazide, 190
—, 2,3,6-trimethyl-, 178
γ-lactone, 178
δ-lactone, 178
phenylhydrazide, 178
—, 3,5,6-trimethyl-, 185
amide, 185
γ-lactone, 185
sodium salt, 185
D-Gluconolactone, 2,3,4-trimethyl-, 175
—, 3,4,6-trimethyl-, 184
phenylhydrazide, 184
—, 2,3,5-trimethyl-, 176
phenylhydrazide, 176
D-Glucono-δ-lactone, 4-methyl-, 156
D-Gluconophenylhydrazide, 2,3-dimethyl-, 163
D-Glucopyranose, *see also* Dextrose, 31, 105
fermentation, 74, 75
—, 2,3,4-trimethyl-, 175
anilide, 175
α-1,6-di-azobenzoate, 175

1,6-dinitrate, 175
β-D-Glucopyranoside, benzyl 2,4,6-trimethyl-, 181
—, methyl, enzymatic synthesis of, 57
α-D-Glucopyranoside, methyl 2,3-dimethyl-, 163
   4-acetyl-6-trityl-, 163
   4-benzoyl-6-trityl-, 163
   4,6-benzylidene-, 163
   4,6-di-azobenzoate, 163
   4,6-furylidene-, 163
   4-tosyl-6-trityl-, 163
   6-trityl-, 163
β-D-Glucopyranoside, methyl 2,3-dimethyl-, 163
   4,6-benzylidene-, 163
   4,6-dibenzenesulfonate, 163
   4,6-dibenzoate, 163
   4,6-ethylidene-, 163
α-D-Glucopyranoside, methyl 2,4-dimethyl-, 164
β-D-Glucopyranoside, methyl 2,4-dimethyl-, 164
   3-acetyl-6-nitrate, 164
   3,6-dinitrate, 164
   6-nitrate, 164
   3-tosylate, 164
   3-tosyl-6-trityl-, 164
α-D-Glucopyranoside, methyl 2,6-dimethyl-, 166
   3,4-di-N-phenylcarbamate, 166
β-D-Glucopyranoside, methyl 2,6-dimethyl-, 166
   3,4-dinitrate, 166
   3,4-ditosylate, 166
—, methyl 3,4-dimethyl-, 168
   2-benzoyl-6-trityl-, 168
   2,6-dinitrate, 168
   6-trityl-, 168
—, methyl 3,6-dimethyl-, 169
   2,4-dibenzoate, 169
   2,4-ditosylate, 169
α-D-Glucopyranoside, methyl 4,6-dimethyl-, 171
   2,3-di-azobenzoate, 171
   2,3-dibenzoate, 171
   2,3-dibenzyl-, 171
   2,3-di-p-nitrobenzoate, 171
   2,3-ditosylate, 171

β-D-Glucopyranoside, methyl 4,6-dimethyl-, 171
   2,3-ditosylate, 171
α-D-Glucopyranoside, methyl 2-methyl-, 150
   3,4,6-triacetate, 150
β-D-Glucopyranoside, methyl 2-methyl-, 150
   4,6-benzylidene-, 150
   4,6-ethylidene-, 150
   3,4,6-triacetate, 150
   3,4,6-tribenzoate, 150
   3,4,6-tritosylate, 150
—, methyl 3-methyl-, 155
   2,4,6-triacetate, 155
   2,4,6-tribenzoate, 155
   2-acetyl-4,6-ethylidene-, 155
   2-tosyl-4,6-ethylidene-, 155
α-D-Glucopyranoside, methyl 3-methyl-, 155
   4,6-benzylidene-, 155
   4,6-benzylidene-2-tosyl-, 155
β-D-Glucopyranoside, methyl 4-methyl-, 156
   2,3,6-triacetate, 156
   2,3,6-tribenzoate, 156
   2,3,6-trinitrate, 156
α-D-Glucopyranoside, methyl 6-methyl-, 161
   2,3,4-tribenzoate, 161
β-D-Glucopyranoside, methyl 6-methyl-, 161
   2,3,4-triacetate, 161
   2,3,4-tribenzoate, 161
α-D-Glucopyranoside, methyl 2,3,4,6-tetramethyl-, 188
β-D-Glucopyranoside, methyl 2,3,4,6-tetramethyl-, 188
—, methyl 2,3,4-trimethyl-, 175
   6-azobenzoate, 175
   6-desoxy-6-bromo-, 175
   6-nitrate, 175
α-D-Glucopyranoside, methyl 2,3,4-trimethyl-, 175
   6-trityl-, 175
—, methyl 2,3,6-trimethyl-, 178
   4-azobenzoate, 178
   4-(3,5-dinitrobenzoate), 178
β-D-Glucopyranoside, methyl 2,3,6-trimethyl-, 178

4-azobenzoate, 178
4-benzenesulfonate, 178
α-D-Glucopyranoside, methyl 2,4,6-trimethyl-, 181
  3-tosylate, 181
β-D-Glucopyranoside, methyl 2,4,6-trimethyl-, 181
  3-tosylate, 181
α-D-Glucopyranoside, methyl 3,4,6-trimethyl-, 184
  2-azobenzoate, 184
β-D-Glucopyranoside, methyl 3,4,6-trimethyl-, 184
  2-azobenzoate, 184
  2-benzyl-, 184
  2-tosylate, 184
—, phenyl 1-thio-, 65
  tetraacetate, 27
—, phenyl 2,4,6-trimethyl-, 181
—, p-tolyl 1-thio-, tetraacetate, 27
D-Glucosaccharamide, 2,3-dimethyl-, 163
—, 2,5-dimethyl-, 165
—, 2,3,5-trimethyl-, 176
D-Glucosaccharic acid, 2,4:3,5-dimethylene-, 217
D-Glucosaccharolactone, 2,3,4-trimethyl-, methyl ester, 175
—, 2,3,5-trimethyl-, methyl ester, 176
D-Glucosamine, diethyl thioacetal, pentaacetate, 11, 26
D-Glucose. *See also* Dextrose.
  alcoholic fermentation, 50
  configuration in sucrose, 39
  from amygdalin, 60
—, formed from starch, by acid hydrolysis, 127–143
  by amylase of *Aspergillus oryzae*, 250–255
  by amylase of malted barley, 255–265
  by pancreatic amylase, 235, 247–250
—, acetobromo-, 202
—, 3,6-anhydro-, 197
—, 6-desoxy-, tetraacetates, 24
—, 6-desoxy-6-iodo-, tetraacetates, 24
—, diethyl thioacetal, pentaacetate, 6, 26
—, 2,3-dimethyl, and derivatives, 160–163
  anilide, 163
  phenylhydrazone, 163

1,4,6-triazobenzoate, 163
—, 2,4-dimethyl and derivatives, 162–164
—, 2,5-dimethyl, derivatives of, 164–165
—, 2,5-dimethyl-3,6-anhydro-, anilide, 165
—, 2,6-dimethyl, and derivatives, 165–167
  1,3,4-triazobenzoate, 166
—, 3,4-dimethyl-, and derivatives, 167–168
  phenylosazone, 168
—, 3,5-dimethyl-, 168
—, 3,6-dimethyl-, and derivatives, 168–169
  1,2-isopropylidene-, 169
—, 4,6-dimethyl-, and derivatives, 170–171
  1,2,3-triazobenzoate, 171
—, 5,6-dimethyl-, and derivatives, 171–172
  p-bromophenylosazone, 172
—, 1,2-isopropylidene-, 172
  3-benzoxymethyl-, 172
  3-benzyl-, 172
  3-carbanilyl-, 172
—, 1,2,3-triazobenzoate, 172
—, 1,2,3-tri-p-nitrobenzoate, 172
—, methyl ethers of, 145–190
  dimethyl ethers, 160–172
  monomethyl ethers, 148–160
  tetramethyl ethers, 186–190
  trimethyl ethers, 172–186
, 2 methyl-, and derivatives, 148–151
  phenylhydrazone, 150
  p-toluidide, 150
  1,3,4,6-tetrabenzoate, 150
—, 2-methyl, diethyl thioacetal tetraacetate, 11, 26
—, 3-methyl-, and derivatives, 151–155
  α-form, 155
  β-form, 155
  anilide, 155
  1,2:5,6-diisopropylidene-, 155
  phenylosazone, 155
  β-tetraacetate, 155
  1,2,4,6-tetraazobenzoate, 155
  β-tetrabenzoate, 155
—, 4-methyl-, and derivatives, 154–157
  phenylosazone, 156
  1,2,3,6-tetraacetate, 156

—, 4-methyl-, dibenzyl thioacetal, 156
   2,3,5,6-tetraacetate, 156
—, 5-methyl-, and derivatives, 157–158
   1,2-isopropylidene-, 158
   1,2-isopropylidene-3,6-diacetyl-, 158
   1,2-isopropylidene-3-tosyl-6-benzoyl-, 158
   phenylosazone, 158
—, 6-methyl-, and derivatives, 158–161
   1,2-isopropylidene-, 161
   phenylosazone, 161
   α-1,2,3,4-tetraacetate, 161
   β-1,2,3,4-tetraacetate, 161
   1,2,3,4-tetraazobenzoate, 161
   α-1,2,3,4-tetrabenzoate, 161
   β-1,2,3,4-tetrabenzoate, 161
—, 1-phosphate, enzymatic reactions with, 33, 35, 45–47, 58, 71
—, 2,3,4,6-tetramethyl-, and derivatives, 31, 44, 186–188
   anilide, 188
   β-azobenzoate, 188
   p-toluidide, 188
—, 2,3,5,6-tetramethyl-, and derivatives, 189–190
—, 1-thio-, β-tetraacetate, 3, 26
—, 2,3,4-trimethyl-, and derivatives, 172–175
   anilide, 175
   α-1,6-di-azobenzoate, 175
   1,6-dinitrate, 175
—, 2,3,5-trimethyl-, and derivatives, 174–176
   1,6-anhydro-, 176
—, 2,3,6-trimethyl-, and derivatives, 176–179
   1-chloro-5-benzoate, 178
   β-1,4-diacetate, 178
   1,4-di-azobenzoate, 178
   dibenzoate, 178
   diethyl thioacetal, 178
—, 2,4,6-trimethyl-, and derivatives, 179–182
   anilide, 181
   3-benzyl-, 181
   1,3-di-azobenzoate, 181
—, 2,5,6-trimethyl-, 182
—, 3,4,6-trimethyl-, and derivatives, 182–184
   α-form, 184

β-form, 184
α-1,2-di-azobenzoate, 184
phenylosazone, 184
—, 3,5,6-trimethyl-, and derivatives, 184–186
   1,2-dichloroethylidene-, 185
   1,2-isopropylidene-, 185
   phenylosazone, 185
   1,2-trichloroethylidene-, 185
α-Glucosidase (maltase), 31, 64
β-Glucosidase, 57, 60–65, 67
β-D-Glucoside, phenyl, enzymatic hydrolysis of, 62–63
—, phenyl 1-thio-, tetraacetate, 16
—, p-tolyl 1-thio-, tetraacetate, 16
"1,4-Glucosido-styracitol," 16
"1,6-Glucosido-styracitol," 16
D-Glucothiose, tetraacetyl-, 202
D-Glucuronic acid, in hemicellulose of cottonseed hulls, cotton stalks, wheat straw and alfalfa hay, 280
—, in xylans of corn cobs and esparto, 280
—, 2,3-dimethyl-, 284
—, presence in xylan, 269, 278
——, 2,3,4-trimethyl-,
   amide of α-methyl uronide, 175
   amide of β-methyl uronide, 175
   β-methyl uronide, 175
β-Glucuronidase, 61
Glutaric acid, (xylo)trihydroxy-, from oxidation of xylan, 285
Glutathione, 54
(levo)-Glyceraldehyde and (dextro)-lactic acid, correlation of configurations, 11
L-(levo)-Glyceraldehyde and (dextro)-alanine, correlation of configurations, 11
Bis(L-Glyceraldehyde, 3-tosyl-) 2,2'-ether, 206
Glycitols, 1,5-anhydro-, 18
—, ω-desoxy-, 9
Glycogen, phosphorolysis of, 34
   hydrolysis by beta amylase, 231, 232
Glycosidases, 55, 59, 61
Glycosides, enzymatic hydrolysis of, 49–78
—, 1-thio-, hydrogenolysis, 14
Glycine soya, emulsins, 63
Grape sugar (dextrose, D-glucose), 128
L-Gulomethylitol, pentaacetate, 6

α-D-Guloside, methyl 2-desoxy-, 21, 28
  4,6-benzylidene-, 21, 28

H

Helix pomatia, action of digestive juices on hemicellulose, 288
Hemicellulose, action of *Helix pomatia* digestive juices on, 288
—, of apple wood, 282
—, of English oak, presence of monomethyl-hexuronic acid in, 280
—, of raw corn cob, 276, 282
—, xylan as a component of, 270
Hemlock, Western, 107, 113, 285
Hemp, 107, 122
—, Manila, xylan percentage in, 270
Hexamethylenediamine, 288
1,4-Hexanediol, 192
Hexanetetrol, 1,6-(*erythro*-3,4), 6
Hexitols, anhydrides of, 191–228
  industrial uses, 222
Hexokinase of yeast, 73
Hexose phosphates, in leaves of sucrose-producing plants, 33
Horse carboxyhemoglobin, dipole moment of, 55
Hexuronic acid, monomethyl-, in hemicellulose of English oak, 280
HMF. See 5-Hydroxymethylfurfural.
Holocellulose, as source of xylan, 274
Hydrogen bonds, between carbohydrates and hydrophilic groupings of proteins of enzymes, 55
Hydrogen peroxide, reaction with peroxidase, 53
Hydrogenolysis, of carbon–sulfur bond, 1
5-Hydroxymethylfurfural (HMF), production and removal in manufacture of dextrose, 141

I

L-Iditol, 2,5-anhydro-, 205, 206, 226
  1,6-didesoxy-1,6-diiodo-, 226
  1,6-di-*p*-tosyl-, 206, 226
  1,3,4,6-tetraacetyl-, 226
  1-*p*-tosyl-, 206, 226
—, 1,4:2,5-dianhydro- (?), 206

—, 1,4:3,6-dianhydro-, (Isoidide), 195, 213, 215–217
  2,5-ditosyl-, 216
—, 1,4:5,6-dianhydro-, 195
L-Idosaccharic acid, 2,4:3,5-dimethylene-, 217
β-D-Idoside, methyl 3-desoxy-, 21
  4,6-benzylidene-, 21, 28
  2-methyl-4,6-benzylidene-, 28
α-D-Idoside, methyl 2-methylthio-4,6-benzylidene-, 20, 28
β-D-Idoside, methyl 3-methylthio-4,6-benzylidene-, 21, 28
  2-methyl-, 28
Inhibition, of phosphorylation and sucrose synthesis by iodoacetate, 33
Intermediate compound theory of enzymatic action, 51
  identification of intermediate compounds, 53
Invertase. See β-D-Fructofuranosidase.
Iodine, colors of products formed from starch by amylases, 261
Iodoacetate, inhibitor of phosphorylation and sucrose synthesis, 33
Iron, removal in manufacture of dextrose, 139
Isoidide. See L-Iditol, 1,4:3,6-dianhydro-.
Isomannide. See D-Mannitol, 1,4:3,6-dianhydro-.
Isosorbide. See D-Sorbitol, 1,4:3,6-dianhydro-.
Isosucrose, 69
  octaacetate, 30
Isothioureas, *S*-substituted, desulfurization with Raney nickel, 24

J

Jute, xylan percentage in, 270
Jute fibers, diffraction pattern, 104

K

Ketoacids, α- and β-, specificity of carboxylase action on, 50
α-Ketoacids, aromatic, 50
α-Ketobutyric acid, 50

α-Ketocaproic acids, 50
α-Ketovaleric acids, 50

L

L-(*dextro*)-Lactic acid, configuration, 11, 12
  O-methyl-, 12
  O-methyl-, *p*-phenylphenacyl ester, 12
Lactitol, 1,5-anhydro-, 18
β-Lactopyranoside, 2'-naphthyl 1-thio-, 18, 27
Langenbeck's formulation of enzymatic glycoside hydrolysis, 58
Laurel, pentosan content of California-, 271
Leuconostoc dextranicum, methylated dextran from action of organism on sucrose as source of 2,3-dimethyl-D-glucose, 70, 161
Leuconostoc mesenteroides, dextran produced from sucrose, 47
  methylated dextran as source of 2,3-dimethyl-D-glucose, 161
Levan, produced from sucrose or raffinose by bacterial enzymes, 47, 48
Lignin, removal in purification of xylan, 274
"Ligno-cellulose," 104
Linters. *See* Cotton linters.
Lithium aluminum hydride, for hydrogenation of 1,2-epoxides, 22
Lithium chloride, influence upon the activity of pancreatic amylase, 237
Lucerne seed, emulsins, 63
D-Lyxomethylitol, tetraacetate, 7
L-Lyxomethylitol, tetraacetate, 7
α-D-Lyxopyranoside, phenyl, 62, 66

M

Madrone, pentosan content of Pacific, 271
Malonic acid, protection of enzyme by, 54
Maltase. *See* α-Glucosidase.
Maltitol, 1,5-anhydro, 18
β-Maltopyranoside, phenyl 1-thio-, heptaacetate, 18, 27
Maltose, from starch by pancreatic amylase, 235, 247–250
  by amylase of *Aspergillus oryzae*, 250–255
  by amylase of malted barley, 255–265
α-Maltose, 1-phosphate, 71
Mandelonitrile, catalytic formation of optically active, 53
Mannan, presence in alpha cellulose, 275
β-Mannide, 217
Manninotriose, 66
Mannitan. *See* D-Mannitol, 1,4-anhydro-.
D-Mannitol, 8, 192, 209
  reaction products with boiling hydrochloric acid, 210
  —, 1,4-anhydro-, (Mannitan), 192, 209, 210, 226
  5,6-benzylidene-, 226
  2,6 (or 3,6)-dibenzoyl-, 226
  2,3-dibenzoyl-5,6-benzylidene-, 226
  2,3:5,6-dibenzylidene-, 226
  2,3,5,6-tetraacetyl-, 226
  —, 1,5-anhydro-, (Styracitol), 15, 16, 27, 198, 200, 218, 226
  dibenzylidene-, (*a*) and (*b*), 226
  diisopropylidene-, 226
  proofs of structure and configuration, 198
  tetraacetyl-, 226
  tetrabenzoyl-, 226
  tetramethanesulfonyl-, 226
  tetramethyl-, 200, 226
  tetranitro-, 226
  transformation to D-fructose, 200
  —, 5,6-anhydro-1,2:3,4-diisopropylidene-, 219, 226
  —, 1-desoxy-, 8, 26
  —, 2-desoxy-, pentaacetate, 6
  —, 6-desoxy-6-amino-1,2:3,4-diisopropylidene-, 219
  —, 1,2:5,6-dianhydro-, 219
  3,4-ethylidene-, 219, 220
  3,4-isopropylidene-, 219
  —, 1,4:3,6-dianhydro-, (Isomannide), 206, 210, 213, 215, 228
  2(5)-acetyl-, 227
  2,5-dibenzoyl-, 209, 228
  2-desoxy-2-chloro-5-methanesulfonyl-, 228
  2-desoxy-2-chloro-5-phenylcarbamyl-, 228

2,5-diacetyl-, 225, 227
2,5-dicrotyl-, 228
2,5-didesoxy-2,5-di-(N^4-acetylsulfanilamido)-, 228
2,5-didesoxy-2,5-diamino-, and derivatives, 215, 225, 228
2,5-didesoxy-2,5-dichloro-, 227
2,5-didesoxy-2,5-diiodo-, 228
2,5-didesoxy-2,5-di-(p-nitrobenzenesulfonamido)-, 228
2,5-didesoxy-2,5-disulfanilamido-, 228
2,5-didesoxy-2,5-imino-, 215–217
2,5-diethyl-, 228
2,5-diformyl-, 227
2,5-dimethacrylyl-, 224
2,5-dimethanesulfonyl-, 210, 228
2,5-dimethyl-, 208
2,5-di-(phenylcarbamyl)-, 227
2,5-ditosyl-, 228
2,5-ditrityl-, 228
2(5)-methyl-, 228
production from sucrose, 213
—, 1,5:3,6-dianhydro-, (Neomannide), 217, 228
2,5-dimethanesulfonyl-, 228
—, 1,6-dibenzoyl-2-tosyl-, 205
—, 1,6-didesoxy-1,6-dibromo-, 211
—, 1,6-didesoxy-1,6-dichloro-, 207, 210, 218
2,5-dimethyl-, 208
3,4-isopropylidene-, 208
—, 1,2:5,6-diisopropylidene-3, 4-ditosyl-, 214
—, tetradesoxy-1,3,5,6-tetrachloro-, 208
D-Manno-D-*gala*-heptitol, 1-desoxy-, 9, 26
L-Manno-L-*gala*-heptitol, 7-desoxy-, 10
α-D-Manno-D-*gala*-heptopyranoside, methyl, 66, 67
D-Manno-D-*gala*-heptose, diethyl thioacetal, 9, 26
D-Manno-L-*manno*-octitol, configuration, 8
D-Manno-L-*manno*-octose, 10
L-Manno-D-*manno*-octose, 8-desoxy-, 11
α-D-Mannopyranose, fermentation, 74, 75
β-D-Mannopyranose, fermentation, 74, 75
α-D-Mannopyranosidase. See α-Mannosidase, 61, 66

β-D-Mannopyranoside, ethyl 1-thio-, tetraacetate, 15, 27
α-D-Mannopyranoside, methyl
  2,3-anhydro-4,6-benzylidene-, 20
  3-desoxy-4,6-benzylidene-, 20, 27
  3-desoxy-2-methyl-4,6-benzylidene-, 20, 27
—, phenyl, 62, 66
—, phenyl 2-desoxy-, 66
β-D-Mannopyranoside, phenyl, 66
D-Mannosaccharic acid, 2,4:3,5-dimethylene-, 217
D-Mannose, 6-desoxy-, 8
D-Mannose, diethyl thioacetal, 7, 26
  pentaacetate, 7, 26
α-D-Mannose, 1-phosphate, 71
α-Mannosidase, 61, 66
Mannosidostreptomycin, 15
Melezitose, 69, 70
Melibiose, 66, 67
—, 1-phosphate, 71
Mercaptals (*see also* Thioacetals), reductive desulfurization of, 5
Methylated dextrins, not affected by pancreatic amylase, 250
O-Methylhydroxylamine hydrochloride, use in estimation of carbonyl groups, 285
Molds, pectin-esterases in, 86, 98
Mucor javanicus, emulsins of, 63
Muscle phosphorylase, 35
Mutarotation, of D-fructose, 32, 75
Mutarotation, alpha, of reacting mixtures of pancreatic amylase and starch, 236
  of amylase of *Aspergillus oryzae* and starch, 251
  of alpha amylase of barley malt and starch, 257

N

Nacconol, for pectin-methylesterase (PM) determination, 94
Neomannide. See D-Mannitol, 1,5:3,6-dianhydro-.
Nettle stalk, pentosan content, 271
Nickel, Raney, reductive desulfurization in the carbohydrate field by, 1–28

SUBJECT INDEX

Nickel-aluminum alloy, for preparation of Raney nickel, 3
Nicotine, 53
Nitrocellulose, degree of polymerization, 106
Nomenclature of pectic enzymes, 80, 82, 85, 92

O

Oak, pentosan content of Tanbark, 271
Oat hulls, pentosan content, 271
—, optical rotation of xylan from, 282
Oats, beta amylase of, 231
Orange, pectin-methylesterases (PM) in flavedo and albedo of, 93, 95, 96

P

Peas, hexose phosphates in, 33
Pectase. See Pectin-methylesterase, 82
Pectic acid, 81
 action of depolymerase on, 82, 92
Pectic enzyme preparations, commercial, 86, 101
Pectic substances, enzymes acting on, 79–102
Pectin, chemistry and nomenclature, 80–82
Pectin-esterase. See Pectin-methyl-esterase, 82
Pectinase. See Pectin-polygalacturonase, 82
Pectinic acids, 81
Pectin-methylesterase (PM), 82, 86, 92–98, 101
Pectinol, 86, 102
Pectin-polygalacturonase (PG), 82, 85–91, 93, 101
"Pecto-cellulose," 104
Pectolase. See Pectin-polygalacturonase, 82
Pentitol anhydrides, 191–228
Pentosans, amount in various natural products, 271
Peptidase, 55
Peroxidase, 53
PG. See Pectin-polygalacturonase, 82
Phosphorolysis, of sucrose, starch and glycogen, 34
 definition, 34

Phosphorylases. See Transglycosidases.
Phosphorylation, sucrose formation and, 33
Pine, pentosan content of ponderosa, 271
PM. See Pectin-methylesterase, 82
Polyaffinity between polar groups of enzyme and substrate, 55
Polygala amara, 198
Poly-α-D-galacturonic acids, of pectins, 80
Polygala senega L., 15
Polygala tenuifolia, 198
Polygala vulgaris, 198
Polygalitol. See D-Sorbitol, 1,5-anhydro-.
Polymerization, degree of, (DP), for celluloses, 106
Polyuronides, pectin class of, 79
Potassium chloride, influence upon the activity of pancreatic amylase, 237
Potato phosphorylase, 35
Potatoes, sweet, beta amylase of, 231, 234
Propionic acid, N-acetyl-2-amino-, 11
Protopectin, 81, 82, 84
Protopectinase, 82, 84
Prunus amygdalus, emulsins, 63
Prunus avium, emulsins, 63
Pseudomonas saccharophila, 32, 33, 36, 39, 41, 43, 46, 70
keto-D-Psicose, pentaacetate, 6
Pyridine, 5-amino-2-butoxy-, 223
Pyruvic acid, yeast carboxylase action on, 50
—, dimethyl-, 50

Q

Quinidine. See also Camphor, 53
 catalyst for synthesis of optically active mandelonitrile, 53
Quinine. See also Camphor, 53

R

Raffinose, constitution as 6-α-D-galactopyranosyl-α-D-glucopyranosyl β-D-fructofuranoside, 68
 hydrolysis by yeast β-D-fructofuranosidase, 68, 69

Ramie, 104, 107, 116, 122–124, 271
  crystallinity of, 116
  non-crystalline cellulose of, 107
  rate of acetylation, 122
Raney nickel. *See* Nickel.
Rayon, 107, 113, 114, 115, 116, 119, 121
Rayons, accessibility of cellulose of, 114, 115, 121
—, degrees of crystallinity, 116
—, yarn properties, 116, 119
Residual affinity, in theory of enzyme action, 54
D-Rhamnitol, 7
"Rhamnoheptose" of Fischer and Piloty, 10
"α-Rhamnohexose," 10
D-Rhamnose, 8
Rhizobium radicicolum, methylated capsular polysaccharide of, as source of 2,3-dimethyl-D-glucose, 160
Ribitol, 1,5-anhydro-, 14, 15, 221, 222, 225
  2,3,4-triacetyl-, 225
  2,3,4-tribenzoyl-, 27, 225
D-Ribonate, ethyl thiol-, tetraacetate, 22, 28
D-Ribonyl chloride, tetraacetate, 22
β-L-Ribopyranoside, methyl 2,3-anhydro-, 22
β-D-Ribopyranoside, 2′-naphthyl 1-thio-, tribenzoate, 15, 27, 222
aldehydo-D-Ribose tetraacetate, 22, 28
β-L-Riboside, methyl 3-desoxy-, 22
Rice, beta amylase of, 231
Rosa canina, emulsins, 63, 64
Rye straw, pentosan content of, 271

S

Saccharase. *See* β-D-Fructofuranosidase.
Saccharomyces fragilis, source of galactokinase, 49
  emulsin of, 63
Salivary amylase, 265
Schardinger dextrins, 266
Sierra juniper, pentosan content of, 271
Sisal, xylan percentage in, 270
Slow rates of change in amylase actions, 241, 253, 267, 268

Sodium chloride, addition compound with D-glucose, 132
Sodium salts, influence upon the activity of pancreatic amylase, 237
Solvents, for reductive desulfurizations, 2
D-Sorbitol (*synonym*, D-Glucitol), 5, 15
  hexaacetate, 23, 28
  2,3,4,5,6-pentaacetate, 23
—, 1,4-anhydro-, (Arlitan), 194, 226
—, 1,4-anhydro-benzylidene-(*a*), 226, (*b*) 226
—, 1,4-anhydro-3,5-benzylidene-, 196, 226
—, 1,4-anhydro-3,5-benzylidene-6-*p*-tosyl-, 226
Bis-(D-Sorbitol, 1,4-anhydro-3,5-benzylidene-6-desoxy-6-chloro-) 2,2′-sulfite, 226
D-Sorbitol, 1,4-anhydro-6-desoxy-6-chloro-, 195, 226
—, 1,4-anhydro-6-desoxy-6-chloro-3,5-benzylidene-, 196, 226
—, 1,4-anhydro-6-desoxy-6-chloro-2,3,5-triacetyl-, 226
—, 1,4-anhydro-6-desoxy-6-iodo-3,5-benzylidene-, 226
—, 1,4-anhydro-2,3,5,6-tetramethanesulfonyl-, 226
—, 1,4-anhydro-2,3,5,6-tetramethyl-, 194, 226
—, 1,4-anhydro-5-tosyl-, 195
—, 1,4-anhydro-6-*p*-tosyl-, 226
—, 1,4-anhydro-2,3,5-tribenzoyl-, 226
—, 1,5-anhydro-, (*synonym*, Polygalitol), 5, 15, 27, 198, 226
  proofs of structure and configuration, 198
  tetraacetate, 5, 23, 26, 28, 202, 226
  tetramethyl-, 200
—, 1,5-anhydro-4-(β-D-galactopyranosyl)-, 18, 27
—, 1,5-anhydro-4-(α-D-glucopyranosyl)-, 16–18, 27
  heptaacetate, 27
—, 1,5-anhydro-4-(β-D-glucopyranosyl)-, 16, 17, 27
  heptaacetate, 27
—, 1,5-anhydro-6-(β-D-glucopyranosyl)-, 27
  heptaacetate, 27

SUBJECT INDEX

—, 2,5-anhydro-, 226
—, 2,5-anhydro-1, 6-dibenzoyl-, 226
—, 2,5-anhydro-1,6-dibenzoyl-3,4-di-*p*-tosyl-, 226
—, 3,6-anhydro-, 194, 197, 226
—, 3,6-anhydro-2,5-dimethyl-, 226
—, 5,6-anhydro-, 226
—, 5,6-anhydro-1,3:2,4-diethylidene-, 219, 226
—, 5,6-anhydro-1-tosyl-2,4-benzylidene-, 206, 226
—, 2,4-benzylidene-1,6-ditosyl-, 205
—, 2,4-benzylidene-1-*p*-tosyl-, 206
—, 6-desoxy-6-amino-1,3:2,4-diethylidene-, 219
—, 1-desoxy-, pentaacetate, 26
—, 1-desoxy-2-methyl-, 11 tetraacetate, 26
—, 2-desoxy-, pentaacetate, 26
—, 1,4:3,6-dianhydro-, (Isosorbide), 195, 197, 211, 227
  differing stabilities of rings, 212
  production from sucrose, 213
—, 1,4:3,6-dianhydro-5-desoxy(?)-5-chloro-, 227
—, 1,4:3,6-dianhydro-*x*-desoxy-*x*-iodo-*x'*-*p*-tosyl-, 227
—, 1,4:3,6-dianhydro-2,5-diacetyl-, 224, 225, 227
—, 1,4:3,6-dianhydro-2,5-diacrylyl-, 224, 227
—, 1,4:3,6-dianhydro-2,5-diallyl-, 224, 227
—, 1,4:3,6-dianhydro-, diamine, 216
—, 1,4:3,6-dianhydro-2,5-dibenzoyl-, 227
—, 1,4:3,6-dianhydro-2,5-didesoxy-2,5-diamino-, 225, 227
  dimethylene-D-glucosaccharate, 227
  hydrochloride, 227
  oxalate, 227
  picrate, 227
  sulfate, 227
—, 1,4:3,6-dianhydro-2,5-didesoxy-2,5-dichloro-, 227
—, 1,4:3,6-dianhydro-2,5-didesoxy-2,5-di-(N⁴-acetylsulfanilamido)-, 227
—, 1,4:3,6-dianhydro-2,5-didesoxy-2,5-di-(*p*-nitrobenzenesulfonamido)-, 227

—, 1,4:3,6-dianhydro-2,5-didesoxy-2,5-disalicylideneamino-, 227
—, 1,4:3,6-dianhydro-2,5-didesoxy-2,5-disulfanilamido-, 227
—, 1,4:3,6-dianhydro-2,5-diethyl-, 227
—, 1,4:3,6-dianhydro-2,5-dimethacrylyl-, 227
—, 1,4:3,6-dianhydro-2,5-dimethallyl-, 227
—, 1,4:3,6-dianhydro-2,5-dimethanesulfonyl-, 227
—, 1,4:3,6-dianhydro-2,5-dimethyl-, 217, 227
—, 1,4:3,6-dianhydro-2,5-dinitro-, 227
—, 1,4:3,6-dianhydro-2,5-di-*p*-tosyl-, 227
—, 1,4:3,6-dianhydro-2,5-ditrityl-, 227
—, 1,2-didesoxy-2-amino-, pentaacetate, 11
—, 1,3:2,4-diethylidene-6-methyl-, 219
α-L-Sorbofuranose, configurational formula, 42
β-L-Sorbofuranose, configurational formula, 42
α-L-Sorbofuranoside, α-D-glucopyranosyl-, 42, 57, 70, 71
L-Sorbose, 41, 43, 47, 57
Soy beans, beta anylase of, 231
Specificity, of enzymes in the domain of carbohydrates, 49–78
Spinner's fluff, accessible cellulose of, 115
—, definition, 115
Spruce, pentosan content of White, 271
Stachyose, 68, 69
Starch, 105
  and cellulose, comparison of chain structures, 105
  commercial acid hydrolysis, 128–143
  hydrolysis by amylases, 229–268
  phosphorolysis of, 34
  soluble, Lintner, action of alpha amylases, 233, 239, 243, 244, 251–254, 257
Stilbene, 4
Straw, as source of D-xylose, 279
Streptobiosamine, didesoxy-dihydro-, tetraacetate, 19, 26
Streptobiosaminide, ethyl thio-, diethyl thioacetal, 12
  hydrochloride, 12
  tetraacetate, 26, 27

—, methyl, dimethyl acetal hydrochloride, 13
—, methyl pentaacetyl-dihydro-, 19
Streptobiose, desoxy-, tetraacetate, 27
Streptomycin, mercaptolysis, 12
—, dihydro-, methanolysis, 19
—, mannosido, 15
Styracitol. See D-Mannitol, 1,5-anhydro-.
"Styracitol, 1,4-glucosido-," 16
"Styracitol, 1,6-glucosido-," 16
Styrax obassia, 15, 198
Succinic acid, protection of enzyme by, 54
—, D(−)-dimethoxy-, 202
—, L(+)-dimethoxy-, 199, 202
Succinic dehydrogenase, 54
Sucrose, 29–39, 67–71
  diagram of postulated union with saccharase, 56
Sucrose phosphorylase, (transglucosidase), 31–36, 47, 48, 59, 70–73
Sweet potato, beta amylase of, 231, 234

T

Taka diastase, 250
D-Talitol, 1,5-anhydro-, 17, 203, 204, 226
  tetraacetate, 226
α-D-Talopyranoside, phenyl, 66
Thallous ethylate, use in methylation of cellulose, 123
Thioacetals, reductive desulfurization of, 5
Thiocyanates, reductive desulfurization of, 24
Thioethers, formation and hydrogenolysis, 19
1-Thioglycosides, reductive desulfurization of, 1, 14
Thiol esters, hydrogenolysis, 22
D-Threaric acid, dimethyl-. See Succinic acid, D(−)-dimethoxy-.
L-Threaric acid, dimethyl-. See Succinic acid, L(+)-dimethoxy-.
Tobacco, pectin-esterases of, 93, 95
Tomato, pectin-esterases of, 85, 93, 97
Torula monosa, for fermentation of only monosaccharides, 38
Transfer reactions between one enzyme and two species of substrate, 57

Transglucosidase (sucrose phosphorylase), 49, 58, 59, 70–73
Transition temperature in dextrose manufacture, 135
Transphosphorylation, 49, 59, 70–73
Tree bark, pentosan content, 271
Trehalose, unaffected by sucrose phosphorylase, 72
"Tween 80," favoring growth of tubercle bacilli, 223

V

Vanillin β-D-glucopyranoside, 64, 65
Vibrio perimastrix, action on cellulose, 122
Viscose, 107, 113, 114, 116, 119

W

Wheat, beta amylase of, 231
Wheat straw, L-arabinose in xylan of, 279
—, pentosan content, 271
—, xylan of, 282
Whey, *Saccharomyces fragilis* grown on, 49
Wood fibers, diffraction pattern, 104
Wood pulp, 106, 113, 115, 121

X

Xanthate, ethyl tetraacetyl-β-D-glucopyranosyl-, 23, 28
—, ethyl triacetyl-D-arabinopyranosyl-, 23
Xanthates, S-glycosyl-, reductive desulfurization to anhydrides of sugar alcohols, 23
X-ray diffraction studies of crystallites in cellulose, 104, 106–108
D-Xylal, 2-hydroxy-, triacetate, 221
Xylan, 269–290
  biological decomposition, 288
  composition and structure, 278–284, 289
  furfural from, 288
  nitric acid oxidation, 285
  optical rotation, 282
  periodate oxidation, 284

—, crystalline, from birchwood, 278
—, dimethyl-, 281
—, esparto, yield of crystalline D-xylose from, 278
—, holocellulose as source of, 274
—, straw, 288
   yield of crystalline D-xylose from, 278
Xylan esters,
   benzoyl ester, 287
   diacetate, 286
   dinitrate, 287
   oleoyl ester, 287
   stearoyl ester, 287
   sulfate, 287
   xanthate, 287
Xylanase, enzyme from *Aspergillus niger*, 288
D,L-Xylitol, 1-desoxy-2,4:3,5-dimethylene-, 25, 28
—, 1-desoxy-1-thiocyano-2,4:3,5-dimethylene-, 24, 25, 28
Xylitol, 1,4:2,5-dianhydro-, 220, 221
D,L-Xylitol, 1-tosyl-2,4:3,5-dimethylene-, 24
Xylobiose, from partial acid-hydrolysis of xylan, 280
   crystalline hexaacetate, 280
Xylobioside, methyl, 280
   crystalline pentaacetate, 280
—, methyl pentamethyl-, 280
D-Xylofuranose, 3,5-anhydro-1,2-isopropylidene, 207

β-D-Xyloketofuranoside, α-D-glucopyranosyl-, 35, 39, 69, 71
D-Xyloketose, 39, 46
γ-D-Xylonolactone, 2,3,5-trimethyl-, 282
D-Xylopyranoside, methyl 2,3-dimethyl-, 281, 289
—, methyl 2-methyl-, 289
—, methyl monomethyl-, 282
—, methyl 2,3,4-trimethyl-, 289
β-D-Xylopyranoside, phenyl, 62, 65
—, phenyl 1-thio-, triacetate, 14, 27, 221
D-Xylose, preparative methods, 279
—, dimethyl-, 284
—, 2,3-dimethyl-, 282, 283, 284
—, 2-methyl-, 283
—, monomethyl-, 284
α-D-Xylose, 1-phosphate, 71
D-Xylose, 2,3,4-trimethyl-, 282, 284
β-L-Xyloside, methyl 3-desoxy-, 22, 28
—, methyl 3-methylthio-, 22, 28
(*Xylo*)trihydroxyglutaric acid, from oxidation of xylan, 285
D-Xyluronic acid, 1,2-isopropylidene-, 197

Y

Yarns, accessibility of regenerated cellulose of, 114, 116
Yeast, carboxylase of, 50
Yeast, α-galactosidase from brewer's, 66
Yeast, hexokinase of, 73
Yew, pentosan content of Pacific, 271

ERRATUM

Volume 2

Page 74. In the systematic naming of chondrosic acid read "2,5-Anhydro-D-talosaccharic acid" for "2,5-Anhydro-D-allosaccharic acid." The alternative common name "Anhydrotalomucic acid" remains unaffected.

ADVANCES IN CARBOHYDRATE CHEMISTRY

Volume 1

C. S. HUDSON, The Fischer Cyanohydrin Synthesis and the Configurations of Higher-carbon Sugars and Alcohols	1
NELSON K. RICHTMYER, The Altrose Group of Substances	37
EUGENE PACSU, Carbohydrate Orthoesters	77
ALBERT L. RAYMOND, Thio- and Seleno-Sugars	129
ROBERT C. ELDERFIELD, The Carbohydrate Components of the Cardiac Glycosides	147
C. JELLEFF CARR and JOHN C. KRANTZ, JR., Metabolism of the Sugar Alcohols and Their Derivatives	175
R. STUART TIPSON, The Chemistry of the Nucleic Acids	193
THOMAS JOHN SCHOCH, The Fractionation of Starch	247
ROY L. WHISTLER, Preparation and Properties of Starch Esters	279
CHARLES R. FORDYCE, Cellulose Esters of Organic Acids	309
ERNEST ANDERSON and LILA SANDS, A Discussion of Methods of Value in Research on Plant Polyuronides	329

Volume 2

C. S. HUDSON, Melezitose and Turanose	1
STANLEY PEAT, The Chemistry of Anhydro Sugars	37
F. SMITH, Analogs of Ascorbic Acid	79
R. LESPIEAU, Synthesis of Hexitols and Pentitols from Unsaturated Polyhydric Alcohols	107
HARRY J. DEUEL, JR. and MARGARET G. MOREHOUSE, The Interrelation of Carbohydrate and Fat Metabolism	119
M. STACEY, The Chemistry of Mucopolysaccharides and Mucoproteins	161
TAYLOR H. EVANS and HAROLD HIBBERT, Bacterial Polysaccharides	203
E. L. HIRST and J. K. N. JONES, The Chemistry of Pectic Materials	235
EMMA J. MCDONALD, The Polyfructosans and Difructose Anhydrides	253
JOSEPH F. HASKINS, Cellulose Ethers of Industrial Significance	279

Volume 3

C. S. HUDSON, Historical Aspects of Emil Fischer's Fundamental Conventions for Writing Stereo-Formulas in a Plane	1
E. G. V. PERCIVAL, The Structure and Reactivity of the Hydrazone and Osazone Derivatives of the Sugars	23
HEWITT G. FLETCHER, JR., The Chemistry and Configuration of the Cyclitols	45
BURCKHARDT HELFERICH, Trityl Ethers of Carbohydrates	79
LOUIS SATTLER, Glucose and the Unfermentable Reducing Substances in Cane Molasses	113

JOHN W. GREEN, The Halogen Oxidation of Simple Carbohydrates, Excluding the Action of Periodic Acid... 129
JACK COMPTON, The Molecular Constitution of Cellulose..................... 185
SAMUEL GURIN, Isotopic Tracers in the Study of Carbohydrate Metabolism.... 229
KARL MYRBÄCK, Products of the Enzymic Degradation of Starch and Glycogen 252
M. STACEY and P. W. KENT, The Polysaccharides of *Mycobacterium tuberculosis* 311
R. U. LEMIEUX and M. L. WOLFROM, The Chemistry of Streptomycin......... 337

Volume 4

IRVING LEVI and CLIFFORD B. PURVES, The Structure and Configuration of Sucrose (Alpha-D-Glucopyranosyl-Beta-D-Fructofuranoside)............... 1
H. G. BRAY and M. STACEY, Blood Group Polysaccharides................... 37
C. S. HUDSON, Apiose and the Glycosides of the Parsley Plant............... 57
CARL NEUBERG, Biochemical Reductions at the Expense of Sugars........... 75
VENANCIO DEULOFEU, The Acylated Nitriles of Aldonic Acids and Their Degradation... 119
ELWIN E. HARRIS, Wood Saccharification.................................. 153
J. BÖESEKEN, The Use of Boric Acid for the Determination of the Configuration of Carbohydrates.. 189
ROLLAND LOHMAR and R. M. GOEPP, JR., The Hexitols and Some of Their Derivatives... 211
J. K. N. JONES and F. SMITH, Plant Gums and Mucilages................... 243
L. F. WIGGINS, The Utilization of Sucrose................................. 293

QD
321.98
A2.95
V5
1950

DATE DUE

MAY 25 1967			
MAY 20 1977			